AN INTRODUC

TO

CONTENTS

	Preface	vi
1	General Introduction	1
2	Horses	33
3	Dairy Cattle	95
4	Beef Cattle	147
5	Goats	167
6	Sheep	185
7	Pigs	217
8	Rabbits	247
9	Dogs	265
10	Cats	311
11	Fowls	339
12	Turkeys, Ducks, and Geese	377
	Index	417

PREFACE

Good animal husbandry is the main essential for success in the management of animals of all species. It is necessary at all stages of life, but is particularly important during the early part. Animals need to be kept contented both with their attendant and their surroundings if they are to live in comfort and develop their potentialities to the full. This requires the provision of adequately balanced rations and comfortable, healthy surroundings, and the avoidance of any form of stress.

Although the subject deals primarily with the welfare of animals it is also concerned with the welfare of their owners. To all livestock keepers losses through disease or unsatisfactory development are a lurking threat to the success of the enterprise. In general, it is found that the better the standard of management the fewer the failures.

Good animal husbandry methods require an understanding of an animal's needs. A vast amount of technical information has been accumulated in the last few years, but there has been a delay in transmitting the findings to those people who are actually tending animals. This is due, to a large extent, to the fact that the results of research investigations are published in scientific journals, and concise summaries of the findings are required so that their application in practice is possible.

The aim of this book is to present for each of the common species of domesticated animal a balanced summary of the basic knowledge available in a clear and simple form. The intention is to meet the needs of the following groups of readers:

(1) Veterinary students starting a course in animal husbandry.
(2) Agricultural students early in their course.
(3) Would-be students seeking information prior to entering a college or university.
(4) Animal attendants, working with stock, who are anxious to obtain information of an elementary technical nature.

(5) People interested in home food production on a small scale or in keeping animals as companions and for pleasure.

The contents of this book have been restricted and essential details only, selected by the writer from his personal experience as likely to be most helpful to a beginner, have been included. The aim has been to provide a background of information on which the reader can build. The writer has gained his experience by the use of imperial measurements and the metric equivalents given are only approximate.

Acknowledgements
The writer is indebted to many friends and colleages for help and advice. He is especially grateful to Mr. R. Ewbank, Dr. F. T. W. Jordan, and Dr. T. L. J. Lawrence who read through various sections of the manuscript and offered constructive criticism. The assistance of a number of stock owners who allowed their animals to be photographed is recorded with gratitude. They include Mr. P. H. Attfield, British United Turkeys, Mrs. R. E. Day, Mr. & Mrs. K. Dick, Mr. J. A. Green, Mr. T. Houghton, Mr. N. G. Job, Mrs. E. A. Jones, Miss S. E. Jones, Miss J. O. Joshua, Mrs. J. C. Lawrie, Mr. & Mrs. R. M. Mather, Mr G. S. Mottershead, Mrs. J. S. M. Noakes, Mrs. M. J. Raven, Frederic Robinson Ltd's Unicorn Brewery, Miss M. Thomas, and Mr. J. Walker.

The writer's thanks are also extended to his wife who gave unstinting assistance during the checking of the final draft, to Mrs. P. Wilton for many hours of willing, painstaking typing, and to Miss S. A. Pleavin who exercised great care in taking a number of the photographs. Miss J. E. Kent also kindly helped by taking one photograph. Finally he would like to thank the publishers and their staff for their help at all stages during the preparation of this book.

February 1977 J. O. L. King

CHAPTER 1
GENERAL INTRODUCTION

Introduction

In each chapter the subject matter has been set out under specific headings and in this introductory chapter general points applicable to all the species covered in the book have been grouped under the same titles.

At the start of each chapter, as an introduction, a brief reference is made to the main values of the species to man in the British Isles, whether for food production, sport or companionship. Where appropriate, reference has been made to the linkage between efficiency of production and profitability.

As it is important to be able to describe animals easily and clearly, certain general terms are commonly employed when reference is made to groups of animals of different ages, sexes, colours and types and the meanings of those in general use are given. Definitions of important terms which are widely used in connection with the housing, feeding and breeding of animals and in the maintenance of health are also set out.

In each of the species considered there are a number of recognised breeds and the principal ones are described. The relative importance of these breeds alter as economic requirements and fashion changes dictate, but an attempt has been made to include those breeds which the keeper of a certain class of animal should be able to identify. References are also made to the systems employed by breeders to improve the quality of their stock.

Methods used for the identification of individual animals are described and details are given of the changes in the incisor teeth which can help in determining the ages of mammals of most of the domesticated species.

It is difficult to decide how best to cover the subject of animal housing because new methods are continually being introduced. Thus, on the one hand, tried and tested systems, which may be considered old fashioned by some, have been described and, on the other, modern, recently introduced developments have been included.

In all forms of livestock keeping it is essential that animals should receive foods of the right kinds in the correct amounts. As food is the largest single item of cost in animal production this is particularly important, economically, in those types of animal kept commercially. The aim of this section in each chapter has been to give a broad outline of the ways in which these requirements can be fulfilled.

For each species the reproductive cycle is outlined and the sequence of events in normal parturition described. Where appropriate a reference is made to artificial insemination.

Efficient animal production methods as well as humanitarian considerations demand that animals shall be maintained in a healthy state. A severely diseased individual can usually be recognised easily, but a chronic slight departure from health may not be noticed. Thus, in each chapter, information is included on the signs of health and it must be appreciated that the more familiar an animal attendant becomes with a species the easier it is to recognise those deviations from normal which indicate disease. The acquisition of such a skill is of value because, obviously, the earlier a disease condition can be detected the better.

The methods used to control animals while being handled are given special consideration and finally a general section covers such subjects as vices, minor operations, marketing, licences and welfare codes.

Definitions of common terms

Sire. The father of an animal.

Dam. The mother of an animal.

Litter. All the young brought forth at one birth in species producing a number of young at a time.

Sibs or full brothers and sisters. Have the same sire and dam.

Half-brothers and half-sisters. In horses, particularly, out of the same dam by different sires. Animals with the same sire but from different dams are, properly, referred to as being 'by the same sire' and not as half-brothers or half-sisters.

Purebred. An animal descended from a line of ancestors, all of which belonged to a particular breed.

Cross-bred. The progeny resulting from the mating of different breeds.

Hybrids. The products of crossing inbred lines or strains. Hybrids are noted for their vigour and the uniformity of their type.

Type. An abstract term used for a standard of perfection which fits an animal for a specified purpose.

Polled. The condition of being naturally hornless.

Dehorned. Having had the horns removed.

Disbudded. Having had the horn buds treated to prevent the horns from growing.

Killing out percentage. The measure of the amount of meat in a carcase by excluding the head, skin, blood and internal organs. It is calculated by comparing the dead weight of the two sides of a carcase, including the kidneys, with the liveweight of the animal.

Food conversion ratio. The number of units of food consumed required to produce one unit of liveweight gain.

Permanent pasture. A field laid down to grass and maintained in good condition for a long period by manuring and careful management, without ploughing and reseeding.

Ley or lea. A temporary pasture sown to grass for a period, usually of from 1 to 4 years' duration, after which it is ploughed. A ley is generally more productive than a permanent pasture.

Electric fence. Consists of easily moved insulated posts about 0·91 m (3 ft) high, carrying a wire through which an electric current, generated by a battery unit, is passed.

Lux (lx). One lumen per sq m. A lumen is the amount of light falling per second on a unit area placed at a unit distance from such a source.

Anthelmintic. An agent which destroys worms in the digestive tract.

Vaccine. A suspension of attenuated or killed micro-organisms (bacteria or viruses) which is injected into animals to prevent, ameliorate or treat infectious diseases.

Toxin. A poisonous substance produced by a micro-organism which causes a particular disease.

Antitoxin. A substance which can neutralise a toxin.

Antibody. A substance in the blood or other body fluid which exerts a specific destructive action on bacteria or neutralises their toxins.

Principal breeds

The improvement of domesticated animals dates back to the time man first tamed animals and tried to improve them so that they more closely satisfied his needs. Animals were developed to produce meat, milk, wool, eggs, or power or to give pleasure, and groups of animals were evolved having certain distinguishable characteristics. These are now known as breeds, and a number of breed associations have been established and standards have been drawn up for the various breeds. These standards list the desirable points for each individual, and, in most cases, undesirable characteristics are also noted. Livestock shows are held where animals can be exhibited on a competitive basis and judged on the breed points. Although showing has been criticised because sufficient attention is not paid to the economic points of breeds developed for utility purposes, shows have undoubtedly helped to improve livestock quality. In this book the important breeds in each species have been described impartially and any advantages and disadvantages have been mentioned for the general information of the reader only.

Breeding systems

The considerable progress which has been made in improving animals by breeding has been achieved mainly by animal breeders who have the ability to select stock of merit and plan a constructive breeding programme. Successful breeders have to possess the skill required to make visual appraisals of body type and the knowledge to use the various tests now available to measure the merits of individual animals. In companion animals kept for show, such as some types of horse and dogs and cats, the ability of the trained judge to evaluate body form is still of fundamental importance. In species which are used for meat production an external examination can determine the proportions of the valuable meat producing areas of the body in relation to the less valuable parts such as the head, neck and lower parts of the legs. In horses bred for racing the visual assessment can be backed by the performance on the racecourse, and in dairy cows and laying poultry by a record of the yield. A measure of the growth rate of meat-producing animals, termed a performance test, can help in the selection of breeding stock, while an evaluation of the performance of the progeny, which is a more efficient guide to breeding value, can be used in some classes of animal, e.g. dairy cattle. This is known as a progeny test.

Most animal breeders plan a constructive breeding programme and the systems most commonly used are briefly summarised.

Inbreeding
This covers the mating of closely related animals, such as brother and sister or parent and offspring, and was the method used by most of the early breeders of improved livestock because their animals were so superior that they did not wish to mate them to inferior animals from other breeders. Inbreeding can result in rapid progress in producing a number of animals with highly desirable qualities, but genetically linked undesirable characters can become a problem. Thus, to use the method satisfactorily, breeders must be prepared to eliminate all animals failing to attain the desired standard, and the culling required can make the system expensive.

Linebreeding
This is the mating of distantly related animals, and aims at obtaining the benefits of inbreeding without incurring the penalties. In practice it generally consists in breeding back to members of a family having certain desirable characteristics, and has been adopted successfully by a number of breeders in the development of particular strains.

Outcrossing
This term means the mating of stock which are unrelated with the object of introducing into the strain some desirable character which is lacking. Care must be exercised for there is always a danger that undesirable characters may be introduced as well as the one for which the cross was made.

Cross-breeding
Animals of different breeds are mated in order to combine the best characteristics of the breeds and obtain the hybrid vigour which results from the mating of unrelated individuals. Females produced by cross-breeding may be served by a male of a third breed, because the hybrid vigour of the dams can increase their mothering ability and the sire may introduce qualities from the third breed. However, indiscriminate cross-breeding must not be practised as it only leads to the production of mongrels having no special qualities.

Hybrid production
As very closely inbred lines produce the best results when crossed, certain lines within breeds are closely bred, and then, when these are breeding fairly true to a particular character, they are crossed to

produce a hybrid. This system is only effective if the inbred lines are culled of all individuals showing undesirable traits. This technique is chiefly employed in the commercial production of hybrid fowls, and fuller details of the method as applied to the breeding of this species are given in Chapter 11.

Identification

In most species it is important that individual animals can be easily and accurately identified. This is necessary to prevent the substitution of one animal by another and to enable production and breeding records and pedigrees to be kept correctly. In some species a written description of each animal is used as the method of identification, while in others the animals are marked in a particular way. Such markings may be permanent or only temporary.

Ageing

All the larger domesticated mammals are provided with two sets of teeth. The first set, called the temporary or milk teeth, are subsequently replaced, at more or less regular intervals, by the second set, known as the permanent teeth. With advancing age these permanent teeth show signs of wear. It is possible to estimate the approximate age of individual animals of most mammalian species by an examination of their teeth, particularly the incisor or front teeth. Although variations occur, the majority of animals cut their teeth at the periods stated in the appropriate chapter, or within a narrow margin before or after. There are, however, certain factors which can influence the age of eruption and the rate of tooth wear. Earlier eruption and greater wear can be induced by heavy feeding on hard concentrate foods and also by precociousness, which may result from breed selection for early maturity. Slower wear may follow the giving of food in a sloppy state.

Housing

Most animals are housed for at least a part of their lives in order to protect them from inclement weather conditions and the greatest impact of the intensification of livestock production seen in recent years has been in housing. It is difficult to lay down standards for animal houses but there are certain basic principles which need to be satisfied if the optimum conditions are to be achieved within a building. Housed animals require a free movement of air without draughts, a fairly even temperature with protection from extreme cold, freedom from excessive humidity and a dry, relatively clean bed.

An adequate supply of fresh air, without draughts, is essential. If the ventilation is poor the atmosphere within the building tends to approximate the air expired by the animals with a resultant shortage of oxygen and accumulations of carbon dioxide and water vapour. Also any infectious agents contained in the exhaled breath are cycled and transfer to uninfected individuals in the house is thus facilitated.

The hot exhaled air is lighter than the colder fresh air of the building and so rises and will pass out of the building if outlets are provided in or near the roof. The resultant air movement will facilitate the admission of fresh air through inlets placed in the external walls of the building. These inlets are usually baffled in some way to reduce the risk of draughts and should not be sited too near the floor to prevent any draughts blowing on the animals. The air outside a building is rarely still as it is almost always being moved by breezes or wind, and this movement helps to force fresh air through the inlets and to draw foul air out by creating a suction force as it passes over the outlets.

Most mature animals are kept in buildings which are heated by the warmth given off from their own bodies, but as warmth is particularly important for young or rapidly growing animals artificial heat may be provided in these cases. Insulation of the floors, roofs and walls of animal buildings is necessary to contain the warmth produced naturally by the animals or artificially by the heaters within the building as well as to prevent big diurnal temperature changes, for, although animals of most species can become acclimatised over longish periods to quite wide temperature ranges they find it difficult to adjust to short-term variations.

Damp conditions, in most species, have a more marked deleterious effect on health than the environmental temperature, but temperature has an effect on humidity. The quantity of water vapour the air can hold increases with a rise in temperature so a fall in temperature in a building with a humid atmosphere can result in condensation on the inside of the building, producing wet living conditions. Humidity is measured as relative humidity (r.h.) which is the percentage of saturation, and, under British conditions during the winter months, the r.h. is usually 80 per cent or more.

Drainage must also be adequate, so that urine and stagnant water can drain away from the bedding and provision must be made for the dung to be removed regularly, incorporated with straw into manure, or broken down by bacteria under a deep-litter system. In addition the building should not have any protruding nails, sharp edges or awkward projections which could cause physical injuries.

Feeding

Foods for animals consist of plants, plant products, animal flesh, and animal products, such as milk and eggs. They are all composed of one or more of a comparatively small number of basic ingredients: proteins, fats and oils, carbohydrates, crude fibre, minerals, and vitamins.

Proteins

Proteins are very complex compounds which form the greater part of the body tissues. For this reason young animals require protein in considerable quantities for growth while adults need a certain amount for the replacement of worn-out tissues. Although the proteins of the muscles, internal organs and skin differ, even in an individual animal, they are all formed from amino acids. There are 24 of these, all being needed to maintain the animal body in health. About ten of them, known as non-essential amino acids, can be formed in the animal body, but the remainder, termed essential,

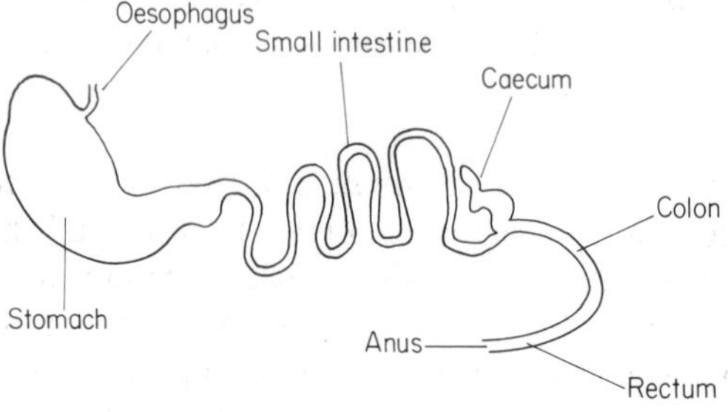

Fig. 1.1. Diagram of the digestive system of the dog.

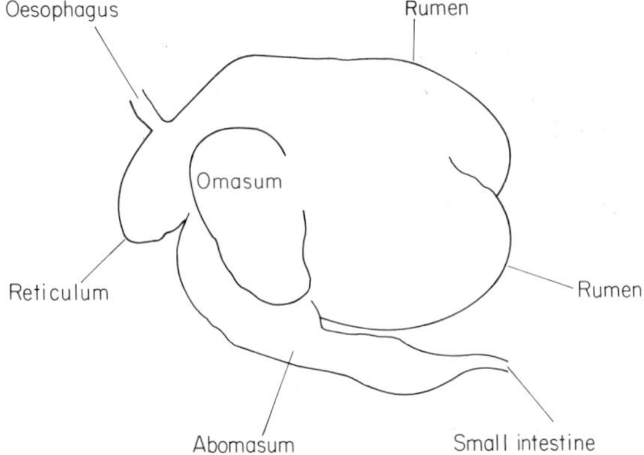

Fig. 1.2. Diagram of the stomachs of an adult ruminant.

must be provided in the diet. The uncertainty about the precise number is that this varies according to the species of animal and the rate of growth.

Protein may be supplied in foods of animal origin, such as meat, fish and fish meal, or in vegetable foods, e.g. soya bean meal, palm kernel, groundnut and linseed cakes, and beans. The animal proteins contain the amino acids in approximately the proportions required for tissue building, but the vegetable proteins tend to be deficient in one or two of the essential amino acids and to contain the others in disproportionate amounts. Thus the animal proteins have a higher biological value as nutrients.

The feeding of rations containing abnormally high proportions of protein is uneconomic and can be harmful, although the undesirable effects of retarded growth and increased mortality may be due to a deficiency of other nutrients created by the high protein content and may not be due directly to the ingestion of excess protein.

Fats and oils

These are similar but fats are normally solid at ordinary temperatures while oils are liquid. A certain amount of fat and oil in the diet is essential for the well-being of all animals. Fats are mainly used to provide energy and, weight for weight, fats and oils will provide from two to two and a half times more energy than carbohydrates. Most animals store fat in the body and this acts as a source of energy when required. The majority of the foods in general use contain some fat but the fat content of most foodstuffs used for animal feeding is low. Because fat and oil are valuable for human consumption and industrial purposes the oil from seeds with a high oil content is extracted and only the residues are left for animal feeding. Although expensive, fats and oils such as high grade tallow and soya bean oil, are sometimes added to diets when there is a special need for a high energy intake, such as rations for young animals. An excessive use of fats and oils can lead to a digestive upset resulting in scouring, or to the laying down of an undesirably oily fat in the carcases of animals being fattened for meat production.

Carbohydrates

These are very important in animal feeding, and provide the chief source of energy in the body. The energy is needed to provide heat to maintain the body temperature and to perform the normal functions of life. These include movement, which may involve working at high speeds, and the production of such commodities as milk, eggs, or wool. Carbohydrates consist of sugars and starch which are broken down into glucose during digestion. The glucose is then absorbed from the digestive tract and some is converted into glycogen which is stored in the liver and, to a certain extent, in muscle tissues, where it provides a readily available source of energy for muscle action. When the need for glycogen is satisfied any remaining glucose is converted into fat and stored in the body in this form. Foods with a high carbohydrate content are maize, barley, oats and wheat.

Crude fibre

The fibre in animal foods consists mainly of cellulose and lignin, and forms an essential part of the diet of all animals except the very young on a milk diet. It aids digestion because its bulk stimulates the action of the alimentary canal and opens up concentrated foods so that the digestive juices can come in close contact with them. However, a very high proportion of fibre in the diet can prevent the consumption of sufficient concentrate food to meet the animal's requirements for essential nutrients. Ruminants require more crude fibre than other herbivores and herbivores need more than omnivores or carnivores. Fibre is digested to some extent but the degree depends on the species. Animals with simple stomachs are only able to use small amounts unless they have voluminous caeca, in which case larger quantities can be digested. In ruminants fibre can be broken down by the rumen bacteria into simple substances which can be used as a source of energy. Hay and straw are examples of foods with a high fibre content.

Minerals

The bodies of animals contain considerable amounts of minerals for

Table 1.1. The physiological functions and chief sources of the minerals and trace elements.

Name	Bodily functions for which required	Dietary sources
MINERALS		
Calcium	The formation and maintenance of bones and teeth, blood clotting	Chalk, bone flour, milk, fish meal
Phosphorous	The formation and maintenance of bones and teeth, protein and carbohydrate metabolism	Bone flour, milk, cereals and oil-seed meals
Sodium	Digestion. Maintenance of body fluid balance	Fish, meat, milk and their by-products
Potassium	Maintenance of health of cells	Most natural foods
Chlorine	Digestion, particularly in the stomach	Linked with sodium and potassium
Magnesium	Bone formation	Natural foods
Sulphur	The formation of horns, claws, feathers, hair and wool	Chiefly obtained from the breakdown of proteins
TRACE ELEMENTS		
Copper	Haemoglobin formation	Liver. Traces in most foodstuffs except milk
Cobalt	In ruminants for the formation of the vitamin cobalamin	Vegetable foods grown on soils with a good cobalt status
Iron	Haemoglobin formation	Most natural foods, except milk
Iodine	Body metabolism, controlled through thyroid gland	Only fishmeal of the common food-stuffs
Manganese	Efficient reproduction, development of fetuses and growth of bone	Most foods, although in small amounts
Zinc	Maintenance of skin health	Normal feeding-stuffs

they form part of the various body tissues and are important in ensuring the efficient functioning of the animal body. The minerals found in the largest proportions are calcium, phosphorus, sodium, potassium, chlorine, magnesium, and sulphur. Other minerals found in smaller amounts are termed trace elements. These are copper, cobalt, iron, iodine, manganese, and zinc, and, although only required in minute quantities, they are essential for the maintenance of health. Other minerals may be found in the bodies of animals but are only of minor importance.

Minerals and trace elements are required for a variety of purposes in the body. These include the maintenance of the rigidity and strength of the skeleton, and the constant composition of the body fluids, including the blood and lymph. They are essential constituents of various soft tissues and are required for enzyme formation. They are also needed for efficient reproduction and the production of meat, milk, eggs and wool. Some of the main physiological functions are set out in Table 1.1.

A deficiency of any of the minerals or trace elements over a length of time will impair health and production. Also, a deficiency of one mineral may affect the utilisation of another and a shortage of certain vitamins may reduce the efficiency with which some minerals are used. Conversely an excess of some trace elements may cause signs of toxicity.

The various foods normally fed contain some minerals but in certain areas the soils may be deficient in a particular trace element and this will be reflected in the composition of the grass, or other crop to be used in animal feeding, grown on the land. Mineral supplements can be used to rectify such deficiencies and also to provide the additional minerals required by animals expected to attain high levels of production.

Vitamins

These are organic substances needed, in minute quantities only, for the maintenance of life and health in man and animals but which do not necessarily contribute to the structural framework or energy requirements of the body. Most of the vitamins have to be supplied in the food but some species of animals can form certain vitamins in their body tissues or within their digestive tracts. The known vitamins are listed in Table 1.2 with brief comments on their functions in the animal body and good dietary sources.

Table 1.2. The physiological functions and chief sources of the vitamins.

Name	Bodily functions for which required	Dietary sources
FAT-SOLUBLE		
Vitamin A	Vision, health of skin and walls of respiratory and urogenital tracts, growth and reproduction	Some fish liver oils. Carotene (pro-vitamin A), fresh green leaves and carrots
Vitamin D	Absorption and deposition of bone, growth and reproduction	Some fish liver oils, irradiated yeast
Vitamin E	Normal reproduction. Muscle health	Wheatgerm, whole cereal grains, some vegetable oils
Vitamin K	Clotting of blood	Liver, egg, fresh green leaves
WATER-SOLUBLE		
Vitamin B group		
B_1 (thiamine)	Carbohydrate metabolism, functioning of central nervous system	Dried yeast, soya bean meal, liver, heart
B_2 (riboflavin)	Oxidation processes in the body	Dried yeast, dried milk, liver, heart
Nicotinic acid	Oxidation processes in the body	Dried yeast, dried milk, liver, heart, wheat and barley
B_6 (pyridoxine)	Amino acid metabolism. Of greatest importance in meat eating species	Cereals, dried yeast, dried milk, liver, egg yolk
Pantothenic acid	Oxidation of fats	Dried liver, dried yeast, dried milk, egg yolk, liver
Biotin	Amino acid metabolism	Wide distribution
Choline	Fat formation, amino acid metabolism	Most foods contain reasonable amounts
B_{12} (cobalamin)	Amino acid metabolism. Normal growth, reproduction and blood formation	Cheese, dried milk, liver, kidney
Vitamin C	Normal digestion and carbohydrate metabolism	Liver, green leaves, some fruits

Animals of all species, other than man, monkeys and guinea pigs, can synthesise sufficient vitamin C in their bodies.

Ruminant and equine animals have enough bacteria in their alimentary canals to enable adequate supplies of the vitamin B group to be produced, but the simple digestive systems of the dog and cat are not capable of containing bacteria in sufficient numbers to permit this to be effected. Animals kept outdoors can make enough vitamin D provided their coats and skins are irradiated by good summer sunshine but weak winter sunlight is unlikely to be strong enough to enable adequate supplies to be produced during the winter months. However, as a surplus of the fat-soluble vitamins can be stored in the body, chiefly in the liver, vitamin D produced in the summer can be utilised during winter.

Although the foods listed in the table are commonly used to supply vitamins most concentrate rations are now fortified with proprietary mixtures of synthetically made vitamins in balanced amounts to meet the particular requirements of the animal.

Classification of feeding-stuffs

In practice the foods fed to animals can be roughly divided into groups according to their bulkiness. In general the bulky foods have a lower feeding value than those which are more concentrated.

Roughages

These are bulky foods with a relatively low feeding value, largely because of their high crude fibre content. Hay and straw are examples.

Succulent foods

Have a relatively low food value because of their high water content. Roots, kale and silage are included in this group.

Concentrate foods

This group contains the seeds, both legumes and cereals. They all have a high food value and a low fibre content, some being high in protein and others high in carbohydrates or fat.

Special foods

These are used in small quantities for particular purposes and include foods rich in minerals and vitamins and additives which can increase the rate of growth.

Descriptions of common feeding-stuffs

Roughages

Hay. Two main types are recognised, meadow hay made from pastures which contain a variety of grass species and some clover, and seeds hay made from leys which consist of one grass species, nearly always Italian rye grass, and red clover. Two subsidiary types are red clover hay and lucerne hay, each comprising the one species of legume only. Both these hays have a relatively high protein content. The feeding quality of hay can vary considerably depending on the material from which it is made, the stage of growth at which the crop was cut, the weather at the time of making and the methods used. The best quality hays are cut early, because there is less indigestible fibre, are dried quickly in good weather, because rain can wash out soluble foodstuffs, are handled as gently as possible, to avoid breaking and losing the leaves which are more brittle but have a higher feeding value than the stems, and are baled and stacked when relatively dry, as moist material goes mouldy and may become overheated and unpalatable. The barn drying of grass, using artificial heat, is valuable in seasons when the weather is bad at haymaking time. Samples of good quality hay are greenish in colour, have a sweet smell, are not mouldy and contain leaves as well as stems.

Dried grass. The grass to be dried is cut at an early stage, preferably before the flowering heads appear. After wilting the cut grass is dried by artificial heat. The methods used vary but in one the grass is placed on a moving belt which passes through a heated chamber. The feeding quality of dried grass varies considerably. Some very leafy samples are equivalent to a concentrate food while others cut in the early flowering stage resemble very good quality hay. When ground it forms a light, dusty meal and should not be fed dry as it may induce coughing in animals and may be wasted by being blown away. It is best mixed with molasses and fed in cube form or added to a wet food such as wet sugar beet pulp.

Straw. The feeding value of straw is limited by the amount of fibre it contains and is normally only used to provide bulk in a ration when hay is expensive or in short supply. Oat and barley straw contain less fibre than wheat straw, are more palatable and are the two varieties most commonly fed. Straw is best supplied long so that the upper, more nutritious portions can be eaten and the lower, more fibrous bases discarded.

Succulent foods

Roots. These supply mainly carbohydrate, but are valuable because of their palatability and their laxative action, which are particularly beneficial during a period when stock are housed and fed on a dry diet. Roots are low in fibre, protein and fat. Mangolds, swedes, and turnips are valuable foods, but are expensive to grow and feed because of the amount of manual labour required. However modern sowing and chemical weed killing techniques could make them popular again. Carrots are liked by horses and may be added with advantage to dry rations. Waste potatoes can be used but if fed whole may become lodged in the oesophagus and cause choking and death, and if fed with green skins or sprouts are toxic.

Kale and the cabbage family. These foods are better balanced in their protein to carbohydrate contents than are roots and are usually rich in carotene. Two types of kale are popular, marrow-stem and thousand-headed, and are grown specially for feeding from the autumn onwards when the grass has ceased to grow. Both can be grazed or cut and carted to housed animals. The marrow-stem variety is used in the autumn before the severe frosts start as this type is susceptible to frost damage and is of particular value for cattle as they can utilise the thick stems. Thousand-headed kale is more resistant to cold winter weather and is useful for sheep in the early spring. Waste leaves from brussels sprouts, cabbages, and cauliflowers grown for human consumption can be fed with advantage.

Sugar beet tops. When sugar beet is harvested root crowns and leaves are cut off and used for feeding. They must be wilted for about a week or be mixed with powdered chalk at the rate of 113·6 g of chalk per 113 kg of leaves (4 oz per 250 lb) before being fed to reduce their oxalic acid content which might otherwise prove poisonous.

Rape and mustard. Rape is related to the turnip but has no root storage of food and white mustard is a cruciferous crop. Both grow rapidly and are often planted between two main crops, being termed catch crops. They are usually grazed, particularly by sheep in the autumn.

Silage. Most silage is made from grass but a variety of green crops, including green maize, sugar beet tops, lucerne, and kale, can be used. The fodder plants to be ensiled are cut, allowed to wilt slightly, and tightly packed in a silo. Silos vary markedly in construction but the most popular are pits, solid-sided barns or towers. The plants respire for a short time and liberate heat which is conserved in the silo so that the temperature of the mass rises. The plants then die and the mass is acted upon by bacteria with the production of organic acids which act as preservatives so that the silage will keep for a long period. Lactic acid is the most desirable, and thorough packing and the reduction of air intake to a minimum are required to promote rapid acidification and the development of the bacteria forming this acid. Chopping the crop into short lengths permits tighter packing and hastens lactic acid fermentation. A highly nitrogenous crop, such as one containing a high proportion of clover, contains insufficient carbohydrate for the multiplication of the desirable type of bacteria. To make this deficiency good molasses is added at the rate of about 9·1 kg per 1,016 kg (20 lb per ton) for grass and clover crops and 13·6 kg per 1,016 kg (30 lb per ton) for lucerne. The molasses is diluted with three to five times its volume of water and is sprinkled evenly over successive layers of plant material. Failure to obtain the conditions required for lactic acid development can result in the production of butyric acid which has an unpleasant rancid odour and results in a loss of energy and palatability. If a crop is very wet when ensiled nutrients are lost in the water which is expressed through pressure. Good silage should be made from quality crops, be yellow-brown in colour, have an acid, fruity smell and just exude moisture when squeezed.

Concentrate foods, protein concentrates and vegetable protein

Oil-seed cakes and meals. These are the residues which remain after the oil has been expressed from the oil-seeds and are very valuable animal foods. They are rich in vegetable protein and contain some residual oil. They are marketed in two forms: undecorticated, with the seed husk left on; and decorticated, with the husk removed. The undecorticated cakes contain less protein, more fibre and are more astringent on the alimentary canal than are the corresponding decorticated cakes. As the oil-seed cakes are fed for their protein contents they can roughly be divided into a high protein group containing decorticated groundnut, decorticated cotton-seed and soya bean cakes, a medium protein group illustrated by linseed cake and a group in which the protein and carbohydrate contents are reasonably balanced which includes coconut and palm kernel cakes. Some varieties of cotton-seed cake contain gossypol in sufficient quantities to be toxic and some samples have been responsible for deaths in calves and fattening pigs. Some consignments of groundnut cake affected with certain strains of a fungus are poisonous, particularly to ducklings and poults, as the result of the production of aflatoxin. Linseed cake is safe to feed under normal systems but if given as a warm mash to hungry animals, such as calves, there is a risk of poisoning by the hydrocyanic acid released in the warm, damp conditions. Boiling linseed cake for 10 minutes renders it safe.

Beans and peas. Field beans and field peas used to be grown fairly extensively for livestock feeding but have lost popularity. They are similar in feeding value, both being relatively rich in protein. Peas can be fed whole to poultry but beans for horses must be broken (kibbled) before use. Both can be added as coarsely ground meals to concentrate rations.

Animal protein

Animal protein concentrates are produced from meat, bone, fish and milk. The meals are obtained by drying the raw materials and low temperature processes produce meals which have a higher nutritive value than do methods requiring higher temperatures. Meat meal and meat and bone meal are obtained from the carcases of animals not being used for human consumption or from meat from slaughter houses which is unfit for human use. The proportion of flesh to bone in meat and bone meal is not specified but a good quality product is a valuable food as it supplies both protein and minerals. White fish meal is obtained from the flesh and bones of white fish, being mainly prepared from parts unsuitable for human use. In colour, white fish meal is yellowish-brown. Herring meal comes from herrings which have a higher oil content. Both are very valuable foods because of their protein and mineral contents. The oil in herring meal may taint flesh and so herring meal should not be used in the last 14 days of a fattening period.

There are several milk by-products which are used in animal feeding. Skim milk has had the fat removed manually and separated milk has had the fat removed mechanically. The latter has a lower fat content because the separation method is more efficient. The powders produced by drying these milk residues are particularly valuable for young stock feeding, being rich in protein and lactose. When reconstituted, dried whole milk is similar in composition to liquid cows' milk and can be used as a milk replacer.

Carbohydrate concentrates

Cereals and cereal by-products. These are mostly starchy foods, rich in carbohydrate and low in protein and are used as a source of energy. The grains are not fed whole but are rolled or crushed or ground into a meal. Oats are rather high in fibre. Sussex ground oats is the term used for whole oats ground to a fine meal, while oatmeal is the ground kernel of the grain, but, because of its cost, oatmeal is only used for special purposes. Wheat is not usually fed to animals, except perhaps to poultry, but the by-products of wheat milling are widely used. The scaly outer part of the kernel is marketed as bran in the form of large flakes and the inner seed coat as a fine meal, termed weatings, middlings, or millers' offals. Barley which is not of the quality required for malting is fed as rolled barley and as a meal. Brewers' grains are a by-product from the brewing of barley and are sold either wet or dry. They contain the fibre and protein of the original grain but much less starch. The wet grains are palatable when fresh but soon become sour and mouldy. The dried grains are more popular because of the difficulty in preserving and transporting the wet grains. Rye is less palatable and is not popular. Maize is palatable, rich in carbohydrate and low in utilisable protein and fibre. As a meal it is widely used in fattening rations and it is also fed flaked. Flaked maize is produced by heating the grains and then squeezing them between hot rollers. The flaking process increases the digestibility of the maize.

Sugar beet pulp. This residue from the extraction of sugar is sold either wet or dry. The wet material is used on premises near sugar beet factories but its high water content makes it difficult to transport and store. Dried sugar beet pulp contains valuable soluble carbohydrate but is low in utilisable protein and high in fibre. Dried sugar beet pulp should not be fed dry as it rapidly absorbs water from the oesophagus which causes it to swell and cause a blockage.

Special foods

Sterilised bone flour. Bones are boiled to remove the fat and are then ground and sterilised by steam under pressure. The resultant powder is an excellent source of calcium and phosphorus.

Fish liver oils. Cod-liver and halibut-liver oils have been used for many years as sources of vitamins A and D. Cod-liver oil contains at least 600 I.U. of vitamin A and 85 I.U. of vitamin D per g. Halibut-liver oil is much more potent, averaging about 30,000 I.U. of vitamin A and 3,000 I.U. of vitamin D per g. Rancid oils can cause extensive destruction of other fat-soluble vitamins.

Growth-promoting food additives. Growth-promoting substances are widely used in the diets fed to growing poultry and are included to some extent in the foods given to young pigs. To be of any economic value the improved performance which results must produce a financial return which is greater than the cost of the additive. Also the substance should have little or no value as a therapeutic agent in human or veterinary medicine and must not reduce the efficacy of other drugs used in the treatment of disease through the development of resistant strains of bacteria. Growth promoters in common use which satisfy these conditions are the antibiotics virginiamycin, flavomycin and zinc bacitracin and the mineral compound copper sulphate.

Antibiotics. These are included in the diet at a nutritional or low level which varies from 2 to 40 g per 1,016 kg (0·07 to 1·41 oz per ton) and animals on a supplemented diet can increase in body weight from 5 to 20 per cent more rapidly than do controls on a normal diet and show an improved food conversion ratio. Another advantage is that groups of animals grow at a more even rate. The younger the animal the greater the response. The responses are also greater when poor quality rations are fed and when the standard of management is low. It is not certain how antibiotics promote animal growth but the beneficial action is probably mainly due to the fact that an antibiotic may inhibit the development of, or destroy, micro-organisms which produce a subclinical infection and so cause a slowing of growth in the host animal. This theory is supported by the fact that growth stimulation is minimal, or even non-existent, in animals kept under disease free conditions. Antibiotics appear to exert a sparing effect on some nutrients, which may, to some extent, be due to an inhibiting action on organisms which compete with the host for available nutrients and may also act by stimulating the appetite. They thin the intestinal wall which may contribute to the growth-promoting effect by facilitating the absorption of essential nutrients.

Copper sulphate. Copper salts are very palatable to pigs and copper sulphate fed at a level of 0·05 per cent of the diet to pigs after weaning will cause improvements in both the rate and efficiency of liveweight gain. The mode of action by which copper sulphate stimulates growth has not yet been fully explained. This compound is not used in other species.

Water

In addition to food, animals require water for the maintenance of life. Water accounts for over 60 per cent of the body weight of an adult domesticated animal and gives the rigidity combined with elasticity to the cells which form the body. Water provides the basis for all the complex physiological processes necessary for life. It facilitates digestion by assisting the passage of food through the alimentary canal and by forming the basis of the enzyme solutions necessary for the breakdown of food. Blood and lymph, which are mainly composed of water, are essential for the transportation of nutrients from the intestinal wall to the tissues where they are needed and of the waste products formed to the excretory organs. Water is also used in the control of body temperature. An animal body is continually producing heat, some of which must be eliminated either through expiring water vapour which can be increased in amount by panting or by sweating. In the form of joint-oil water is needed to lubricate the joints of the body and so facilitate movement.

Animals obtain water mainly by drinking or from the food they eat. All foods contain water, but the proportions vary greatly. Dry foods, such as hay, contain about 15 per cent, while succulent foods, e.g. kale, contain 85 per cent of water or more. Animals eating large amounts of food of this latter type ingest appreciable quantities of water, and this water is as useful as that obtained by drinking. There is a third source of water utilised by animals. This is the water, known as metabolic water, which is released by the various chemical reactions which are continually occurring in the animal body. The quantity of water which animals need to drink varies according to the climate and their size and activity, as well as the moisture content of the diet. The safest method of watering animals is to allow them to have access to water regularly and to drink as much as they desire.

Breeding

Efficiency in reproduction is one of the main essentials in profitable animal production and good fertility and high survival rates in young animals are the results of good husbandry. Each animal species has a reproductive cycle, but the details vary from one species to another. In male animals of the species considered in this book there are no clearly defined cycles. In female mammals there are a number of oestrous cycles, each consisting of a period of

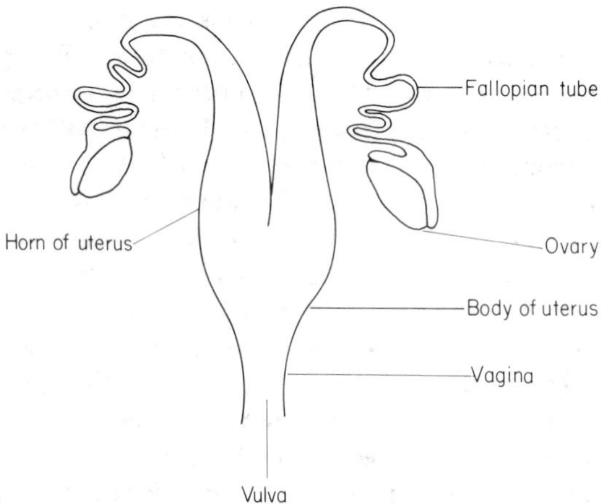

Fig. 1.3. Diagram of the reproductive system of a female mammal.

sexual activity, known as oestrus or 'heat', followed by a period of sexual inactivity known as anoestrus. If the female does not become pregnant these events occur throughout the breeding cycle. During oestrus ovulation occurs, at which time the eggs are shed by the ovary and pass down the Fallopian tube into the uterus. The number of eggs shed at ovulation depends upon the species, and, to a smaller extent, upon the individual animal.

Females of the species of poultry described do not have clearly-defined periodic sexual cycles, although they do lay eggs in clutches. This means that they lay on a series of successive days, and then have a break of 1 or 2 days before starting to lay the next clutch. The clutch lengths vary greatly, some lasting for 2 days only while others may last for 50 days or more. Obviously the most efficient producers are those with long clutch lengths.

Breeding efficiency can be reduced by poor standards of husbandry. Various factors are involved but the most important is the plane of nutrition. Undernutrition is the most common factor depressing fertility in female animals, but an unbalanced diet may also be responsible. Deficiencies of protein or certain minerals, which lead to dietary imbalances, are the most common.

The act of mating male and female animals is generally known as service. As puberty occurs when animals are still young and actively growing, domesticated animals are not usually mated until some time after attaining puberty. The time taken for service varies considerably according to the species. At service the male sperms are deposited in the female tract in or near the uterus and may have to travel some distance through the uterus and up the Fallopian tube before the eggs can be fertilised. This journey can take several hours. There are species variations but in general sperms will only remain alive in the female tract for from about 24 to 36 hours, while the life of an egg after it is shed is shorter. Therefore, there is only a limited time for fertilisation, and so serving an animal too early in an oestrous period may mean that the sperms are dead before ovulation has begun, while serving too late may mean that the egg is dead before the sperms have completed the journey to the point of fertilisation. In the case of the poultry species the female stores semen after mating and can lay several fertile eggs over a period of a week or even longer without being mated again.

After service it is sound practice to keep a female mammal quiet for a few hours. If frightened or exercised violently there is a danger that some of the semen introduced into the genital tract may be expelled.

The time of pregnancy or gestation in mammals is fairly constant for each species, although slight variations can occur depending on the breed and the individual. A fetus grows relatively slowly at first but during the last quarter of pregnancy it develops very rapidly, often doubling in size in a short time towards the end. This places a strain on the mother, and her feeding during this time is very important in ensuring that the offspring are strong at birth.

Parturition can be divided into three stages. The first is characterised by contractions of the muscles of both the abdominal wall and the uterus. At first these contractions are mild and infrequent but later become stronger and more frequent, reaching a maximum at the time of giving birth. This straining presses the fetus and the fluid-filled membranes surrounding it against the narrow opening of the uterus (the cervix), causing it to widen. The front parts of the membranes then appear at the vulva, forming what is known as the water bag. The second stage commences with the rupture of the water bag which releases some of the pressure within the uterus. Part of the fetus can now be seen, this usually comprising the forelegs with the head between them. The contractions become stronger and force the chest and shoulders, the largest part of the fetus, through the vulva. The rest of the fetus usually follows with less effort. Some young animals are born with the hind quarters first. The third stage consists of the expulsion of the fetal membranes, usually termed the afterbirth. Animals which give birth to several young at a time usually pass the afterbirths after each fetus. The times taken over each stage vary according to the species.

The umbilical cord, which forms the main connection between the unborn fetus and its mother, generally breaks at birth or is bitten through by the mother. If it does not it should be broken, leaving a length attached to the navel of the newly-born animal. This will dry and fall off in a short time. After the young are born the mother licks them dry and, in so doing, removes any fetal membranes round the mouth and nose, thus facilitating breathing. Shortly after this the young start to suckle, and their chances of survival are much greater once this has been accomplished.

Artificial Insemination (A.I.)

This technique comprises the introduction of spermatozoa collected from a male animal into the female reproductive tract by means of instruments so that direct contact between the male and female animals is unnecessary. There are several advantages which arise from the use of this system of breeding. The main one is that each semen ejaculate contains many more spermatozoa than are required for conception and so, by semen dilution, it is possible to increase greatly the number of young which can be sired by one male, thereby permitting the careful selection of sires of high quality for widescale use. Also, in some species, semen can be stored in a frozen state for long periods and so females can be inseminated years after the semen sample was collected which enables young to be sired by deceased male animals. The fact that semen samples can be so easily transported means that artificial matings between animals kept long distances apart can be conveniently arranged. One great advantage, in certain areas, is that the spread of venereal diseases is prevented as there is no sexual contact. A minor point is that there are some females which, because of a deformity of the reproductive tract, cannot be mated normally but can be successfully inseminated. Disadvantages are that the preservation of frozen semen is expensive, and that some male animals produce semen which will not freeze satisfactorily. Also, in certain breeds, the heavy utilisation of frozen semen might limit the number of sires bred from and so lead to inbreeding.

Lactation

With the exception of poultry all the animals mentioned in this book are mammals which suckle their young. Milk is the ideal food for young mammals and the rate of their early growth can be slowed if they do not receive a good supply of milk. Thus, breeding females should be good milk producers.

The mammary tissue enlarges in late pregnancy and begins to secrete fluid as parturition approaches. The first secretion produced in each lactation is colostrum. This is a yellow, sticky liquid which is laxative and rich in protein and vitamin A but its chief importance is that it contains antibodies which are absorbed from the intestines of the newly-born animal and fortify its resistance to disease. The antibodies from the dam will afford protection against the strains of organisms present in the dam's environment but, if an animal is moved to new premises shortly before giving birth, or if the offspring is moved shortly after birth, the different surroundings may contain pathogenic organisms against which the dam's colostrum gives no protection.

During the first 3 or 4 days of lactation the colostrum changes rapidly into milk of normal composition for the species and, provided the milk is removed, the quantity produced increases for a period, the duration of which varies with the species. If any teat is not sucked or milked the mammary tissue supplying it becomes distended for 2 or 3 days but production rapidly ceases and the swollen gland contracts. Milk production usually terminates when the young are weaned or milking stops but will, in any event, cease after a period of time which differs between species.

Health

A healthy animal is one whose body processes function properly so that it can lead an active life, grow steadily, reproduce and attain the maximum level of production of which it is genetically capable. Health is commonly defined as 'freedom from disease' and disease can be defined as any disturbance of the normal body processes which affects an animal adversely. Such an upset can be caused by physical injuries, bacteria, viruses, parasites, fungi or poisons, or by dietetic errors, metabolic disturbances or hereditary defects. Every effort must be made to support health and so avoid disease in order to prevent losses through deaths or reduced production levels.

There are a number of clinical signs which an animal attendant can look for which can give an indication of the health status of his charges. An animal which carries its head high and is interested in its surroundings, and which has a clear eye and a dry, clean coat with a sheen, shows signs of well-being. Nasal discharges, inflamed running eyes, a staring, scurfy coat and a dull, dejected appearance are indications that all is not well with an animal. Healthy animals have good appetites and a first sign of illness is a refusal to eat. Ruminant animals should also chew the cud normally. Healthy adult animals should maintain, and young animals increase, their body weights, so an animal showing an obvious loss of body weight is probably nnwell.

The appearance of the faeces indicates the state of the digestive tract and constipation and diarrhoea are signs of digestive disturbances. Normal urine is a pale straw-coloured liquid with a distinct smell and alterations in colour and smell may be caused by disease conditions. One of the first signs of disease in lactating animals is a drop in milk yield. The posture and movement of an animal should be carefully noted. Lameness may be due to a variety of causes and arched backs and an inability to rise and stand are indications of abnormality. Breathing should not be too rapid or erratic and it should be noiseless. Continuous or intermittent coughing shows that there is an irritation in the respiratory tract, and groans, grunts and grinding of the teeth are evidences of pain. The reproductive tracts of healthy animals function normally and a failure to breed is a sign of disease.

The internal body temperatures of healthy animals remain fairly constant at levels which vary according to the species. Temperatures may be raised by exertion, particularly in hot weather, and by fear, but disease is the principal cause of a rise in temperature. A fall below the normal level may occur just before death or in animals in a comatose condition. The temperature of an animal is taken by inserting a half-minute clinical thermometer into the rectum and retaining it there for the prescribed length of time. The thermometer should be held in position and directed towards the rectal wall and not buried in a mass of faeces. In small animals, or in cases where insertion is difficult, the thermometer can be wetted with water or lubricated with soap or vaseline before use as this facilitates its entry into the rectum. A stub-nosed thermometer should be obtained as the long-nosed thermometers normally placed under the tongue in human beings may break and cause injury.

The walls of the arteries are muscular and elastic and expand when there is a thrust of blood from the heart and then contract on to the blood to reinforce the heart's initial push. In certain places in the body this movement can be detected by touch and is known as the pulse. The optimal sites for detecting these pulsations vary with the species. Feeling the pulse enables its frequency, regularity and strength to be assessed. A raised pulse rate is generally an indication of disease but may be caused by severe exertion or excitement. In animals of all species the number of pulsations during a period of 30 seconds, timed by a watch, should be counted while the animal is in a quiet state. In the case of nervous individuals it may be necessary to wait for the animal to become calm before starting to make the count.

As a thriving animal has a higher resistance to disease than one in poor bodily condition stock should be well housed, well fed and well managed. These principles are particularly important in the case of young animals. Regular daily exercise is important for the maintenance of health and should be provided for all animals except those which are fattening rapidly, such as pigs, when it should be limited. Exercise stimulates the activities of the internal organs, such as the heart and lungs, prevents the joints and muscles from becoming stiff, and keeps the feet from becoming overgrown.

Common diseases and their prevention

Methods of disease prevention will vary according to the particular causal agent and, in some cases, the species of animal but there are a few points which have a general application. As far as possible stock owners should prevent animals from injuring themselves, for physical injuries can cause considerable discomfort and possibly, in farm animals, a fall in production. Nails or pieces of wire in the food can cause injury to one of the stomachs in ruminants and sharp posts or badly fitted strands of barbed wire in fences can result in skin cuts and limb injuries in grazing animals. Dogs and cats allowed to stray on roads are frequently injured by vehicles. Injuries at parturition can often be averted if animals giving birth are closely supervised so that expert assistance can be obtained if necessary.

Many diseases are caused by infectious or contagious agents. An infectious disease is one which can be transmitted from one animal to another without direct contact and is usually caused by a bacterium or virus which can be carried in the air and breathed or swallowed. A contagious disease is one which is spread by contact and examples are some external parasites and fungi. As far as possible animals should be kept away from sources of infection, or contagion, and in this connection the separation of young and old stock, where possible, is a wise precaution. Adult animals may harbour infectious agents to which they have acquired a resistance and which do not, therefore, cause clinical signs in them, but which can be transmitted to susceptible young stock which may develop the disease.

Certain chemical compounds and plants are poisonous and may cause clinical signs if ingested. Toxic materials should be kept away from animals and not allowed to contaminate the foods they eat. Wild plants known to be harmful to livestock should be eradicated from grazing land. Animals grazing on good pastures tend to avoid poisonous plants, but if on bare fields with a choice between hunger and the consumption of a poisonous plant they often adopt the latter course.

One disease-causing dietetic error is sheer undernutrition, particularly seen in animals entirely dependent on grazing bare pastures; another is a deficiency of some essential item in the animal's diet, while imbalances of nutrients, e.g. minerals, may also have a depressing influence on health. Metabolic disturbances are caused by a deficiency within the animal itself, due to some physiological failure in adjustment, and not by an inadequate supply of some nutrient in the diet.

Health may also be affected by heredity. Some breeds or strains may be particularly susceptible to certain infectious agents, while inherited defects are recognised which may be transmitted by parents to some of their offspring. However hereditary diseases are not of major importance in animal husbandry.

Handling

It is necessary to control the movements of animals in order to manage them under the husbandry conditions being practised. The methods employed will vary according to the degree of control which is required. Unrestricted movement can be prevented by the use of hurdles or fences. Individual animals need to be restrained while being examined or being used for a specific purpose such as riding, milking or shearing, which will vary according to the species. In most cases physical force will be needed to some degree, but with some trained animals verbal commands may be sufficient to check unrequired movements.

Handling should be a gentle, but firm, process which does not cause the animal unnecessary pain. According to the particular requirement the restraint may be general, or applied to a particular body area only. Alternatively it may comprise some way of preventing movement by diverting the animal's attention. The methods used will also vary according to the species and to the temperament of the individual, but the one selected should always be the least harsh method which will achieve the required objective. Gentle animals may be excited into rebellion by the use of unnecessarily severe methods.

When a decision has been reached as to the method to be employed it should be applied quickly and smoothly. For this, experience and the manual dexterity which is acquired by practice are needed. It is also important that the handler's actions should be calm and deliberate so as to encourage the animal's confidence in man. Nervousness or uncertainty on the part of a handler are sure to arouse an animal's apprehension, which may lead to difficulties which could have been avoided.

Administration of medicines

For each species the common methods by which medicaments can be administered are briefly described, and advice given on the most satisfactory procedure to be adopted. Medicines can be given by mouth in a liquid form, commonly termed a drench in the larger animals, as pills, tablets, capsules or boluses, or as electuaries, which are semi-fluid mixtures with a treacle base prepared for application to the back of the tongue. Increasingly, medicaments are being administered by injection using a sterile hypodermic syringe and needle and various sites are employed. Subcutaneous injections are made by inserting the fluid under the skin, in intramuscular injections the preparation is placed in a muscle, while intravenous injections are made directly into the blood flowing in a vein. Intramammary injections are made through a teat canal into an udder quarter, generally by using a specially prepared tube of the drug prescribed.

General

Vices

Animals of several species acquire bad habits, commonly termed vices, which can have deleterious effects on their health and performance or on the well-being of other members of the group. Some of these habits are quickly copied by other animals. The most important vices found in each species are described.

Minor operations and manipulations

In most species certain routine operations are commonly performed to facilitate the control of individuals, to improve their well-being or to reduce the risk of damage to other animals or property These are noted where appropriate.

Species control and product marketing

Some regulatory bodies have been formed to govern and control the use of animals kept for sport or pleasure and comments are made on the objects and powers of some of the most important. The various animal products are marketed through different channels and in some instances brief descriptions of the marketing authorities arranging the sale of the commodities are given.

Licences and welfare codes

In a few cases licences have to be obtained before animals of a specified age or sex can be kept and the main provisions of the controlling legislation are outlined. Welfare codes making recommendations for the management of a few species have been drawn up and notes on the principal clauses have been included.

CHAPTER 2
HORSES

Introduction

A great interest is being shown in riding horses of all types at the present time. Riding is a popular pastime and the number of horses and ponies kept for pleasure is steadily increasing. Some people own the horses they ride, while others hire them from a riding school. Pony riding can be a great benefit to children. They obtain exercise in the open air and develop a sense of responsibility and self-control by managing a living creature. Pony show competitions teach them to accept both successes and disappointments. Riding holidays are attracting more people and riding centres are being established in country areas noted for their natural beauty. Disabled persons are finding that horse riding is one of the few sports they can enjoy and the Riding for the Disabled Association has active groups in most parts of the United Kingdom. Expert riders enjoy the thrills of competition sports of various types, including show jumping, gymkhanas, dressage and three-day events, which include dressage, riding across country and show jumping.

Many horse lovers who do not ride themselves enjoy watching horse racing, both on the flat and over jumps, as well as trotting races. Show jumping competitions are also popular with spectators, and both racing and show jumping are popular with television viewers.

Light horse driving is at present experiencing a resurgence of interest and equitation schools specialising in driving courses are being established. There instruction is given on driving horses as singles, pairs and four-in-hands.

The number of heavy draught horses kept in the British Isles has declined in recent years because of the increased use of mechanical transport but a few heavy horses are still used in towns, particularly by brewery companies, for pulling carts on short journeys. In addition to the advertising value of a well turned out pair of horses the claim is made that the costs of running and depreciation are less than for a motor vehicle. Similar claims are made by those farmers who still use horses for field cultivations.

Definitions of common terms

Ages

Thoroughbred horses are all aged from 1 January, irrespective of the actual date of birth.

Foal. Up to the end of the year in which it was born.

Yearling. During the whole of the year after that in which it was born.

2-*year-old*. The year after the yearling stage. Then, 3-year-old and so on.

Aged horse. Over 8 years.

Sexes

Males

Colt foal. Uncastrated male in first year of life.

Colt. Uncastrated male, 1–4 years of age inclusive.

Stallion, entire or horse. Uncastrated male, 5 years and upwards.

Gelding. Castrated male horse of any age.

Cryptorchid or Rig

unilateral—one testicle in scrotum, the other in the abdominal cavity;

bilateral—both testicles in the abdomen.

All cryptorchids have the instincts of a stallion, and are unreliable. Bilateral cryptorchids are also infertile.

Monorchid. One testicle has not developed.

Females

Filly foal. In the first year of life.

Filly. 1–4 years of age inclusive.

Mare. 5 years and upwards.

Brood mare. Mare kept for breeding.

Spayed mare. A mare with the ovaries removed. This operation is rarely performed.

Horse × donkey crosses

Mule. Sire, donkey. Dam, mare.

Hinny. Sire, stallion. Dam, donkey.

Both these hybrids are hardy and sterile. They take their size from their dams, so mules are considerably larger than hinnys. The mule has long ears and a donkey's tail and the hinny short ears and a horse's tail.

General terms

Near. The left-hand side of a horse, derived from the fact that a riding horse is mounted from the left side.

Off. The right-hand side of a horse.

Hand. Horses are measured from the ground to the withers in hands and inches. A hand equals 10·16 cm (4 in).

Lop ears. Ears which droop and are not held upright in the normal manner.

Roman nose. When viewed from the side has a convex appearance.

Dished face. Presents a concave appearance from the side.

Crest. The upper border of the neck which is normally arched. It is always larger and more clearly arched in stallions than in mares or geldings.

Ewe neck. There is no arch to the neck, which has a concave and not a convex appearance.

Chestnuts. Horny growths on the insides of the legs situated above the knees on the forelegs and below the hocks on the hind legs.

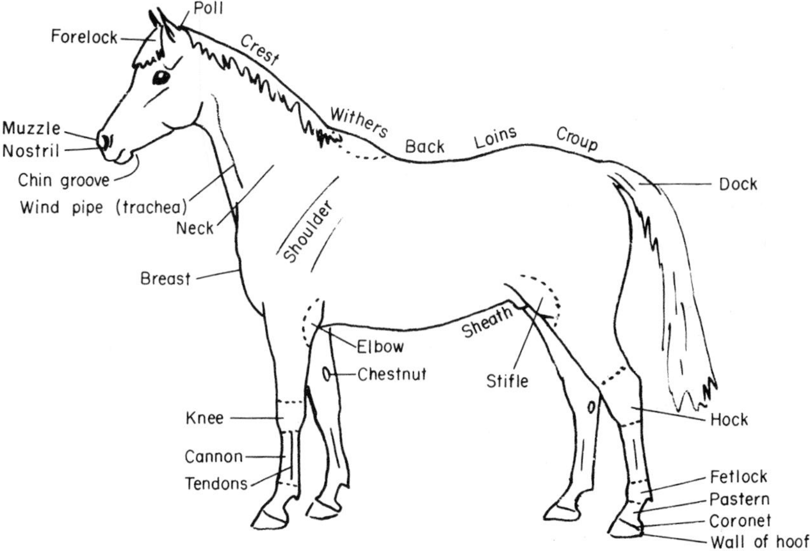

Fig. 2.1. Points of the horse.

Colours and markings

The following list of body colours and markings is based on a Royal College of Veterinary Surgeons Report (1977).

Body colours

Section A

Colours acceptable to the thoroughbred authorities.

Black. Black over the body, limbs, mane, and tail.

Brown. A mixture of black and brown pigments over the body with black limbs, mane, and tail.

Bay. The body colour varies considerably from a reddish-brown to a yellowish colour, but the limbs, mane, and tail are always black.

Chestnut. The colour over the body, limbs, mane, and tail varies from yellow to a reddish shade. The mane and tail may be darker or lighter than the body.

Grey. The coat colour is a mixture of black and white hairs. When young, most greys are nearly black in colour but they lighten with advancing age until most become nearly white.

Blue Roan. The body colour is a mixture of black and white hairs giving a blue coloration. The limbs, mane, and tail are black.

Bay Roan. The body colour is a mixture of bay and white hairs with black limbs, mane, and tail.

Chestnut Roan. The body colour is chestnut with a mixture of white hairs.

Section B

Colours, additional to the above, accepted by non-thoroughbred authorities.

Blue Dun. Dilute black body colour with a black mane and tail.

Yellow Dun. Yellow body colour with black pigmentation on the limbs, mane, and tail.

Cream. A cream body colour with blue eyes. The skin is pink in colour.

Piebald. The coat consists of large irregular patches of black and white, with lines of demarcation between the two colours usually well defined.

Skewbald. The coat consists of large irregular patches of white and of any definite colour except black. The line of demarcation between the colours is generally well defined.

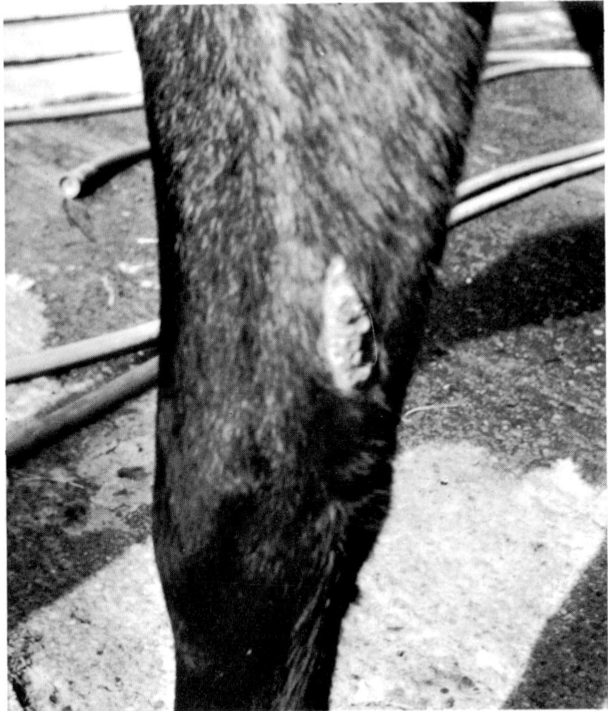

Fig. 2.2. Chestnut on the inside aspect of the foreleg of a horse.

Odd coloured. The coat consists of large irregular patches of more than two colours which may merge into each other.

Palomino. Newly minted gold coin colour (lighter or darker shades are permissible) with a white mane and tail.

Appaloosian. The body colour is grey, covered with a mosaic of black or brown spots.

Markings

The term 'whole coloured' is used where there are no hairs of any other colour on the body, head, or limbs. The majority of horses, however, have markings which are divided into the following groupings.

Head

Star. Any white mark on the forehead.

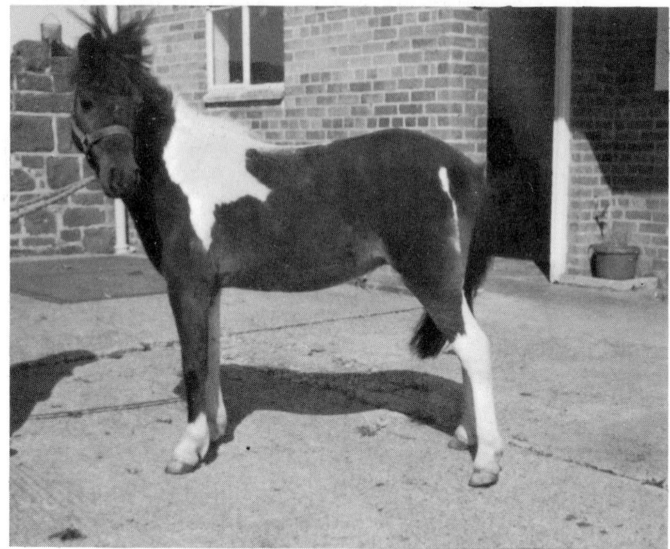

Fig. 2.3. Skewbald pony.

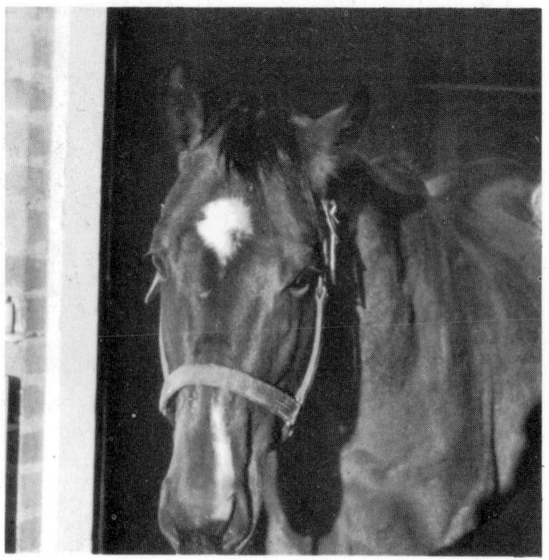

Fig. 2.4. Star with a separate stripe below running into the left nostril.

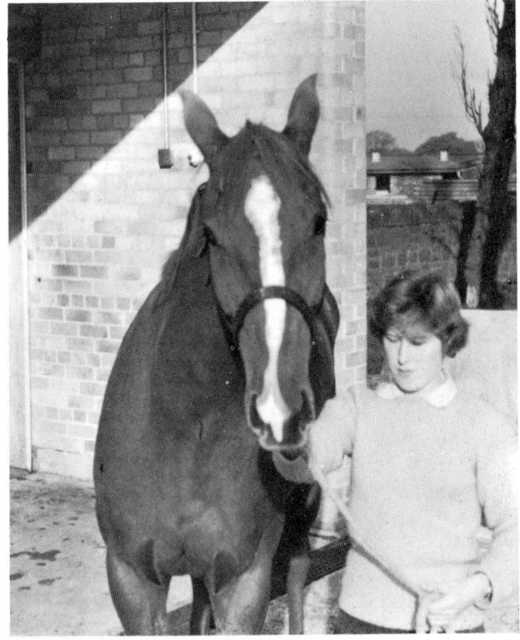

Fig. 2.5. Star and stripe conjoined bordered in the upper and lower sections.

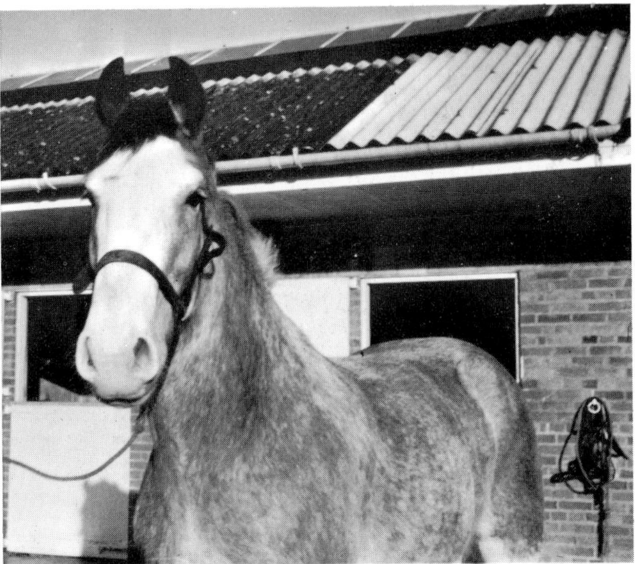

Fig. 2.6. Blaze on a roan-coloured horse.

Fig. 2.7. Wall eye.

Stripe. Narrow white marking down the face not wider than the flat anterior surface of the nasal bones.

Blaze. A white marking covering almost the whole of the forehead and extending beyond the width of the nasal bones and usually to the muzzle.

White face. The white covers the forehead and front of the face, extending laterally towards the mouth.

Snip. An isolated white marking in the region of the nostrils.

Wall-eye. A lack of pigment, either partial or complete, giving a pinkish or bluish appearance to the eye.

Body

Spot. Small collection of hairs differing from the general body colour.

Patch. Large well-defined irregular area differing from the general body colour.

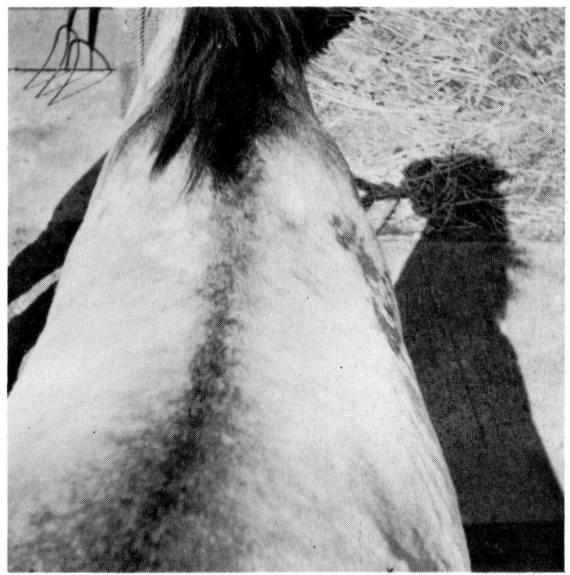

Fig. 2.8. Dorsal band on the back of a dun-coloured pony.

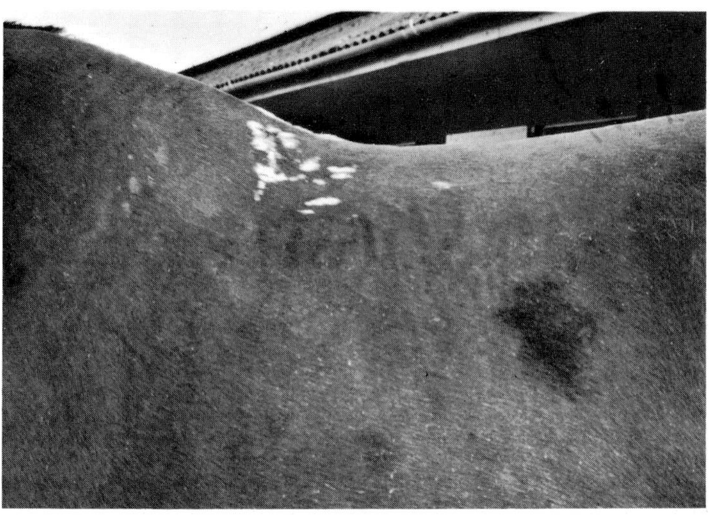

Fig. 2.9. Acquired saddle marks on a horse's back.

Dorsal band. Dark stripe down the centre of the back, often seen in duns.

Zebra marks. Striping on the limbs, neck, withers, or quarters.

Limbs

White markings. Should be defined according to their position anatomically.

Hoofs. Any variation in the colour of the hoofs should be noted.

General

Mixed. A white marking which contains varying amounts of hairs of the general body colour.

Bordered. White markings circumscribed by a mixed border.

Flesh marks. Patches where the pigment of the skin is absent.

Acquired marks. Permanent marks not present at birth but acquired during life, such as saddle marks, branding marks or surgical scars.

Whorls. Consist of changes in the hair pattern. The anatomical position of any on the head, neck, and body should be described.

Principal breeds

Heavy draught

Shire

The largest and heaviest of the British breeds, with stallions up to 17 hands in height and 1,016 kg (1 ton) in weight. Slow in pace but powerful pullers. Long hairs are found round the sides and at the back of the legs, these being said to form the 'feather'. All colours except chestnut and cream are found.

Clydesdale

A more active breed with lighter bodies and longer legs. 'Feather' is present but is not as marked as in Shires. Stallions stand at about 16 hands 2 in. Most are browns or bays and often have prominent white markings on the head and legs.

Suffolk

A heavy bodied breed with short legs without 'feather'. Slow but docile with a willingness to work. The colour must be some shade of

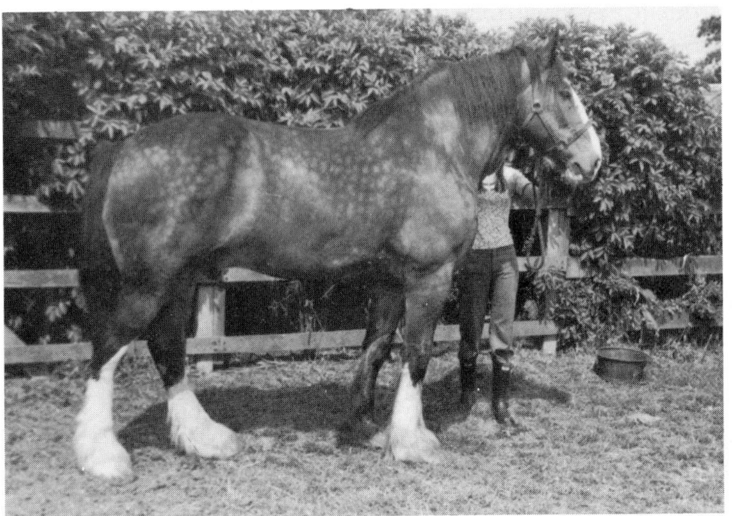

Fig. 2.10. Shire horse.

chestnut and there are usually few white markings. The stallions are about 16 hands 3 in high.

Percheron
A French breed, rather like a Suffolk in conformation but lighter and more active. The stallions are about 16 hands 3 in high and grey and black are the accepted colours.

Light draught

Cleveland Bay
A carriage horse, bay in colour with few white markings, standing about 16 hands high.

Yorkshire coach horse
Taller than the Cleveland Bay, reaching 16 hands 2 in high, with a freer action. Bay and brown are the colours which are found.

Hackney
Used for show because of their exaggerated leg action. At a trot the knee is raised high and vigorously snapped in the air and the hocks

are well flexed. The height varies from 15 hands to 16 hands 2 in. They can be any colour but bays, browns, and chestnuts predominate.

Hackney pony
Similar to above but under 14 hands.

Trotters
Harnessed to light carts and used for racing. Trotters must not break into a canter. Most are about 15 hands 2 in high and tend to have short legs and long bodies. Have great powers of endurance. Colour is not important.

Pacers
Are used for racing like trotters but have a lateral leg movement known as pacing as contrasted with the diagonal leg movement of trotting.

Riding

Thoroughbred
This term is reserved for racehorses which are registered with the firm of Weatherby. In order to be registered both their parents must have been similarly registered. There tend to be two types, the first bred for flat racing over relatively short distances at 2, 3, and 4 years of age principally, and the second raced as older horses over longer distances including the jumping of hurdles or fences.

Thoroughbreds have been selected for generations for their ability to gallop. The average height is about 16 hands and chestnut, bay, and brown are the most common colours with some greys and blacks.

Arab
Bred for many centuries by the Arabs and selected for intelligence, courage, endurance and sure-footedness. They are nearly all under 15 hands in height. Most are grey, bay, brown, or chestnut in colour.

Hunter
Hunters are a type of horse used for riding after hounds and are not a pure breed. Compact, fairly short legged, well-balanced horses somewhere about the height range from 15 hands 3 in to 16 hands

2 in which are able to trot as well as jump and gallop, are required. They must have a strong constitution to be able to withstand the fatigue of prolonged and severe exertion. Most have a considerable proportion of thoroughbred blood balanced by crosses with hackneys, ponies, Arabs, or possibly cart-horses such as Suffolks or Percherons to give a quieter temperament and, in the case of the last two, greater body size.

Polo pony

Similarly a type and not a breed. Speed combined with docility, hardiness, good balance when turning, weight-carrying power and courage are required. Good depth of body with powerful loins and quarters are conformation points which assist selection.

Hack

Again, a type and not a breed. Hacks are a refined type of riding horse, generally with a large proportion of thoroughbred blood. They do not usually exceed 15 hands 3 in in height and are chosen for good body conformation and true and level movements.

Hunters, polo ponies and hacks can be of any colour.

Ponies

Welsh

(a) Section A of the Stud Book, up to 12 hands, often known as Welsh Mountain ponies. Usually live on mountains or moorlands.

(b) Section B of the Stud Book, up to 13 hands 2 in. Very popular for show and as riding ponies.

Ponies in both classes have small attractive heads and a gay tail carriage and are free and fast movers with pluck, endurance, and dash.

(c) Section C of the Stud Book. Welsh pony (cob type). A more stoutly built type than the above.

(d) Section D of the Stud Book. Welsh cob, usually between 13 hands 2 in and 15 hands. These are stockier in type with strong deep bodies and very powerful hind quarters and are capable of carrying heavier weights. Most have some silky feather on the legs.

In all four sections animals are found in all colours except piebald or skewbald.

New Forest

Type A, 12 hands 2 in to 13 hands 2 in and type B from 13 hands 2 in to 14 hands 2 in. Can be any colour except piebald or skewbald. Weight-carrying ponies with a free action which make excellent riding ponies.

Dartmoor

Most are brown or dark bay in colour although other colours, except skewbald and piebald, are accepted. Up to 12 hands 2 in in height and hardy and enduring.

Exmoor

Similar to the above but many of the bays and browns have mealy-coloured noses and distinct and wide mealy rims running round the eyes.

Highland

These ponies come from the highlands of Scotland and are divided into three types which vary in height from 12 hands 2 in to 14 hands 2 in. They have been used for a long time to carry deer carcases and are strong, docile, sure-footed ponies which make good riding animals. Black, brown, dun, and grey are the main colours.

Connemara

This breed originated in Ireland and most of the ponies are within the height range of 13 hands to 14 hands 2 in. Their docility and hardiness make them excellent riding ponies. Grey is the commonest colour, but blacks, browns, and bays are also found.

Fell

At one time these hardy ponies were used as pack animals but are now valued for riding and driving. In height they should not exceed 14 hands and black, brown, bay, and grey are the common colours. Fine hair is present at the heels.

Dales

Strong, miniature cart horses of an active kind, originally bred to carry heavy lead ore from the Yorkshire mines. Up to 14 hands 2 in in height, most are brown or bay, but other colours except chestnut, skewbald, or piebald are seen. Feather is grown on the fetlocks.

Shetland
The smallest of the British breeds, registered stock must not exceed 1·07 m (42 in) at 4 years of age and over. Blacks and dark browns with few white marks are the commonest colours, but skewbalds are becoming popular and chestnuts and greys are found. They are intelligent, hardy and enduring, but can be self willed.

Identification

Horses are generally identified individually by a detailed description of their colours and markings. Coat colour at birth is not always stable and so a horse should be examined for colour when 9 months old or over, by which time the foal coat has been shed and the permanent coat has become established. However there is a compulsory registration scheme for thoroughbred horses and the thoroughbred authorities now call for a foal to be identified by colour and markings shortly before it attains 4 months of age, or earlier if it is to be exported. This is difficult, for not only is there uncertainty about the final body colour, but white markings are not

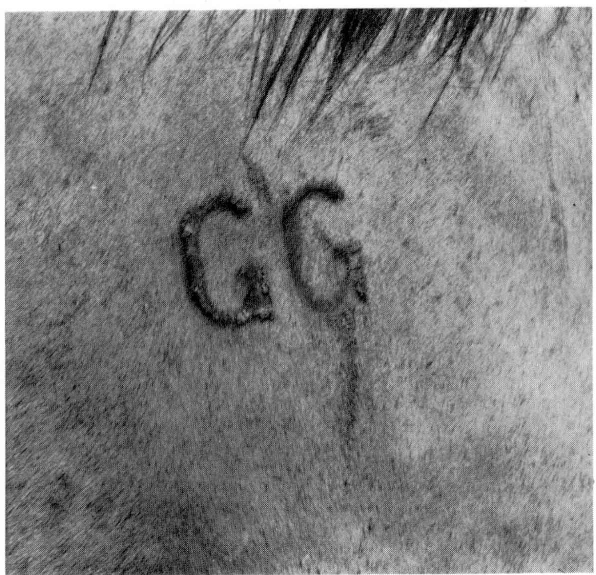

Fig. 2.11. A brand mark on a pony's shoulder.

always obvious. The Stud Book Authority now requires the head and neck whorls to be recorded on a foal identification certificate as these changes in hair pattern are the easiest to distinguish early in life.

Any marks acquired by a horse during its life which are permanent, and congenital abnormalities, such as incisor teeth which do not meet correctly, can also help in the identification of some animals.

In certain countries horses are branded with numbers on the neck or have numbers tattoed on the inner surface of their upper lips to identify individuals. Close up photographs of the chestnuts on the fore and hind legs can be used as there are about eight different basic type patterns. A difficulty is that the shape is not set until an animal is 18 months old and so the method cannot be used for foals. A few horses have no chestnuts, or only vestigial ones, on their hind legs.

Ponies running on moors can be branded on the withers or quarters with their owner's mark, and some native ponies have clip marks made in their ears for the same purpose. These methods denote ownership but do not identify individual animals.

Ageing

Ages are estimated, almost entirely, by reference to the incisor teeth. The horse has six incisor teeth in each jaw, set in the front of the mouth. The pairs nearest the centre of the mouth are known as the centrals, those on either side of the centrals as the laterals and the outside teeth as the corners. Temporary incisor teeth possess a definite neck and are rounded where they meet the gum so that a small triangle of gum is seen between adjacent teeth. Permanent incisor teeth are more square-cut and little or no gum is seen between adjacent teeth. When both temporary and permanent teeth are present in the same mouth the permanent teeth can be seen to be larger. When the dentition is completed there are also twenty-four

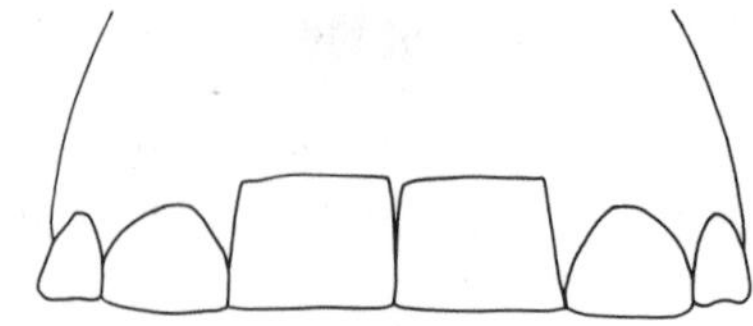

Fig. 2.12. Incisor teeth of a 3-year-old horse showing two central permanent and four temporary teeth.

Table 2.1. Incisor tooth eruption.

Age	Teeth
Birth to 1 week	2 temporary centrals
2–4 weeks	2 temporary laterals
7–9 months	2 temporary corners
2 years 6 months	2 permanent centrals
3 years 6 months	2 permanent laterals
4 years 6 months	2 permanent corners

molar teeth, six on each side of both jaws. The first three in each of the rows are preceded by temporary teeth and, therefore, are properly known as pre-molars. In male animals there are, in addition, four canine teeth, commonly known as tusks or tushes, one on each side of the upper and lower jaws between the corner incisors and the molar teeth. Some mares may have small or rudimentary canine teeth.

The teeth do not come into wear until about 3 months after eruption. The canine teeth in male animals erupt at about 4 years of age.

The following anatomical characteristics of the permanent incisor teeth are taken into consideration when ageing.

The infundibulum or mark. A cavity on the upper surface formed by an infolding of the outer enamel layer. This grows shallower and eventually disappears with wear.

Pulp cavity or dental star. A brownish linear streak in front of the infundibulum which appears with wear.

Galvayne's groove. A groove running down the labial surface of the upper corner incisor.

Appearance of incisor teeth from 5 years onwards

5 years. The corner incisors have met their opposites on the other jaw along the anterior edges but the posterior corners are rounded and unworn.

6 years. The corner teeth are in wear. In the central incisors the infundibulum is almost worn out, but the rings of enamel are distinct.

7 years. The infundibulum in each corner tooth is shallow. The outlines of the central teeth are becoming triangular. A '7-year-old

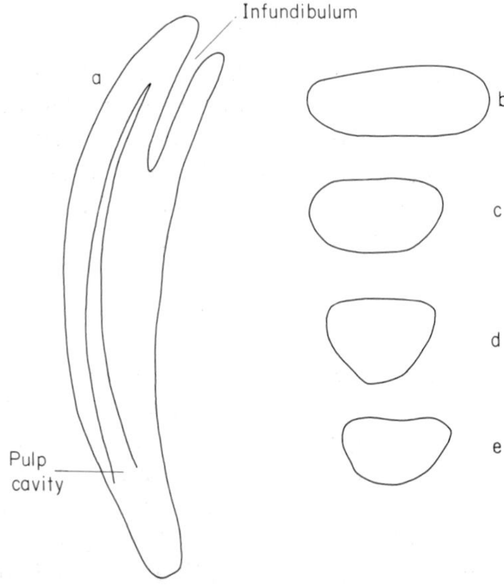

Fig. 2.13. (a) Longitudinal section of a horse's tooth showing the infundibulum and pulp cavity. Cross sections at (b) 5 years, (c) 9 years, (d) 11 years, and (e) 20 years of age.

hook' often develops on the posterior edges of the upper corner teeth.

8 years. The infundibulae in the centrals and laterals are nearly obliterated but rings of enamel can be seen. The pulp cavities appear in the centrals. The 7-year-old hook has worn out.

Beyond 8 years the teeth vary to such an extent that only a rough estimate is possible. The appearance of Galvayne's groove, changes in the shape of the cross-sections of the teeth, and the angle at which the teeth meet are practically the only guides.

Galvayne's groove. Appears near the gum at 10 years, is half way down at 15, extends the whole length at 20, has grown half way out at 25 and has disappeared at 30.

Cross-section of teeth. Changes from an oval (5 years) to a triangle (11 years) and eventually to a quadrilateral (20 years). These ages apply to the shapes of all incisor teeth in the mouth.

Angle at which the teeth meet. A profile view shows the upper and lower incisors meeting at about 180° until about 7 years, but this angle becomes less until at 20 years it is about 90° only.

Housing

It is probably true to say that insufficient attention has been paid to the housing of horses and that, on many establishments, elaborate buildings have been erected mainly to impress visitors.

Heavy draught horses can be kept loose in yards, the advantage being that they obtain a certain amount of exercise which helps to maintain body fitness. A few stalls are needed into which the animals can be taken for harnessing.

Stalls

Heavy draught horses are usually, and hunters and hacks occasionally, kept in stalls. The floor space allocation is usually 4 × 2 m (13 ft × 6 ft 6 in) for heavy draught horses and 3 × 1·8 m (10 ft × 6 ft) for hunters and hacks. The stall divisions are 2·14 m (7 ft) high at the front sloping to 1·37 m (4 ft 6 in) at the back, the length varying according to the stall sizes given above. A manger and a hay rack are fitted at the front of each stall with a small hole bored in the metal between them through which the rope from the headcollar is passed. This ends in a block of wood weighing about 1·81 kg (4 lb) and the length of the rope is such that when the horse is standing at the manger the wood block is just clear of the floor, keeping the rope taut. When the horse lies down the block is raised, but its weight still keeps the rope tight. This avoidance of any slackness in the rope minimises the risk of a horse injuring its leg on the rope.

The passage behind a single row of stalls should be 2·4 m (8 ft) wide and behind a double row 4·2 m (14 ft). The passage must be wide enough to back a horse right out of a stall and swing its head clear of the heel post.

Loose boxes

Loose boxes are now being used to a greater extent to house light horses of all types. They allow increased freedom and avoid the dangers inherent in leaving horses tied up for long periods. Most loose boxes are designed so that each one opens on to a yard but some are situated within a building, each opening on to an internal passage which gives access to an outside yard. The size varies according to the type of horse being accommodated but most are about 3·7 × 3·7 m (12 ft × 12 ft). A manger and a hay rack are provided plus either an automatic water bowl or a bucket ring in

Fig. 2.14. Loose boxes showing doors in two halves, the upper section being open and the lower closed. There are grids which can be closed across the top opening to prevent the horse from putting his head out.

which a water bucket can be placed. Tying rings should be fixed at the side of the manger and in the back wall.

At studs where thoroughbred horses are bred a series of foaling boxes should be provided adjacent to an attendant's room with small viewing windows so that each mare can be watched at intervals during both day and night without the disturbance which would be caused by the entry of a groom. These boxes are usually 4·9 × 4·9 m (16 ft × 16 ft) and should be provided with some form of artificial heating. The floors are frequently made of grooved rubber to provide a surface which is softer than concrete. After foaling the mares and foals are moved to boxes of a similar size, some of which are fitted with rollers on the sides of the doorway, so that a foal passing through at the same time as its dam does not become jammed against a solid wall.

Stallions can, with advantage, be housed individually in loose boxes with paddocks attached into which they can run at will as this opportunity for regular exercise helps to maintain physical fitness. The paddock fences must be strongly constructed, free from projections which might cause injury, and at least 1·83 m (6 ft) high.

Stables of all types need to be well ventilated, but the horses should be protected from draughts. Good lighting is desirable, but windows should not be fitted just over the heads of tied horses to avoid bright sunlight shining on their eyes. If fitted in loose boxes windows should be out of the reach of the horses. The floors of stables should be durable, to stand up to the wear caused by iron shod hoofs, and must not be slippery. Grooving the floor has been found to be an effective method of enabling horses to obtain a grip. The ceiling height should be between 3·3 and 4·6 m (11 and 15 ft). The former figure is high enough to prevent a horse hitting its head and the latter figure sufficiently low to avoid down-draughts. Stable doors must be 2·44 m (8 ft) high and at least 1·22 m (4 ft) wide to minimise the risk of a horse rearing up and striking its head against the lintel or knocking its flanks against the door posts when being led through the aperture. Frequent use is made of the half door principle, in which the lower half can be kept shut to confine the horse while the upper half can be left open during the day to improve ventilation.

Mangers are generally about 0·76 m (30 in) long, 0·36 m (14 in) wide and 0·31 m (12 in) deep and are fitted at about 1·07 m (3 ft 6 in) from the ground. They should be made of metal with broad, round edges to reduce the chances of crib biting. Often there is a bar across the centre from front to back, or internal ridges at the sides to prevent the horse scooping the food out. The bottom of the hay rack should be level with the top of the manger. In some old stables the hay racks were placed higher but hay seeds became tangled in the horses' manes or entered their eyes, while attempts to fit the racks lower have not been very successful because unless a movable grid is placed on top of the hay the horses tend to pull the hay out and either waste it or eat it off the floor, which can lead to the consumption of bedding. An alternative method, growing in popularity, is to feed the hay in nets which can be filled and hung up. On some establishments horses, particularly those kept in stalls, are still led out from the stables at intervals and watered from troughs in a yard, but it is far better to provide individual water buckets, supported in bucket rings attached to the loose-box wall, or, possibly, automatic drinking bowls.

Bedding materials

When horses are housed in stalls or loose boxes they must be supplied with some form of bedding to enable them to lie down in comfort.

Bedding provides a warm resting place and reduces the risk of injury on a hard floor. By encouraging horses to rest the strain on their legs is eased, and by preventing them from coming in contact with their own urine they are easier to keep clean. Cereal straws are the best material as they are not dusty and either absorb urine or allow it to pass through and drain away. Wheat straw is the best as it is tough and so provides a firm bed which is not eaten. Oat straw is not as good as it is softer and more palatable and so may be consumed. Barley straw is also soft and has sharp stiff awns which may scratch and irritate the skin. Wood shavings are considerably cheaper but tend to be dusty and are not as comfortable. Peat moss is a good absorbent for urine but is dusty and because of its brownish colour looks dirty. Both wood shavings and peat moss tend to block drainage systems and have a lower value when the soiled material is used as a manure. Shredded paper is now being used. It is light to handle, makes a clean comfortable bed and rots down into compost fairly rapidly. Some owners use a deep litter system by placing straw over peat and sawdust and only cleaning the building out once a year. Fresh straw is placed over any soiled material when necessary. The advantages claimed are warmth and economy.

Harness room

A harness room should always be built and harness should not be hung on the back wall of a passage. A food store is also required, and it is best to have movable galvanised iron storage bins as fixed concrete bins are difficult to clean. It is possible that a chaff cutter and an oat crusher will be installed and a few stables have machinery to extract nails and other foreign bodies from the food.

Feeding

General

Although horses have been used for riding and driving in many countries for centuries comparatively little research has been conducted into the scientific aspects of equine nutrition. Much of the information available has been deduced by analogy from other species and by trial and error. The special difficulties in calculating the nutrient requirements of horses scientifically are that horses differ more in temperament and in individual food requirements than do other classes of stock and it is very difficult to measure work done.

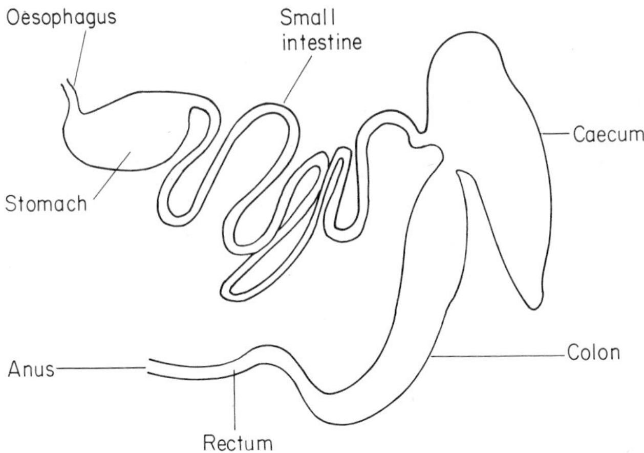

Fig. 2.15. Diagram of the digestive system of the horse.

The horse has a small, simple stomach and a large caecum and colon. The caecum and colon have a population of micro-organisms which are critical to the digestion of fibrous material. As horses have comparatively small stomachs they must be fed relatively frequently. Three feeds a day is the usual and minimal allowance and four feeds are probably better. Horses should be allowed about 1 hour for each meal and should always be fed at least $1\frac{1}{2}$ hours before doing fast or hard work. This is particularly important in the case of riding horses required to work at fast paces. The foods fed must be palatable as horses may refuse to eat foods with an offensive odour or taste, or, if very hungry, may consume a small amount only. Changes in the diet should always be made slowly because some of the digestion is bacterial and modifications of the intestinal flora are often needed to digest a new food. If changes are introduced suddenly colic or scouring may follow.

The horse can utilise starches, sugars and fats as energy sources. Most of the soluble carbohydrates are digested in the small intestine to produce glucose and other simple sugars and these are absorbed in the small intestine. Dietary proteins are digested in the stomach and small intestine and are subject to microbial attack in the large intestine, although the extent to which protein is broken down in the caecum and colon to release amino acids available for absorption is uncertain. Rations fed to horses must always include some roughage to aid digestion and reduce the risk of constipation and colic. However, horses cannot consume large amounts of bulky roughage because of their comparatively small stomachs.

To provide the protein necessary for growth or the replacement of worn-out tissues and the energy needed for work, concentrate foods are fed, oats being the cereal most generally used. Small amounts of beans or oil-seed cakes, such as linseed and soya bean, may be given to increase the protein content of a diet. Other concentrate foods which can be used to replace up to one-third, or even one-half, of the oats in a ration are bran, dried brewers' grains, weatings, barley and flaked maize. For the most efficient feed use grains need to be rolled or cracked to ensure that the digestible inner parts are exposed to the digestive juices. Molassine meal, a sweet tasting by-product of sugar refining, is sometimes used to supplement other concentrate foods because of its palatability. The favoured roughages are hay in the winter and grass in the summer, but oat or barley straw or silage can be used to supply part of the roughage requirements.

Proprietary compound nuts and cubes are also used extensively and are gaining in popularity. They are palatable, consistent in composition, relatively dust free and the labour needed for the mixing of a variety of foodstuffs is avoided. The formulae include cereals, oil-seed cakes, minerals and vitamins. Some also contain a high quality roughage which reduces the hay intake, while others are prepared to form a complete diet without the need to feed hay. Different formulations are prepared to suit the needs of equine animals of different types. Many firms prepare cubes suitable for (a) show jumpers, hunters and ponies, (b) racehorses, with a higher protein and energy content and a lower fibre percentage, and (c) stallions, mares, foals and yearlings on breeding stud farms.

Working horses which are receiving a ration containing a high proportion of concentrate foods must have the concentrate part of the ration reduced and the roughage element increased if, for any reason, they are left standing in a stable without exercise for a few days. If this is not done the horses become difficult to handle when worked and it is possible that diseases of the leg may develop.

Draught horses

Stabled heavy draught horses usually receive a simple diet such as oats 6·4 kg (14 lb) and hay 8·2 kg (18 lb) daily when working but in some stables animals doing heavy work may be given a more varied diet as shown in Table 2.2.

Table 2.2. Daily ration for a draught horse.

Food	In full work		On light work	
Oats	4·0 kg	9 lb	2·3 kg	5lb
Beans or linseed cake	0·9 kg	2 lb	—	—
Flaked maize	1·4 kg	3 lb	—	—
Bran	0·9 kg	2 lb	0·9 kg	2lb
Mineral mixture	28·4–56·8 g	1–2 oz	28·4–56·8 g	1–2 oz
Hay	8·2 kg	18 lb	9·1 kg	20 lb

During the summer months farm horses can be turned out to grass when not actually working. The grass will provide sufficient nutrients for light or moderate work but needs to be supplemented by from 2·3 kg to 3·2 kg (5–7 lb) of oats a day if the animals are being worked hard.

As straw, silage and roots are available on farms they may be used for cart-horse feeding. Oat and barley straw may be substituted for from one-third to one-half of the roughage ration fed to horses which are only doing light or occasional work. Some of the straw can, with advantage, be fed chopped and mixed with the concentrate ration. Silage can be used in similar proportions but must be introduced into the ration gradually. The rate of substitution is 1·35 kg (3 lb) of wet silage for 0·45 kg (1 lb) of hay. Roots can be used to replace some of the concentrate food about 2·7 kg (6 lb) a day being a reasonable level, although this amount is often exceeded. Turnips can be fed whole but carrots should be sliced lengthwise to prevent choking. Dried sugar beet pulp must be soaked before feeding otherwise it will swell, as it absorbs moisture while passing down the oesophagus, and cause a blockage in the throat.

Riding horses

Riding horses are required to work at faster paces than draught horses and their intestines must not be allowed to become too distended with roughage when working. During the hunting and polo seasons medium-sized hunters and polo ponies receive rations along

Table 2.3. Daily ration for a hunter.

Food	In full work		In light work		Out of work	
Oats	5·4–6·4 kg	12–14 lb	3·2 kg	7 lb	1·4 kg	3 lb
Beans or linseed cake	0·5 kg	1 lb	—	—	—	—
Bran	—	—	0·9 kg	2 lb	1·4 kg	3 lb
Mineral mixture	28·4–56·8 g	1–2 oz	28·4–56·8 g	1–2 oz	28·4–56·8 g	1–2 oz
Hay	4·5–5·4 kg	10–12 lb	6·4–7·3 kg	14–16 lb	7·3–8·2 kg	16–18 lb

the lines shown in Table 2.3 according to the amount of work being performed.

The feeding of racehorses in training is an art which requires a considerable amount of skill, and great attention must be paid to the idiosyncrasies of individual animals. The general aim is to supply the required nutrients in increasing quantities as the date of the race approaches and this is attained by including less and less of the bulky foods. Care must be taken to see that the alimentary tract continues to function properly, as demonstrated by the evacuation of faeces of normal consistency. It is essential that only foods of the highest quality are fed.

Table 2.4. Daily ration for a racehorse.

Best oats	6·4–7·3–8·2 and even up to 9·5 kg	14–16–18 and even up to 21 lb
Beans or linseed cake	0·9–1·4 kg	2–3 lb
Dried grass meal	0·9–1·4 kg	2–3 lb
Dried separated milk powder	0·5–0·9 kg	1–2 lb
Mineral mixture	56·8–85·2 g	2–3 oz
Vitamin concentrate	28·4 g	1 oz
Sliced carrots	3·2 kg	7 lb
Hay	3·6–2·7–1·8 kg	8–6–4 lb

Trainers vary in the foods they use, but the ration given in Table 2.4 is a fairly typical one for racehorses while in training. As the quantities of oats are increased, as shown in the first line, the amounts of hay are reduced as indicated in the last line.

Two and three-year-old horses, in particular, cannot stand such a concentrated ration for long periods at a time and they receive up to 7·3 kg (16 lb), or exceptionally 8·2 kg (18 lb), of oats for short periods before a race only. It is sometimes difficult to induce horses to eat these large quantities and sliced eating apples, carrots, molassine meal or glucose may be mixed with the food to increase palatability and encourage maximum consumption.

On the day of a race a small feed of corn is allowed in the morning but no hay is fed and the horse is muzzled to prevent it eating the straw bedding. A mash and long hay are given after the race.

Ponies

It is very difficult to suggest rations for children's riding ponies as they vary so markedly in size and temperament. The majority are of the native breeds which have, for generations, had to subsist on poor grazing and these, particularly, must not be fed large amounts of concentrate food. If they are, they are likely to become unmanageable or develop digestive disturbances or an inflammation of the hoofs.

A rough guide is to allow about 0·9 kg (2 lb) of medium quality meadow hay and approximately 0·23 kg (0·5 lb) of oats daily for each 45·4 kg (100 lb) liveweight, increasing the concentrate allowance during periods of hard work. It is difficult to give guides to pony weights but a 13-hand pony should weigh about 225 kg (500 lb) and a 14-hand pony about 315 kg (700 lb). Some well-bred ponies standing at about 14 hands 2 in are given approximately three-quarters as much as the hunter allowance given in Table 2.3.

Because of the high cost of hay and cereals pony owners tend to use substitute foods such as those described for draught horses kept on farms.

Rest period

All hard-working light horses, which have been kept in a fit bodily condition, benefit from a period of rest. Hunters and racehorses used for National Hunt racing are turned out to grass during the summer months. During this time the body can store up minerals and vitamins which can be used during the following period of work. Polo ponies and flat racing horses are rested during the winter and so are given roughages, supplemented with only small amounts of concentrate food, during this time. After a period of rest horses must be brought back into working condition gradually.

Brood mares

Brood mares at the time of service should be in good bodily condition, but obesity should be avoided. If mares are over-fed and under-exercised their ovaries can become embedded in fat which interferes with normal ovarian functioning and can reduce fertility. When horses are running out at grass their bodily condition varies with the quantity and quality of the grazing and so they lose weight during the winter and gain condition during the spring as the natural breeding season approaches. Mares, particularly thoroughbreds, which have to be housed and fed artificial foods during the winter and are served early in the year can, with advantage, be managed on a similar system and be 'let down' in condition during the winter and 'flushed', or raised in condition, as the time for service draws near. Care must be taken to supervise the individual animals, as some will stay fat on a ration which will make others become too thin. The condition of mares under 5 years of age, which are still growing, must be observed with particular care.

The nutritional requirements of pregnant mares differ markedly between the first two-thirds and the last third of pregnancy, due mainly to the rate of fetal growth. During the first period little more than normal maintenance is required as the fetus develops slowly and makes little demand on the dam. Fetal development is rapid during the last third and so additional nutrients have to be supplied. Good quality grass is an excellent food as it contains protein, vitamins, and minerals, but, as part, or most, of this period occurs during the winter months concentrate feeding is necessary. The quantity of food varies but an example is hay *ad lib* and between 0·34 kg and 0·57 kg (0·75 and 1·25 lb) per day, for each 45·4 kg (100 lb) liveweight, of a concentrate ration similar to that given below. When a mare is close to foaling she will probably eat less food due to pressure on the alimentary canal from the developing fetus. Should this be observed the quality of the ration should be increased to provide the required nutrients in a smaller bulk. Shortly before foaling the grain allowance should be decreased and bran mashes or other laxative foods given to prevent the mare becoming constipated.

After foaling mares are generally turned out to grass and this

Table 2.5. Daily ration for a brood mare.

	During pregnancy		While suckling	
Oats	0·9 kg	2 lb	1·4–2·3 kg	3–5 lb
Beans or linseed cake	—	—	0·2–0·5 kg	½–1 lb
Bran	0·5 kg	1 lb	0·9–1·8 kg	2–4 lb
Grass meal	1·4 kg	3 lb	—	—
Mineral mixture	28·4–56·8 g	1–2 oz	28·4–56·8 g	1–2 oz
Vitamin concentrate	28·4–56·8 g	1–2 oz	28·4–56·8 g	1–2 oz
Hay	Quantity varies according to body size and amount of grass available			

food is excellent for milk production. It normally provides sufficient nutrients without any addition during the second half of April, the whole of June and the first half of July. Before or after this period the mare usually receives a concentrate ration. The quantity varies according to the quality of the grass but is generally between 0·45 and 0·68 kg (1 and 1·5 lb) for each 45·4 kg (100 lb) liveweight. Suitable mixtures for brood mares are given in Table 2.5.

Stallions

Stallions must be well fed as nutrition has a direct bearing on the quantity and quality of the semen produced. A stallion's nutritional requirements during the breeding season approximate those of a lactating mare and the ration should be fortified with protein. The protein can be provided by peas or beans, or possibly by decorticated ground-nut cake. Occasionally 0·23 kg (0·5 lb) of fish meal or 0·45 kg (1 lb) of dried skim milk are added to the ration each day. Grazing is beneficial, and has the added advantage that it encourages exercise.

Foals

It is essential that the suckling mare receives a ration which the foal can digest as it will commence to nibble at the mare's food at about 2 weeks of age. At the age of 2 months it will be eating about 0·9 kg (2 lb) of the mixture and at 5 months about 3·6 kg (8 lb) daily. As nursing mares tend to eat all the concentrate food they can reach adequate trough space must be provided so that the foal can eat without being driven away by the mare. As foals tend to stay near their mothers the provision of a separate area with a special foal ration is not very satisfactory as the foals do not readily go there without the mare.

Orphan foals

The rearing of an orphan foal is always a problem and fostering is extensively employed, mares whose foals have died being used to suckle orphans. Help in finding foster mothers can be obtained by applying to the National Foaling Bank, Meretown Stud, Newport, Salop. Occasionally, a heavy milking cart-horse mare will be found which is capable of rearing two foals. If a suitable foster mother cannot be found a foal can be reared artificially. It is very important that it should have a companion; a quiet pony is best, but a calf or sheep will do. Some commercial calf foods are suitable for feeding to orphan foals and at least one powder is manufactured specially for foals. In an emergency cows' milk can be used as the basis of the diet but its composition must be modified because mares' milk is higher in sugar and lower in fat content than cows' milk. A suitable mixture can be prepared by dissolving 14·2 g (0·5 oz) of sugar (preferably lactose or glucose) in a little warm water, adding 42·6 or 56·8 ml (1·5 or 2 fl oz) of lime water to help break up the curd which would otherwise form in the foal's stomach, and then making up the total volume to 0·57 l (1 pt) with fresh cows' milk, preferably of a low fat content.

The mare's milk substitute is best fed from a bottle with a large teat and it must be given at blood heat. At first the foal should be fed 0·14 l (0·25 pt) each hour of the day and night, but after a few days the intervals between feeds can be lengthened and the quantity increased to appetite so that at 2 weeks it receives about six feeds a day, and later only four. Alternatively it is possible to use an automatic calf rearing machine. At 4 weeks the foal can be put out to nibble grass. At 6 weeks skim milk can be substituted for the mare's milk replacer, and at about 2 months the foal will be eating solid food such as bran and crushed oats. By this time the foal should drink from a bucket, and it can be weaned off milk at 3 months of age.

Young horses

A very important stage in the growth of a young horse is the period of about 12 months' duration from weaning to 18 months of age. The diet should contain adequate quantities of protein for muscle development. Excessive amounts of fattening foods, such as maize or barley, should not be given as excess fat in the body imposes heavy burdens on the developing muscles, tendons, bones and joints and predisposes to injuries of these tissues.

When weaned in the autumn foals may be given the ration suggested previously for pregnant mares. In the case of thoroughbred animals, when rapid growth is required, the addition of 0·45 kg (1 lb) of dried separated milk powder greatly improves the quality of the ration. Good quality hay should be provided *ad lib.* After the first winter the yearlings will run at grass and good quality pasture is the best food for them. Except in the case of thoroughbreds, they will probably be allowed nothing except grass, but thoroughbreds will continue to be fed concentrate foods in a mixture similar to that recommended for weaned foals until they go into training. During the second winter hunter and cart-horse type yearlings will receive hay to supplement the grass and about 0·9 kg (2 lb) of crushed oats and 0·5 kg (1 lb) of bran per head per day. This style of feeding will be continued until the animals are broken for work when rations, on the lines previously advised for working horses, varied according to the amount of work performed, will be given.

Mineral requirements

Horses presumably require dietary sources of the same mineral elements as other animals, although the calcium, phosphorus, magnesium and sodium chloride requirements are the only aspects studied in detail. The dietary quantities depend upon the liveweight of the animal, the amount of work done and the loss in sweat of each mineral. Ordinary foods of good quality generally contain the minerals required by adult animals in sufficient quantities, but salt may need to be supplied as a supplement. In some stables salt licks are provided as rectangular blocks fitted into holders attached to the wall, while in others a lump of rock salt is placed in the manger. Both methods enable a horse to take the quantity it requires.

The diet of a brood mare must contain adequate amounts of calcium and phosphorus to provide for the bone development of the fetus, and these minerals must be supplied to young growing horses for the same reason.

Vitamin requirements

Insufficient experimental work has been conducted on the vitamin requirements of horses to allow precise estimates to be given. The average horse ration usually meets the vitamin needs of mature horses, but supplementation for growing and breeding animals is desirable. Pregnant mares must be provided with vitamin A at a level of about 24,000 I.U. per day, as a deficiency before mating may lower fertility and during pregnancy can result in the birth of a dead or weak foal. This vitamin is best supplied by adding a synthetic preparation to the concentrate ration. On some stud farms wheat germ is included in the ration of mares before service to provide additional vitamin E, as infertility due to a deficiency of this vitamin has been recorded. The level of vitamin D in the diet of young horses needs special consideration to ensure good bone development. It has been estimated that 300 I.U. per 45 kg (100 lb) of body weight are required.

Water requirements

Horses will drink from 36–45 l (8–10 gal) or more during a normal day, but their requirements are increased by dry food, sweating and high environmental temperatures. Lactating mares obviously need additional amounts for milk production. If water is not constantly available horses should always be watered before feeding, or not less than 1½ hours afterwards, to avoid partially digested food particles being washed from the stomach into the intestines as this can cause violent attacks of colic. Excessive drinking after performing heavy work may be followed by signs of indigestion and so it is usual for tired, thirsty horses to be given only a small quantity of fluid at first, being allowed to drink their fill later.

Breeding

Fillies attain puberty at about 18 months of age but are not normally mated until they are 3 years old, although ponies may be served at 2 years. The natural breeding season is during spring and summer, that is during the period of increasing hours of daylight. However, thoroughbred mares tend to be mated as soon after 15 February as possible so that the foals are born early in the year. The reason is that, as the official birthday of all thoroughbreds is 1 January, early foals meet late foals on equal terms in races, although there may in actual fact be some months difference in their ages.

When in oestrus a mare stands with its hind legs slightly apart, raises its tail and shows some of the mucous membrane lining the vagina. It may eject small squirts of urine. It stands still when a stallion mounts. The average length of an oestral period is 5 days, with a normal range from 3 to 7 days, and the interval between the end of one period and the beginning of the next is 16 days. After foaling a mare usually comes into oestrus at between the fourth and seventh days.

Colts reach puberty at 1 year of age but are not normally used at stud until 4 years old. At this age the number of services per season is restricted, but from 5 years of age onwards about 120 services per annum are allowed. As, on average, thoroughbred mares require three services per conception this means that about forty mares are allowed per stallion each season. In ponies and cart-horses, where the conception rate is higher, more mares can be allocated to each stallion. Artificial insemination can be used in horse breeding, but is rarely practised in Great Britain.

Before a mare is mated it is generally tried to check that it will stand for service. The mare is led to one side of a gate 1·52 m (5 ft) high and the stallion is brought up on the other side. In the case of thoroughbred horses it is customary to cover the gate with coconut matting on both sides to prevent injury if either horse kicks. If the mare is not in a receptive state it will lay its ears back and kick, but if it is ready for service it will stand quietly with its tail raised. When valuable stallions are kept at a stud a stallion of low value known as a 'teaser' is often used for 'trying' mares so as to avoid injury to the valuable horse by kicks from a mare not ready for service. Before service, padded boots are usually strapped to the hind hoofs of a mare to minimise the damage caused to the stallion if it should kick. For mating the stallion and mare are led into a covered yard or a field.

The duration of pregnancy is normally about 336 days, but mares due to foal early in the year usually have longer-than-average pregnancy lengths, while those due to foal late in the season generally give birth earlier than expected.

In order to produce as high a percentage of foals as possible it is necessary to carry out tests for pregnancy so that barren mares can be remated as soon as possible. Careful observation from 16 days after mating onwards will determine whether the mare comes into oestrus. However, this is not always a reliable guide for occasionally in-foal mares show signs of oestrus, while barren mares may not. A rectal examination can be carried out by a skilled veterinary surgeon from about the fortieth day after service and a fetus in the uterus can be felt through the wall of the rectum. Methods of testing which do not involve a rectal examination are a blood test which can be carried out between the forty-fifth and the hundredth day after mating, and a urine test which is reliable from the one hundred and thirtieth day of pregnancy onwards. Later on in pregnancy, after about the seventh month, it is possible to see or feel the movement of a living fetus through the abdominal wall of the mare.

The first sign of approaching parturition is a distended udder, which may be observed from 2 to 6 weeks before foaling. About a week before foaling a drooping of the abdomen will be observed. During the last few days of pregnancy a mare's teats show a waxy secretion and the vulva becomes full and loose. Just before foaling milk will drip from the teats and the mare will become restless and may start to sweat.

It is best to leave a mare alone when foaling, but a watch should be kept in case difficulties are experienced. The mare lies down before the birth and the outer fetal membrane ruptures, followed by the escape of a large volume of fluid. The uterine contractions then become stronger and the foal is usually born within 10–30 minutes of the appearance of the forelegs outside the vagina. The umbilical cord is normally ruptured either by the mare getting to her feet or by the foal's struggles in its attempts to stand. The fetal membranes are usually not expelled when the mare rises, and should be knotted to prevent the foal or mare treading on them and pulling on the wall of the uterus. They are normally expelled within 1–6 hours of the birth, and should not be retained for more than 12 hours. Most mares lick their foals soon after birth, but if a mare fails to do this the foal should be dried and thoroughly rubbed with a warm towel. A healthy foal will attempt to suckle within 3 hours of birth. If a mare will not allow its foal to suck it should be restrained by holding up a foreleg and the foal should be helped to suck. After a few feeds the mare usually accepts the foal.

Foals are generally weaned when from 4 to 6 months of age, by which time they should be eating concentrate food readily. The best method is to shut two or three foals together in a loose box and move their dams away to premises where they cannot be seen, heard or smelt by the foals. If only one foal is to be weaned it should be provided with a quiet pony, donkey, goat or sheep as a companion, and should not be left on its own. The mares and foals soon forget each other, and after being confined in stables for about a week can be turned out to grass, but the mares should obviously be kept apart from the foals. When a number of mares and foals are running together at grass it may be possible to remove one mare at a time until the foals are on their own, without having to confine the foals in a loose box.

Health

Healthy horses are interested in their surroundings and show this by turning the head and constantly moving their ears backwards and forwards. The hair of the coat should be glossy, although animals living out will have long hair which gives a ragged appearance. The eyes should be bright and clear and there should be no nasal discharges. The horse should be keen on its food and the faeces should be voided in the form of loose balls varying in colour from light brown to green. The presence of whole grains or large pieces of grain in the manure indicates that the horse is not chewing properly, either because it has defective teeth or is greedily bolting its grain ration. Sound horses may rest one hind leg with the toe on the ground but do not normally point a foreleg in this way.

The body temperature should be in the range 37·8–38·3°C (100·0–101·0°F). When inserting a thermometer into the rectum it may be necessary to restrain a horse by asking an assistant to hold a foreleg up in a manner which will prevent it from balancing itself and kicking with a hind leg, but the animal must not be unduly upset or its temperature will rise.

The normal pulse rate is between 36 and 42 beats per minute and is best felt in an artery which runs inside the jaw bone below the point where it hinges with the skull. Another suitable artery can be found high up on the inside of a foreleg. The number of pulsations over a period of at least 30 seconds' duration should be counted. Like the body temperature the pulse rate can be raised if a horse, particularly an excitable nervous individual, is upset by being handled.

Exercise

A state of bodily health which makes a horse capable of sustained work can only be attained by exercise. The transformation of fat, flabby flesh into hard, tough muscle is a gradual process which can only be effected by a regular course of graduated exercise. When a horse is brought in after a period of rest out at grass it should be given plenty of walking and a moderate amount of trotting exercise. A minimum period of 2 hours daily is desirable. Fast work should be avoided initially as it does not harden the muscles in the same way, but, as the horse becomes fitter, faster work can be started. Even when the horse is being cantered and galloped the walking and trotting should be continued.

Stabled horses require to be exercised each day in order to keep them fit for work. Racehorses are usually ridden at a walk and canter and hunters and ponies at a walk and trot for 2 or 3 hours each day, with short bursts of faster work at intervals. Thoroughbred stallions kept for stud are led at a walk and trotted and cantered round on a lungeing rein. A lungeing rein is from 6·1 to 9·15 m (20–30 ft) long and is attached to a headcollar. The horse is driven round the circumference of a circle, first one way and then the other, by a handler standing in the centre of the circle. Stallions of other breeds may be sufficiently quiet to be ridden at exercise.

Common diseases and their prevention

Horses sometimes become *injured* while working. Minor cuts generally heal fairly rapidly if kept clean. During warm weather a fly-repellent substance placed round an unprotected wound may prevent some of the irritation which can be caused by flies landing on a sore. Veterinary attention should be sought in the case of serious injuries. There is always a risk that *tetanus* or lockjaw may develop as the causal organism occurs in some soils and horse faeces and can enter the body through a wound and produce a toxin which affects the nervous system. A short-term immunity can be conferred immediately by the injection of tetanus antitoxin. In areas where the disease is prevalent it may be advisable to obtain a long term immunity by vaccinating horses with tetanus toxoid and giving booster doses at intervals subsequently.

Harness galls occur when badly fitting pieces of harness rub an area of skin and cause a painful lesion, saddle and girth galls being the sores most commonly seen. Prevention is by checking that all the items of harness fit properly.

If a horse *quids its food*, that is allows food to drop from its mouth when only half chewed, it is probably due to irregularities in the molar teeth. The grinding surfaces of these teeth are set at an angle, the ones in the upper jaw being longer on the outside and shorter on the inside and the lower teeth the opposite. During the grinding of food sharp points may be formed at the edges which can rub on the sensitive membranes of the mouth and tongue and cause injuries. A veterinary surgeon should be asked to examine the teeth and if necessary use a rasp to remove the sharp points.

A variety of *respiratory conditions* are encountered which are manifested by a high body temperature, loss of appetite, rapid breathing, a harsh cough, and discharges from the nostrils and eyes. They may be caused by either bacteria or viruses, and are spread by the inhalation or ingestion of nasal discharges. Preventive measures consist of avoiding contacts between healthy animals and either infected horses or contaminated foods or premises. Particular care must be taken in the case of young horses, as serious outbreaks of respiratory disease can occur when numbers of young animals are brought together.

Parasitic worms living in the intestines can kill horses but, generally, they reduce bodily condition which results in a lowered work performance. Animals under 4 years of age are most severely affected, and the most noticeable clinical signs are retarded growth, rough hair coats and digestive disturbances. There are a number of species of worms with differing life histories, but the faeces of infested horses which contain worm eggs or larvae are the primary source of spread. Prevention is by keeping the pastures clean, by treating infested animals before allowing them out to grass, by removing the faeces from the paddocks soon after they are voided and by not spreading fresh horse manure on the grazing fields. The provision of pastures temporarily sown down to grass for from 2 to 5 years and then ploughed and used for an arable crop is less likely to cause a build up of parasites than the grazing of permanent pastures. Various drugs can be used for treatment and these are best mixed with the food at intervals during the year.

Colic is a general term describing any abdominal pain, and the common cause is a digestive disturbance. As horses cannot vomit any food consumed must pass through the alimentary canal, and they appear to be particularly sensitive to digestive disturbances following a sudden change in the diet or the ingestion of unsuitable, mouldy or toxic foodstuffs. In acute cases the horse alternately stands and lies and will roll in an effort to relieve the pain. Violent rolling should be checked, for the horse could easily twist or rupture a part of the intestine. In subacute cases the animal becomes listless, kicks at the abdomen, paws the ground and may lie on its side. In both types of colic walking slowly may aid the release of gas from the intestine and so relieve abdominal distension. Veterinary advice should be sought early as death may be the sequel in untreated cases.

Ringworm is a contagious parasitic disease affecting the outer layers of the skin caused by certain microscopic fungi. Although it may appear among animals at pasture it is more common in stabled horses. The clinical signs are round, scaly areas, almost devoid of hair, which appear mainly in the vicinity of the eyes, ears, neck and tail and cause mild irritation. As the fungi are spread by contaminated stables, rugs, harness, and grooming equipment, prevention consists of disinfecting everything which has been in contact with infested animals. When handling infested animals it must be realised that the disease is communicable to man.

Handling

Horses are usually not difficult to handle but, because of their size, strength, and speed can be dangerous. They appreciate being treated gently, but are quick to sense, and take advantage of, nervousness in a handler. It is very desirable that horses should be well handled as foals as this accustoms them to contact with human beings and makes them easier to manage in later life.

Horses running loose in fields can be trained to come when called or when they hear the rattle of a food bucket. Some horses are very inquisitive and will approach a person who enters the field of their own accord. Horses which will not come should be driven into a corner of the field or into a shed or yard. It is a help if horses other than the one to be caught are removed from the field.

A horse should always be approached on the near (left) side and spoken to before it is handled. It should be touched on the head or neck first and sudden movements should be avoided. A fold of skin on the left side of the neck can be grasped with the left hand, as this quietens the animal while the handler carries out a manipulation with his right hand. If the horse does not know the handler, and does not have a headcollar on, either a headcollar or a halter should be

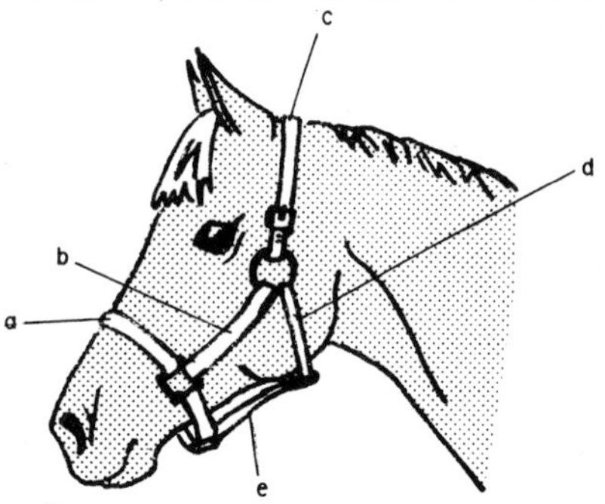

Fig. 2.16. Headcollar. (a) Noseband, (b) cheek piece, (c) poll piece, (d) throat lash, (e) jowl piece.

applied before the body is touched, as horses are not always quiet with strangers.

A halter is made of rope or webbing and consists of a poll piece, two cheek pieces, and a noseband made in one continuous length so that, particularly in the case of a rope halter, it is possible to adjust the length of each section to fit a horse or pony of any size. Headcollars are made of leather or nylon straps in various sizes to suit individual animals. A rope attached to a ring at the back of the noseband is used for leading or tying up the horse. To apply either a halter or a headcollar the poll piece should be held in the left hand with the loose end of the rope over the arm. The right hand can be placed on the horse's head and the left hand on its nose. Then the left hand can be moved up so that the nose band slips over the nose and the poll piece can be grasped by the right hand and pulled over the ears. Then the buckle on a headcollar can be fastened, and a halter can be adjusted for size and fastened with a hitch at the side to prevent it tightening or loosening. If a horse is to be tied up the knot should be one which can be undone easily. When dealing with unhandled, wild horses in a yard it is possible to insert a long stick under the poll piece of a halter prepared with a wide noseband and a long lead and manipulate it quietly over the head and into the required position. The lead can then be pulled so that the parts of the halter adjust.

When a horse is led out of a stable care must be taken to walk it straight to avoid injuring it against the side of the door. The person leading it should then place himself about 0·46 m (1 ft 6 in) from the left side of the animal's head, and hold the halter rope or bridle reins close to its mouth with his right hand. His left hand should hold the loose ends of the rope or reins so that he can still control the animal should it jerk its head free from his right-hand grasp. If the horse is bridled it is easiest to pass the reins over its head and hold them at the side. Should it be necessary to turn a horse it should always be turned to the right, away from the handler. If a horse is drawn towards the handler its hoofs may tread on the handler's feet. When turning a horse loose in a field the handler should bring its head towards the gate before removing the headcollar, and step back himself towards the gate. Horses usually kick when released and can injure a man standing at their side as they gallop to freedom.

To lift a forefoot the handler stands close to the leg, facing the hindquarters, runs his inside hand down the back of the leg, gently pushes in with his shoulder, and lifts the leg by grasping the pastern. The leg can then be gripped between the handler's legs if a prolonged manipulation such as shoeing is being undertaken. A hind leg can be raised by standing looking towards the rear of the horse, running the inside hand down the leg to the middle of the cannon and drawing the leg upwards and forwards. The front of the hoof can then be grasped and the hoof supported on the thigh just above the knee. On no account must a hind leg be held between the handler's legs.

The holding up of a foreleg by an assistant can be a simple method of restraint to prevent kicking when it is necessary for a handler to work near the hind quarters of a horse, e.g., in taking an animal's temperature. The foreleg on the side on which the handler is working should be the one which is raised. If a leg has to be held up for some time it is possible to apply a stirrup leather in the form of a figure of eight, with one loop round the top of the leg and the other round the fetlock joint. The loop round the fetlock should be applied first, and then the leg should be flexed and the other loop made fast. This method is only advised for reasonably quiet horses and the animal should be stood on soft ground or on a thick straw bed in case it falls, and a knee cap should be applied to protect this joint from injury.

Applying a twitch

It may be necessary to apply a twitch to an unruly horse in order to control it, but a twitch should not be used unnecessarily and should be applied for as short a time as possible. A twitch consists of a short pole about 0·91 m (3 ft) long with a hole about 1·27 cm ($\frac{1}{2}$ in) in diameter about 1·27 cm ($\frac{1}{2}$ in) from one end. Through this hole is passed a piece of soft cord about 0·46 m (18 in) long which is tied to form a loop. The horse is haltered and backed into a corner and the operator, standing by its right shoulder, grasps the upper lip and applies the loop of the twitch which he is holding in the fingers of his left hand. With his right hand he twists the pole so that the loop grips the lip tightly. By causing pressure on a sensitive area of the upper lip the twitch diverts the animal's attention from the manipulation to which it is objecting.

Measuring

A measuring stick used to determine the height of a horse consists of an upright pole graduated into inches and hands up to about 19 hands, with a cross rod carrying a spirit level which slides up and down the upright pole. The horse is stood on a level hard surface and is measured from the near side to the highest point of the withers. Some horses show fear when the pole is placed in position and in such cases the attendant holding the animal's head can help by placing a hand over the horse's near-side eye. If the horse is shod some societies allow 12·7 mm ($\frac{1}{2}$ in) to be subtracted from the height recorded.

The circumference of the leg bone below the knee is also measured on occasions. The right foreleg is held up and a measuring tape placed round the left forecannon bone about 50·8 mm (2 in) below the knee. The tape should be drawn tight, without using force.

Grooming

Working horses will not thrive unless they are efficiently groomed. Work causes sweating, although this may not be visible if the horse does not become overheated, and dry sweat blocks the skin pores and prevents them from removing waste products from the tissues and the blood. Regular grooming is therefore required to remove this

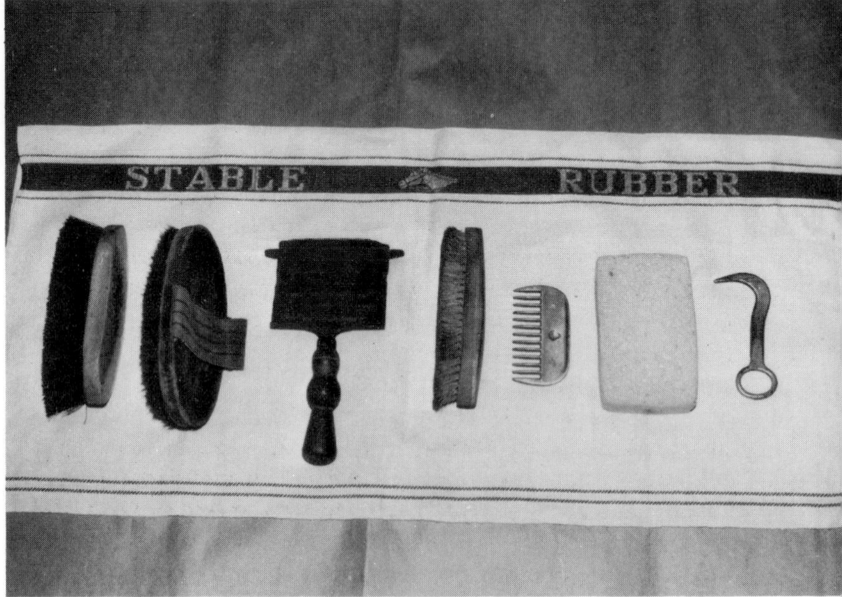

Fig. 2.17. Grooming equipment. From left to right, dandy brush, body brush, curry comb, water brush, tail comb, sponge, and hoof-pick. Set out on a stable rubber.

dandruff and so prevent it from closing the pores and impeding sweating. In addition, grooming improves the condition of the skin by applying friction and massage, and makes the horse more attractive in appearance.

Stabled horses should be groomed at least once a day. A dandy brush is used first, if necessary, to clean away any mud. This is made of stiff fibres which are about 50·8 mm (2 in) in length. Then a body brush is swept along the hair in the direction of its growth to remove the scales of dandruff. This is the principal tool used for grooming and has a broad hand loop of webbing across the back to prevent the brush from slipping from the hand. The body brush is cleaned periodically by pulling the bristles through the teeth of a curry comb. A curry comb is usually made of metal, but can be of rubber. It has a series of fine teeth on its lower surfaces. A water brush which is similar to, but smaller than, a dandy brush is employed for the mane and tail, and a damp sponge is used to clean the eyes, nose, and dock. A cloth, known as a stable rubber, is rubbed over the coat to give a final polish, and the soles of the feet are cleaned out with a hoof-pick. 'Strapping' is the term used for a thorough grooming as described above, which may take as much as an hour to complete efficiently, while 'quartering' is the name given to a quick brushing to improve a horse's appearance before taking it out.

Mechanical groomers are available with a motor unit and a dust bag suspended over the horse. A power-driven rotating brush grooms the coat and a vacuum draws the dislodged dirt and hair into the dust bag. The time taken to groom a horse is halved.

Horses are not groomed when running out at grass. Grooming is not necessary because horses do not exert themselves under these conditions and grooming would also reduce the efficiency of the coat as a protection against cold and wet.

Clipping

Each autumn horses grow a thick winter coat which is shed the following spring, being replaced by a short summer coat. Although a horse running out at grass benefits from the warmth and protection of a thick coat during the winter months, when worked a winter coat causes profuse sweating. This leads to exhaustion which may cause debility, and makes it difficult to dry the horse after work. Horses which are worked during the winter months are therefore clipped as soon as the winter coat has grown. The coat continues to grow during the winter and so the horse must be clipped as often as is necessary. Horses may be clipped all over, known as clipping out, but very often an area of hair under the saddle and the legs of hunters are left unclipped. The saddle patch gives some protection against rubbing, but the chief advantage is to prevent short hair stumps being forced into a horse's back by the weight of a saddle and causing irritation. The leg hair gives protection against minor abrasions. Sometimes horses which are being ridden while being kept out at grass are clipped trace high. The hair is only removed from the lower part of the body, below the position where a trace would lie in a harnessed cart-horse, as this is the area where sweat collects. The head, neck, upper part of the body and the legs are left unclipped to give protection from the weather.

The hair of the forelock, mane, and tail is useful to a horse, chiefly as a defence against attacks by flies. Manes, however, may become too thick and can be trimmed by having the longest hairs pulled out, a few at a time. This improves the appearance of the mane and makes it easier to keep clean. Polo ponies and hunters frequently have their manes cut short, this process being known as 'hogging'. It is effected by cutting the mane along the top of the crest, and then on each side, by the use of a clipping machine. The forelock can be left, as can a lock of hair at the base of the mane which may be grasped by a rider when mounting.

Tails are usually 'banged', that is cut level with the point of the hock, in order to improve their appearance. An assistant places his hand under the horse's dock and holds the tail in the position it would assume if the horse were walking. Then the operator cuts the hair at the desired length with a strong pair of scissors. Thick tails may also be thinned by pulling out some of the hairs which grow from the rear sides of the dock. Tails which are left uncut are known as 'swish' tails.

The clothing of horses

When standing in a stable clipped horses must have their bodies covered with rugs to maintain body warmth. Horse-rugs used for this purpose may be made of wool, but are more usually made of hemp or jute lined with wool. They are designed to cover the body, and are held in position by a strap in front fastened across the breast and a surcingle passing round the body behind the forelegs. The surcingle is made of webbing and can be sewn to the horse-rug, or may be separate. Occasionally a roller is used instead of a surcingle. A roller is always separate from a rug and has padding on each side of the horse's spine. When putting a rug on it should be thrown on the horse's back in a position forward of that where it will eventually rest. The breast strap is then buckled, the rug is pulled back into its correct position and the surcingle or roller is fastened. Some horse-rugs also have a fillet string attached to the lower hind corners which passes round the horse's quarters to prevent the rug being blown forwards. Additional clothing items sometimes used are a short hood, which covers the head down to the nostrils, including the ears, with holes for the eyes, which may be worn in the stable but is more commonly used when at exercise, and a long hood which is similar except that it covers the neck as well, which is worn when a horse is being transported. The hood is placed quietly over the head and the ears are adjusted into the ear covers. Then the tapes are tied in bow-knots under the head and neck. Care should be taken to see that the eyes are not covered, that the ears can be freely moved, and that the horse can feed easily.

New Zealand waterproof rugs can be used for clipped horses which are turned out to grass. They are made of proofed canvas lined with wool and padded at the withers. They are kept in position by straps

Fig. 2.18. Stable bandages on a horse's legs.

passing between the hind legs which will keep the rug in place even when the horse lies down and rolls.

A tail saver is fitted during transport and may be made of either felt or leather with buckles or laces along the part which will cover the outside of the tail. Most have a strap which passes forwards along the back and is buckled to the roller or surcingle. Care should be taken when applying a tail saver to see that all the tail hairs are grasped from the root down and fitted under it. Sometimes a woollen bandage is wound round the tail first, while some owners use such a bandage alone.

Stable bandages are made of wool or a wool and cotton mixture and are used to provide warmth in the stable, protection during travel and to prevent thickening of the legs during periods of inactivity. A stable bandage is about 76·2 mm (3 in) wide and 2·14 m (7 ft) long and has a tape sewn on for fastening purposes. Stable bandages are applied moderately tightly just below the knee or hock in a spiral form round the leg, each turn overlapping the previous one by about two-thirds of its breadth. They end below the fetlock and are then brought up to just below the knee or hock where the tapes are tied in a bow on the outside and the ends are tucked in. Stable bandages should be removed twice a day and readjusted.

Exercising bandages are made of stockinette or a similar elastic material and are applied to give support to the leg tendons during exercise, steeplechasing, hunting, or show jumping. They are usually about 1·82 m (6 ft) long by 101·6 mm (4 in) wide and are applied tightly round the legs between the knee or hock and the fetlock joints. Occasionally, particularly for steeplechasing, a layer of cotton wool is inserted under the bandage which should then be fastened even more tightly. Exercising bandages must be kept clear of both joints, so as not to impede movement, and must not shift or become loose. Thus they should be carefully tied and safety-pins may be inserted on the outside or, on occasions, the tapes are stitched in place.

Knee caps may be fitted when horses are being transported, or when an animal is being exercised on a slippery road at a slow pace, and hock boots can be used during transport. Both are made of leather with a felt lining or of felt with a leather pad at the point most likely to be injured and are shaped to fit the respective joints. Each is provided with an upper and a lower strap; the upper straps are padded and are fastened fairly tightly, first, so as to hold the article in position, while the lower, unpadded straps are tightened just enough to prevent flapping. The strap buckles are always on the outside of the leg.

Harnessing

Riding horses

Saddles used in the United Kingdom are made of leather over a frame or 'tree', usually metal, which forms a central arch over the backbone. The areas which rest on the muscles on either side of the spine are lined with padding. They are of a light type of construction and certain modifications are made to adapt them for use in racing, hunting, and jumping. The saddle should fit on the weight-bearing surface of the back, from behind the shoulder to a point not farther than the last rib. All the weight should be taken by muscle, and the saddle should not touch the bony prominences of the spinal column. Before saddling a horse the stirrup irons should be slipped up the stirrup leathers and the girth left attached to the off side and laid across the seat of the saddle. The saddle should then be placed about 76·2 mm (3 in) in front of the correct position and moved gently back to the right place to ensure that the body hairs lie smoothly. The girth should then be pushed down, grasped under the horse's body, and tightened gently. After the horse has been ridden for a few minutes the girth should always be re-examined and tightened if necessary.

Bridles may be either single or double. A single bridle is fitted with one bit and either one or two sets of reins, whereas a double bridle has two bits and two pairs of reins. When applying the bridle the reins should first be slipped over the horse's head. The head-collar is then removed, the reins round the horse's neck being used to control it if necessary. The bridle is then passed up the front of the horse's face, holding the poll piece in the right hand, and the bit gently slipped into the horse's mouth with the left hand. Most animals will open their mouths as soon as the bit is pressed gently against the incisor teeth, but if a horse will not respond in this way a thumb should be inserted into the corner of the mouth between the incisor and molar teeth. This is usually effective. The poll piece is then raised, pulling the bit into the mouth, and then slipped over the horse's ears on to the head.

There are numerous types of bits, but the commonest is the snaffle, which is the least severe of those in general use. A snaffle is formed of a single bar, which can be straight or curved, with a

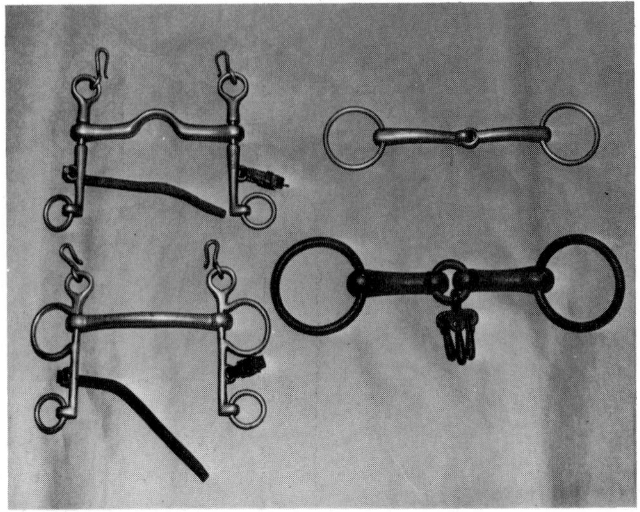

Fig. 2.19. Bits. *Top left*, Weymouth, with a port. *Top right*, plain, jointed snaffle. *Bottom left*, Pelham, without a port. *Bottom right*, mouthing bit.

surface which can be plain or twisted. Most are jointed in the centre. At either end a ring is fitted which serves as an attachment for the cheek pieces of the bridle and the single set of reins. A Pelham usually has an unjointed mouthpiece with an elevated half circle in the centre known as a port. The sides connect with the cheek pieces, each side having one ring at the level of the mouthpiece and another at the lower end. There are two sets of reins, each set being attached to one of the pairs of rings. The rein attached to the rings level with the mouthpiece does not have as severe an action as does that fitted to the lower ends of the cheek pieces. A Weymouth bit is similar to a Pelham, but has rein rings at the bottoms of the cheek pieces only. It is normally used in conjunction with a snaffle to form a double bridle. The snaffle rein should be used as much as possible to give mild restraint, and the Weymouth rein can be employed when greater control is required. Pelham and Weymouth bits are both used with curb chains which pass behind the lower lip and lie in the chin groove. Most bits of these types are provided with a small ring about half-way down each cheek piece to take a lip strap which passes through an extra link on the curb chain to keep it in place.

The proper adjustment of a bit is essential, irrespective of its type. It should rest easily in the mouth, being wide enough to avoid pinching the corners of the mouth and lie about 50·8 mm (2 in) above the corner incisor tooth. In a double bridle the snaffle bit should lie slightly above and behind the Weymouth. All bits should be supplied with rings at the ends large enough to prevent them slipping into the mouth when either rein is pulled when turning.

Bitless bridles are now being used by a few riders, the main advantage being that they cannot damage a horse's mouth because pressure is applied to the nose and chin and not to the bars of the mouth. One type, known as a hackamore, consists of two long metal cheek pieces attached to leather straps which go round the chin and nose. A hackamore must be fitted correctly so that the nose strap rests at least 101·6 mm (4 in) above the nostrils and thus does not interfere with the horse's breathing.

There are three types of martingale in common use. The simplest is the Irish martingale which is a short strap about 76·2 mm (3 in) long with rings at each end through which the snaffle reins pass. It is normally seen only on racehorses and keeps the reins together and may prevent them flying over the horse's head should the jockey lose his hold on the reins when jumping. A standing martingale comprises a strap with a loop at one end attached to the lowest part of the girth and a second loop fitted to the back of the noseband of the bridle. A neck strap passing round the horse's neck is attached to this strap near the middle to prevent the horse catching its leg on it when jumping. The action of a standing martingale is to prevent a horse throwing its head too high, and possibly hitting the rider in the face, or holding its head high, as this renders it more difficult to control because the bit pulls on the corners of the mouth instead of on the bars. A running martingale is similar except that at the upper end the strap is divided and each side ends in a ring through which the snaffle reins are passed. Should a horse throw or hold its head high the running martingale will ensure that the pull on the bit is still on the bars and not on the corners of the mouth.

Driving horses

Driving bridles nearly always have blinkers which cover the sides of the eyes and limit the horse's range of vision. Most collars are of a closed formation, but some open at the top, the two sides being kept together by a strap. The hames are separate bars which are nearly always made of metal and are fitted on either side of the collar, being kept in position by straps above and below. Each carries a ring through which the driving reins pass and a tug to which the traces are attached. A harness saddle carries two terret rings through which the reins are threaded, and on either side are the shaft tugs which are leather loops through which the shafts pass. The breaching consists of a strap passing round the buttocks at a slightly higher level than the shafts, and attached to the shafts on either side, which is used when backing or to check the cart when going downhill. The crupper is a continuation of a strap running back from the saddle which passes round the dock and prevents the saddle moving forwards.

In the case of heavy cart-horses being used for slow work chains are fitted instead of the leather traces and breeching straps and there is a chain which passes from the shafts over the saddle in place of the shaft tugs. In double harness for horses working side by side there are chains or straps which pass from the front end of the centre pole to rings on the fronts of the collars to steer the vehicle and take its weight when stopping or going downhill. The reins split towards the front, and the two halves are attached to the same side of the bits so that a pull on a rein will affect both horses equally.

When harnessing a horse for driving the headcollar should be removed from the head, although the poll piece may be slipped round the neck to leave the head free, but yet give some control over the horse. Closed collars are turned upside down so that the widest part of the collar can pass over the upper, and widest, part of the head. When over the head, the collar is immediately turned towards the side on which the mane falls and then pushed down the neck to rest evenly on the shoulders. The harness saddle is then placed behind the correct position so that the breeching can be applied and the tail drawn through the crupper. The saddle is then moved forward to the correct site and the girth tightened. The bridle is applied and the horse led to the front of the cart and the shafts slipped through the shaft tugs. The traces are then fastened and the driving reins are run through the terret and collar rings and attached to the bridle.

Breaking

If young horses are handled from an early age the training necessary to teach them to be ridden or driven can be carried out comparatively easily. The ages at which horses are broken in to work vary considerably according to the type of horse. Flat-racing thoroughbreds are broken in as yearlings in the autumn so as to be ready for racing the following spring as 2-year-olds. They only carry light weights for relatively short distances at this age. Racehorses for National Hunt racing and hunters are first schooled at about 3 years of age, but horses of the latter type are not hunted until they are 4 or 5 years old. Children's riding ponies similarly are often broken in when 3 years old. The training of heavy draught horses nearly always starts in the autumn when they are $2\frac{1}{2}$ years old. At first they are worked in chains on the land and are not harnessed to carts with shafts until they are about 4 years of age.

Young horses must be handled firmly, but gently, the work load should only be increased slowly and care taken to see that their bodily condition is maintained. A mouthing bit, which has three small pieces of metal attached to the centre, should be placed in the mouth first to encourage the horse to chew and relax the jaw before an ordinary type of bit is fitted. Relaxation of the jaw, and thus of the head and neck, is essential for correct training in whatever capacity the horse is to be used.

The harness should be applied in the stable first and left in position for varying periods so that the horse's skin can become adjusted to it and the incidence of rubbing minimised.

Shoeing

In the British Isles it is necessary to shoe horses to protect the horn of the hoof from excessive wear on hard surfaces. The hoof consists of the wall, the sole, and the frog. The wall is composed of insensitive horn which can be pared, burnt or nailed without causing pain. The sole is a plate of horny tissue which forms the base of the hoof and the frog is a triangular raised structure divided by a central cleft set at the back of the sole which acts as an anti-concussion and anti-slipping device. Both these are also insensitive. Insensitive laminae on the inside of the wall dovetail with sensitive laminae on the outside of the sensitive structures of the foot forming a firm union which can be seen as a white line dividing the sole and wall. At the heels the wall turns forwards and inwards to form the bars, which join with the frog.

Before shoeing, any surplus horn on the sole and frog should be trimmed, but excessive paring should not be practised as this weakens the structures. As the horn of the wall grows continually this also needs trimming. The aim is to provide a level bearing surface on which the shoe can rest. This consists of the lower edge of the wall, the white line, a narrow margin of the sole and a small portion of

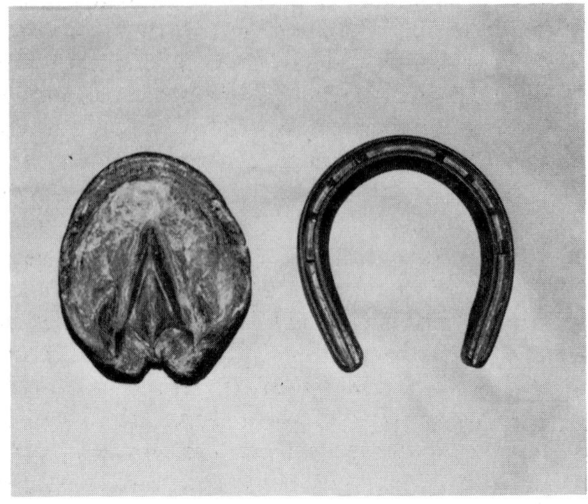

Fig. 2.20. *Left*, Hoof, showing the wall, sole, and frog. *Right*, a fullered shoe.

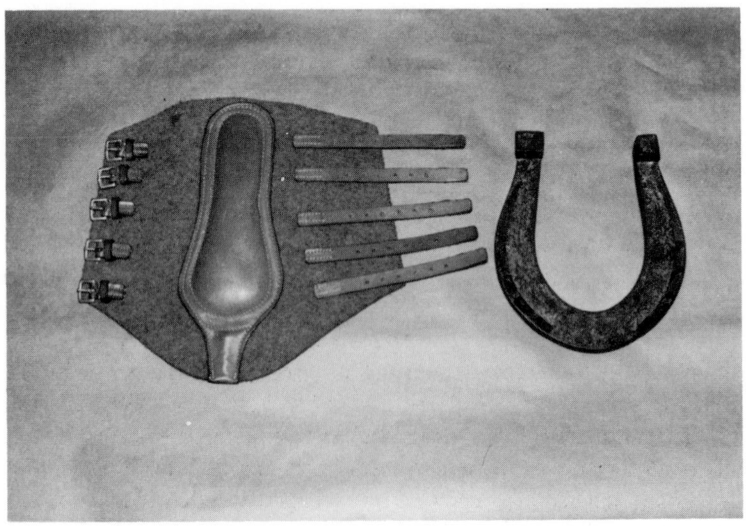

Fig. 2.21. *Right*, Shoe with calkins. The right inside calkin is narrower than the one on the outer branch. *Left*, a brushing boot.

each of the bars. After the hoof has been prepared it should be placed on the ground and viewed from the front, which will show that the inside and outside walls are the same height, from the side, which will show that the angle from the front to the back is correct, and from the rear, which will indicate that the heels are of equal height and that the frog is just touching the ground.

Common faults in the preparation of a foot which may be made are an excessive lowering of the wall, which usually results when a farrier cuts too deeply in one position and has to remove an equivalent amount of horn all round to leave the bearing surface level, an uneven bearing surface following the removal of an excessive amount of wall in one area, rounding off the toe to make the hoof fit the shoe, and over-lowering of the heels.

The front of a horseshoe is known as the toe, the two sides as the branches and the hind ends as the heels. The surface in contact with the foot is known as the foot or bearing surface and the other as the ground surface. Shoes should be as wide as the natural bearing surface of the foot. Many shoes are made with clips which are raised or drawn from the outer edge of the shoe and fit the horn of the hoof closely. In foreshoes there is usually only one, at the toe, known as the toe clip, while in hind shoes there may be two, one on each side of the toe, known as lateral clips.

It is important that the ground surface of a shoe is not so smooth as to be slippery. The most common method of giving a grip is to make a groove, known as a 'fuller' down the centre of the shoe. The edges of the groove cut into soft earth, while on hard surfaces the groove tends to pick up particles of stone which give a grip. The heel ends of the hind shoes are frequently turned down to form 'calkins'. These bite easily into soft ground and to some extent are forced into hard surfaces. Alternatively studs or cogs may be fitted into specially prepared holes in the shoes. One popular type is the 'Mordax' cog, which is made of a relatively soft metal round a very hard projecting metal centre. As the surrounding metal wears more rapidly than the central core, a hard projection is always left in contact with the ground. Studs are not allowed in steeplechasing because they might be a danger to fallen riders or horses.

The sizes and weights of shoes vary considerably, but they should be as light as is compatible with the strength required to stand up to the amount of wear likely to be incurred during a period of 4 or 5 weeks. After the expiration of this period the shoes should be renewed or, if not worn down, removed and refitted. This is necessitated by the horn growth during the period. Most shoes are made of wrought iron, but steeplechasing shoes are made of mild bar steel and flat-racing shoes of aluminium. These racing shoes are known as 'plates' and are fitted for short periods only.

Manufactured shoes can be bought ready made and applied cold, but it is more satisfactory if the shoes are specially made by a farrier for a particular animal and applied to the hoof while still hot to check the fit. The scorching of the horn indicates places where there is not a true fit and makes adjustments possible.

Faults made in fitting shoes are excessive burning of the horn, which follows holding the hot shoe in contact with the foot too long, and the application of shoes which are too wide, which can cause injury to the opposite leg, too long, which means the heels can be trodden on and the shoe torn off, or too short, which can lead to bruising of the rear part of the hoof.

Horseshoes are held in place by nails driven through holes in the shoe into the wall of the hoof. Horseshoe nails are specially made of a type of iron which is sufficiently solid to prevent splitting and buckling, but is ductile enough to prevent snapping when the tips are wrung off. Each nail has a side which is straight from the head to the tip, while the opposite side has a bevel on the head and the tip. The other two sides have bevels on the head only. Nails are made in several sizes, indicated by numbers from 2 to 16; the higher the number the larger the nail. Nails can be selected according to the size of the horse and the thickness of the horn of the hoof. The farrier must be careful to drive the nails into the horn so that they do not penetrate or rub against the white line and the sensitive structures beneath. This is facilitated by the bevelled nail tip. The bevel is placed on the inside so that when hit with the hammer the tip tends to turn outwards and emerge from the wall. The ends which appear are twisted off with the claw of a driving hammer, and the short portion left is hammered down to form a clench and help keep the nail firmly in place. As far as possible the nails should be evenly spaced, but they are usually closer together at the toe, where the horn is thicker, than at the heels where the horn becomes thinner. For small pony shoes five nails are sufficient, but most shoes require seven nails, four on the outside of the hoof and three on the inside. Aluminium racing plates usually have twelve nails as this light metal is liable to twist unless firmly held in position.

Common faults in nailing are driving the nail into the sensitive structures, which causes pain and can lead to infection through the wound left when the nail is withdrawn, driving the nail close to the sensitive structures, which can lead to lameness about 2 or 3 days after shoeing due to the rubbing of the nail on a sensitive area, and driving nails into old holes in the horn in which they only fit loosely so that the shoe may be pulled off.

When horses are turned out to grass their shoes should be removed to minimise the risk of injury to other horses through kicking. While at grass their hoofs should be regularly examined and any overgrown horn removed. If this is not done cracks can start in the excess horn which may rise up the hoof wall and cause lameness.

Under the provisions of the Farriers (Registration) Act 1975 it is unlawful for any person who is not registered with the Farriers Registration Council to carry out farriery, unless he is undergoing training in shoeing, is rendering first aid in an emergency, or is a veterinary surgeon. This Act has been designed to prevent suffering arising from the shoeing of horses by unskilled persons and to promote the proper shoeing of horses.

Injuries due to faulty action

Horses with faulty actions can hurt themselves in a variety of ways, and the severity of the injuries can be reduced to some extent by remedial shoeing. One of the commonest is brushing or cutting, when injuries to the lower part of one leg are caused through being struck by the hoof and shoe of the opposite foot. This fault is more common in young, inexperienced, weak, or tired horses, and training, conditioning, and working reasonably should be carried out as appropriate. The injuries can be aggravated by parts of a shoe which project on the inside of the hoof, and this fault should be avoided when shoeing. Hind shoes with calkins should be fitted with special care in such cases. Specially designed brushing boots or rubber pastern rings can be used to protect from injury the legs of horses which continually brush.

Another injury is overreaching which occurs at fast paces and when jumping and causes tissue damage in two ways. The first is when the tendons or the back of the fetlock joint of a foreleg are struck by the front lower edge of the toe of the hind shoe on the same side, and the second is when the bulbs of a foreheel are cut by the posterior lower edge of the toe of the hind shoe. The remedies are to set the hind shoes well back and to round the front edge in the first case and the posterior edge in the second case.

Speedy cutting only occurs in the forelimbs and results when the upper inner part of the forecannon or lower inner part of the knee is struck with some part of the shoe of the opposite foot. It happens when the horses are travelling fast and on account of the pace the violence of the blow causes the horse to fall. The front of the inner branches of the foreshoes can be made straight so that they can be set well under the hoofs at these sites and special boots designed to cover the part of the leg struck can be applied.

Administration of medicines

Some medicaments can be mixed with the concentrate ration but horses are usually suspicious when drugs are mixed with their food. At one time liquid medicines were administered to horses by holding the head high and pouring the fluid down the animal's throat in small quantities at a time. Now, however, a stomach tube passed down the oesophagus into the stomach is used as there is less wastage of the medicament through spillage, but skill is required in its passage.

Boluses or pills are not now administered to horses, but electuaries may be given. An assistant stands on one side with one hand on the poll and the other on the nose, while the operator grasps the tongue and draws it down and to one side. The electuary can then be smeared on the back of the tongue and the tongue released. Some proprietary drug preparations are now available in the form of a paste suitable for squeezing on to the tongue in a similar manner.

Inhalations may be beneficial in the treatment of horses with respiratory diseases. A bucket containing a little hay is placed in a large sack in the manger and boiling water and the medicament prescribed poured on. The mouth of the sack is held gently round the horse's muzzle so that it inhales the steam.

Injections are being used increasingly as the most convenient method of drug administration, although some horse owners feel that horses are more prone to undesirable reactions than are animals of other species. Subcutaneous injections are usually made in the neck or shoulders, intramuscular injections in the hindquarters or neck, and intravenous injections into the jugular vein in the neck.

General

Vices

The majority of mares and geldings are kindly animals and do not intentionally attempt to injure human beings, although there are exceptions and mares need to be watched when in oestrus. Stallions and rigs are always potentially dangerous, stallions particularly so during the breeding season. However, an excessively high plane of nutrition combined with insufficient exercise can make normally quiet horses difficult to manage and may lead to kicking and biting, both in and out of the stable, and rearing and shying when ridden. These undesirable habits may develop into persistent vices in certain animals and render them dangerous to handle. Chronic boredom can also lead to vices, such as weaving, crib biting, and windsucking, which are principally practised by animals while stabled. Any of these can cause a loss of condition in the animals which practise them. As they are imitative, animals developing any of these vices should be kept away from other horses, except when at work. Other habits shown by individual horses are eating faeces and tearing clothing.

Kicking

Young horses may kick because they are unused to being handled and fear that they may be hurt. Such animals can often be quietened by stroking them gently, particularly over the hind quarters and hind legs, with the end of a pole so that they learn that they have nothing to fear. A few horses persistently kick when being groomed or saddled, and it may be necessary to have a foreleg held up while the hind legs are being brushed or the girth is being tightened. It is not safe to strap a foreleg up as the horse may fall. Some horses kept in stalls kick at the heel posts or the rear parts of the stall divisions in order to make a noise. The noise nuisance can be alleviated by covering the back parts of the stall divisions with matting and placing padding round the heel posts. While out at grass some horses will resent and kick at a strange horse, and the shoes should be removed from grazing animals to prevent serious injuries through kicks. Biting flies or other insects will upset horses and the kicks aimed at disturbing the flies can cause injuries to men or other horses. In heavily fly-infested areas it may be best to stable the horses during the day, when the flies are active, and turn them out to graze at night. Horses that are liable to lash out at other horses when being hunted or hacked are sometimes marked by having a piece of red ribbon tied round the root of the tail.

Biting

Horses usually indicate that they are going to bite by laying their ears tightly back, showing the white of the eye and partially opening their mouths. Biting often starts as play but, if not checked, can become a dangerous habit, particularly during grooming. To prevent the groom being injured while handling a biting horse in a stable the animal can be tied up short or a side-stick can be fitted. This consists of a strong pole of an appropriate length which is attached to a roller fitted round the belly at one end and to the noseband of a headcollar at the other. The horse can move its head vertically but cannot make sideways movements. Muzzles can also be used, but have to be removed at feeding times. When out with other horses the rider of a biting animal must keep a careful control over the head of the horse on which he is mounted at all times.

Rearing

Rearing is often a reaction to bad handling. A horse with a tender mouth may rear if a heavy-handed rider jerks on the bit, or if an unnecessarily severe bit is fitted. A change to a rider with light hands or a gentle bit, or a bitless bridle, may effect a cure. A badly-fitting, or an incorrectly applied, saddle which gives pain when the rider mounts may also cause a thin-skinned horse to buck and rear. Horses which persistently rear may be tried with a circular bit attached to additional cheek pieces on the bridle at the sides and to a tightly-fitted martingale at the back, the other end of which is fitted to the saddle girth. This checks the upward swing of the head which helps the horse to rear. In the past other crude methods were tried, but these were not very effective.

Shying

Horses, particularly young inexperienced animals, will shy at strange objects because they are afraid of them, although shying is not as common now as it was in the past. If at all possible a horse should be led up to the object at which it shied and allowed to smell it and touch it with its muzzle. Such treatment helps the horse to overcome its fear of strange objects. Defective eyesight may also cause shying, and the eyes of any horse which shies frequently should be carefully examined for abnormal lesions.

Weaving

The horse rocks to and fro, often for long periods at a time, possibly lifting its forefeet in turn as the body is swayed to the opposite side. The amount of energy expended in weaving may lead to a marked reduction in body condition. The habit can be checked by tying the horse's headcollar to side rings so as to limit sideways movement, but this will prevent the horse lying down to sleep at night. As a horse often weaves with its head over the bottom half of a stable door it is possible to hang weights from cords attached to the top of the door frame which bump its neck as it moves.

Crib biting

The horse grasps the edge of a manger or other fixed object with its incisor teeth and swallows air. This can lead to tympany of the stomach and intestines, possibly causing indigestion and a loss of body condition. The teeth of a crib biter become worn down and broken by the frequent grasping of solid objects. Crib biting may be prevented by tying the horse up so that there is no fixed object within reach which it can bite, or by housing it in a loose-box with no fittings and introducing a movable trough at feeding times only. However, the horse may learn to suck wind without grasping anything with its teeth.

Wind sucking

Wind is sucked in without the aid of a fixed object to grip. To prevent this a gullet strap may be applied to the upper part of the neck so that it causes pain when the neck is arched to suck wind, or a flute bit, which consists of a thick hollow bit perforated with several holes which can be kept continuously in the mouth, except when the horse is eating, can be attached to the headcollar. Such a bit prevents the horse from sucking in sufficient air to swallow. A more permanent method of prevention is by the performance of an operation for the severing of the neck muscles involved in holding the neck in the required position for sucking wind. This operation does not reduce the work capability of the horse.

Eating faeces

The eating of faeces is an undesirable habit which horses may develop. It may start through a dietetic error or while a horse is suffering from indigestion, but once horses become addicted to this vice they continue, even when the diet has been improved or the digestive upset has been cured. One possible causal dietetic error is a deficiency of hay and some horses receiving a ration low in hay stop eating faeces when the hay intake increases. Another initiating cause can be a mineral deficiency. Some horses discontinue the habit when run at grass after a period of housing.

When stabled a horse can be kept tied so that it cannot reach its faeces, or can be kept muzzled except at feeding times. However some owners prefer to allow their horses to continue this undesirable habit, providing that their bodily condition is not impaired.

Tearing clothing

Some horses bite at the rug applied to keep them warm in the stable and tear it. Such horses can be left without rugs but an effective method of preventing this destructive habit is to apply a piece of leather, known as a bib, to the back and sides of the nose band of the headcollar. This bib projects below the level of the lips and comes round the sides of the mouth, so stopping the horse from seizing the rug, but allowing it to eat normally.

Minor operations

Male horses being used for flat racing are generally left entire, but those used for other types of work are nearly always castrated. The operation is usually performed when the colts are about 1 year of age although some owners, particularly of steeplechasers, prefer to leave them entire for another year on the grounds that they develop more muscle. The spring is the favoured operation time because the horses can be run on clean pastures after the operation, which reduces the risk of infection, and flies, which might be attracted to the surgical wound, are not as active as they would be in the summer.

Species control

The Jockey Club requires all horses which are raced under their rules of racing to be registered through their agents, Messrs. Weatherby. The owner has to supply details of the animal's name and breeding and a veterinary surgeon has to provide a description of its markings.

Licences

Under the Horse Breeding Act 1958, as amended under the Agriculture (Miscellaneous Provisions) Act 1963, no one may keep a stallion over 2 years of age without a licence or permit unless it is a thoroughbred entered or eligible for entry in the General Stud Book, or a Shetland pony, or a pony of a specified breed kept in a specified district. A licence will not be granted if the stallion is permanently affected with any contagious or infectious disease or any of a list of prescribed diseases, or has defective genital organs. A licence will not be given to a stallion with defective or inferior conformation or physique. If a licence has been refused and the owner wishes to keep the stallion entire he may apply for a permit which will be issued subject to any condition the Minister of Agriculture may think fit to impose.

Reference

ROYAL COLLEGE OF VETERINARY SURGEONS (1977) *Report on Colours and Markings of Horses.*

CHAPTER 3
DAIRY CATTLE

Introduction

Milk, and the products made from it, are important sources of food for people in most countries. Milk for human consumption is usually defined as the secretion obtained from the mammary glands of healthy cows. Cows' milk contains, on average, about 12·4 per cent of solids and 87·6 per cent of water. The solids comprise 3·75 per cent of butter fat and 8·65 per cent of solids-not-fat (S.N.F.). These consist of lactose, proteins, and minerals.

Success in dairy farming on any scale requires a planned programme based upon a realistic standard of performance which is executed skilfully. The profitability of the enterprise will depend in large measure upon the ability of the manager and upon the soundness of his judgment in making decisions relating to all aspects of the breeding, feeding, and management of dairy cattle. In recent years profits have been more difficult to make and so the majority of dairy farmers are becoming increasingly business minded. Changes have therefore been introduced designed to raise profits, particularly by reducing production costs.

Milk production is an intensive type of farming and the labour requirement is high. The wages paid to cowmen have risen over recent years and, in order to be able to pay the higher wages, employers have had to require a rise in production per man. This has largely been attained by increasing the number of cows tended and milked by each worker. Milk production may be combined with one or more farming enterprises, but herd sizes are increasing and there is a greater tendency to specialise. Herd sizes are rising because farmers with large herds have greater opportunities for increasing profits as certain investment items can be shared between a larger number of cows and labour-saving methods can more easily be introduced. About 80 per cent of the work involved in keeping cows is conducted in and around the buildings. Working speed can be gained by lessening the amount of walking and handling of milk and foodstuffs. This is one of the reasons why cowshed housing has been largely replaced by the yard and milking parlour system, because, in the latter type of housing, the cows come to the cowman for milking and concentrate feeding, the food being delivered from storage hoppers to the troughs in the parlour and the milk passing direct from the cows' udder to the adjoining dairy.

Definitions of common terms

Calf. An animal under 1 year. Sexes are described by using the terms bull calf for a male and heifer calf for a female.

Heifer. Usually a female over 1 year which has not calved. An unmated animal is known as a maiden heifer and a pregnant one as an in-calf heifer. In some areas the term first-calf heifer is used until the birth of a second calf.

Cow. A female after the production of one or two calves according to local custom.

Down-calver or springer. A female animal nearly ready to give birth to its calf.

Bull. An uncastrated male over 1 year.

Free-martin. A heifer calf born as a twin to a male. Most of these heifers are found to be sterile when they become adult.

Principal breeds

Cows of the pure dairy breeds have similar points of conformation. They should have a lean appearance, without surplus fat. The general outline of the body should be wedge-shaped when viewed either from the side or from above, the front part of the body being narrower than the hind quarters. The abdomen needs to be wide in order to accommodate the capacious rumen required for the digestion of large amounts of fibrous food. The hindquarters should be wide so that the hind legs can be set well apart to support a large udder, which extends well back and well up between the thighs. The front of the udder should be carried well forward, and the base of the udder should be level with the under side of the abdomen, and not be excessively pendulous. The skin of the udder should be soft and pliable. Capacity is gained by a lack of fatty and connective tissue in the udder and so a good udder shrinks after milking. The

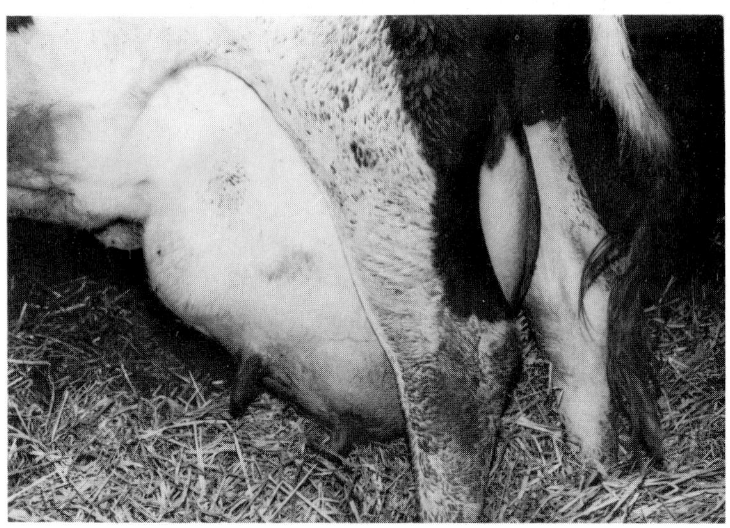

Fig. 3.1. Undesirable pendulous udder.

Fig. 3.2. Friesian cow.

teats should be even in size and sufficiently large to allow them to be milked easily. They should be situated in the centre of each quarter and not on the outside.

Dairy breeds

Friesian

This is the most popular breed of dairy cattle in England and Wales, and is the second largest, with mature cows averaging about 590 kg (1,300 lb) in weight. The breed originated in the Netherlands, and black and white is the only colour accepted for registration in the United Kingdom. Recessive red and white animals do occur and a separate breed society has been formed to cater for them. The average milk yield is about 4,500 l (1,000 gal) per lactation. At one time the compositional quality of the milk was relatively low but this has been improved by selective breeding and the average milk fat percentage is about 3·6 and the S.N.F. percentage about 8·7.

British Canadian Holstein

Animals of this breed are being developed in the British Isles from Canadian importations. Although developed from the same Dutch ancestors as the Friesian they are of a more pronounced dairy type and have a slightly higher body weight. Their reputation is based upon their milk producing capacity. The average milk yield is about 5,000 l (1,100 gal) per lactation, which is higher than that of any other breed, with a fat percentage of about 3·7. In colour they are black and white and although most animals have large black patches on a white background some have a number of small irregular spots. They are not particularly good grazing animals, but can digest large amounts of dried roughage.

Ayrshire

A Scottish breed in which brown and white is the common colour. Cows which have not been polled have longish horns which curve upwards and slightly backwards. On average Ayrshire cows weigh about 450 kg (1,000 lb). The udder formation is excellent with short, regularly spaced teats of uniform size. Lactation yields average about 4,000 l (880 gal) of milk with approximately 3·9 per cent of fat and 8·8 per cent of S.N.F. The fat globules are small, which means that the milk is particularly suitable for cheese making.

Jersey

The breed was developed on the island which bears its name and is the smallest of the important dairy breeds, an average cow weighing between 360 and 400 kg (800 and 900 lb). A variety of shades of yellow are accepted with or without white markings, but all animals have a grey ring of hair behind a black muzzle. The average lactation yield is about 3,300 l (720 gal) of milk containing 5 per cent of fat,

Fig. 3.3. Jersey cow.

and it is for this character of high fat production that the breed is famed. The S.N.F. percentage is about 9·2. On account of its high compositional quality a special price is usually paid for Jersey milk. When quietly handled Jersey cows become very tame, but they tend to have a nervous temperament and are more easily upset by unusual conditions than are cows of the other dairy breeds.

Guernsey

This breed originated on the Island of Guernsey, and the cattle are about the same size as Ayrshires. They are fawn in colour and most have white markings. An average yield of 3,460 l (760 gal) of milk with a fat percentage of 4·5 and a S.N.F. percentage of 9·2 can be expected. The milk has a rich cream colour and its quality means that it fetches a higher price as in the case of Jersey milk.

Dual purpose breeds

Some breeds used for milk production are termed dual purpose. In these breeds an attempt has been made to combine both milk and beef production qualities in the same animal. In appearance they are not as markedly wedge shaped but have well developed frames on which flesh can be laid down when they are not milking. The male calves should develop into quick maturing high quality beef animals. Dual purpose breeds are not as popular now as they have been, as it is generally considered to be more profitable to keep types developed either for milk or beef qualities.

Dairy Shorthorn

The name of the breed is taken from the characteristic short horns, and red, white, red roan, and red and white are the recognised colours. When mature, Shorthorn cows usually weigh between 540 and 590 kg (1,200 and 1,300 lb). The average yield is 3,800 l (825 gal) of milk containing 3·6 per cent of fat. This breed is losing popularity, and there has been a marked reduction in the number kept over the past few years.

Red Poll

As indicated by the name the cattle are red in colour and without horns. The breed was developed in Norfolk and Suffolk. The cows are about the same size as Dairy Shorthorns. The average milk yield is about 3,550 l (780 gal) of milk at 3·7 per cent fat. The breed is a hardy one but is not now widely kept.

Cross-breds

Deliberate cross-breeding is not normally practised in the breeding of dairy cows. One cross which has been tried is the Jersey × Friesian, sometimes known as the Jersian. It has been reported that the milk yield may be nearly as high as that of the Friesian and the fat content may approximate to that of the Jersey. The hybrid vigour resulting from the cross is shown by uniformly good production levels under varied environmental conditions and management systems.

Breeding systems

The aim of all dairy farmers is to breed cows which will produce high yields of good quality milk, utilise the food they are given efficiently, breed regularly, remain healthy for several lactations, and are quiet to handle. Tangible evidence of a cow's producing ability in terms of milk yield can be determined by some system of milk recording. This can best be effected through the National Milk Records Scheme, brief details of which are given later under the heading of milk marketing. Milk samples taken at intervals throughout the lactation should be analysed for their fat and total solids contents and attention is now being paid to the protein component. Milk composition figures are important for often the highest yielders produce the poorest quality milk. A study of the quantity of food fed, particularly the amount of the concentrate ration, enables the efficiency with which the food nutrients are utilised to be assessed. Cows must calve regularly and the aim is for each animal to produce a calf annually. The cows in a herd should wear well and remain productive for several years. The conformation of the cow is important in this respect and special attention should be paid to the udder and the legs and feet. In addition a farmer desires to have cows which are not nervous and are easy to milk. These particulars enable the breeder to decide on the cows whose female progeny should be reared as herd replacements and allow records to be compiled which can be of great assistance in the selection of breeding bulls. In addition to studying the records of an individual cow it is often helpful to consider particulars of a cow's close relatives as these give a further indication of her breeding value.

Progeny testing is the most efficient method of assessing the breeding value of a dairy bull, and entails an examination of the performance of at least 10, and preferably 20 or more, of his daughters. Some breeders use a daughter–dam comparison by comparing the average production of the daughters with the average production of their dams in a similar lactation, usually the first. The hope is that the daughters' yields will be better than those of their dams, both in terms of quantity and quality. The weakness of this method is that the feeding and management conditions are unlikely to have been constant during the lactations of the dams and the daughters and so environmental factors may have been responsible for any changes.

An alternative method, often known as a herdmate comparison, made possible by the widescale use of artificial insemination, is to use the average production of daughters in several herds compared with other animals of a similar age in the herds sired by different bulls. Obviously all the heifers should be out of cows of approximately equal merit. This means that a determination can be made as to whether a bull's daughters have a better or a poorer than average production record compared with other animals in the herds, and this eliminates differences caused by the feeding levels and management standards.

Identification

In the case of marked breeds, such as Friesians, a map card can be

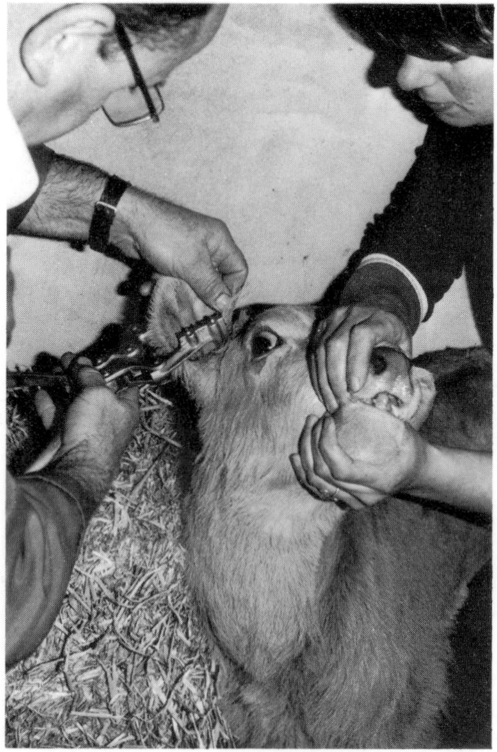

Fig. 3.4. Tattooing the ear of a calf.

Fig. 3.5. Freeze-branded number.

prepared illustrating the distribution of the markings. In other breeds calves can be permanently tattooed along the long axis of the inner surface of one ear, preferably during the first month of life, or numbered ear tags can be fixed to one of the ears. Freeze-branding is becoming popular. In this method the pigment cells in the skin of coloured animals are destroyed through the application of a very cold branding iron, with the result that the new hair grows white. The procedure is generally carried out by trained operators who cool heavy copper branding irons in a mixture of dry ice and alcohol to a temperature of −70°C (−94°F) and hold them on a clipped area of coloured skin, which has just been soaked with alcohol, for 40–50 seconds. The method is comparatively painless and the numbers can be easily read. The best results are obtained when the animals are marked at between 3 and 12 months of age, although older cows can be branded in the same way.

It is very important that milking cows should be quickly identifiable by the cowman so that service and calving dates and milk yields can be recorded accurately. Freeze-branded numbers are easy to read, but animals which were ear-tattooed or fitted with small ear tags as calves need additional identification devices which can be more easily read. Marker straps with numbers can be fitted round their necks, forelimbs, or tails, or large tags with hanging number plates which can be read from front or rear may be inserted in their ears. An alternative is to tattoo the cows on their udders.

Ageing

The development of the incisor teeth can be used to give an indication of the age of cattle. There are four pairs of incisor teeth, situated on the lower jaw only, which contact a hard mass of fibrous tissue known as the dental pad on the upper jaw. Starting from the middle of the jaw the pairs are known as centrals, medials, laterals, and corners. The cutting edges of the temporary incisor teeth show through the gum at birth, or are easily palpable below the gum surface, and are quite prominent at 1 month of age. Thus, they are not of much practical use in ageing but the times at which the temporary teeth are shed and replaced by the permanents are of significance. The temporary incisors are much smaller than the permanent teeth and can easily be differentiated from them.

The centrals usually appear at about 1 year 9 months and the medials at about 2 years 3 months. Until recently it was generally accepted that the laterals are cut at about 2 years 9 months and the corners at approximately 3 years 3 months, but a recent survey by Andrews (1973) indicates that 3 years 2 months and 3 years 9 months respectively may be more accurate figures. Individual animals will, however, be met whose teeth are cut considerably earlier or later than the ages stated. The breed and strain of cow, the methods of management and the type of diet can all exert an influence. At about 5 years of age the incisor teeth show signs of wear along the cutting edges, and the crown wears down in subsequent years until at about 13 years of age only the stumps of the teeth remain. There are no canine teeth in cattle.

Housing

Calves

When calving, cows should be individually accommodated in loose-boxes and not allowed to calve in cowsheds or yards. Calves of the dairy breeds are best housed in individual pens until they reach 90 kg (approx. 200 lb) liveweight, a minimum space of 2 m^2 (20 sq ft) being allowed per head. The partitions between pens should be at least 1·15 m (3 ft 6 in) high, and may be either solid or railed. Rails allow the calves to see each other easily and facilitate the free circulation of fresh air, while solid divisions lessen the risk of draughts and the spread of disease. Young calves can be kept in small groups but should be tied up at feeding time and for an hour afterwards as this reduces the risk of the calves sucking each other's ears, navels or teats.

After they are 3 months of age calves are best housed in small groups, 3 m^2 (30 sq ft) being allowed per calf at this age with the area increasing to about 4 m^2 (40 sq ft) at 6 months. At the latter age they can be transferred to semi-covered yards. Older heifers are kept out at grass when the weather is suitable or in yards.

Cows

When lactating, cows may be kept and milked in cowsheds or kept in yards or some similar type of accommodation and milked in milking parlours or milking bails. Provisions made under the Milk and Dairies (General) Regulations 1959, require buildings used for milking to have a sufficient number of openings to ensure that the air is kept fresh, to be so lit that the work may be conducted in good light and to have those parts of the walls liable to soiling by cows made impervious and easily cleansed. In addition, except in the case of milking bails, there must be an impervious floor so made that the cows are not soiled with dung and so sloped that liquid runs in gutters to drain outside the building. Buildings in which cows are housed must have adequate light and ventilation, must have a clean approach and be kept in such a condition that the cows are not avoidably soiled.

Cowsheds

Under the cowshed system the cows are milked in the cowshed throughout the year, and are kept tied in stalls during the winter

Fig. 3.6. Single row cowshed with individual stalls.

months. During the summer months the cows are turned out to grass between milkings. For small herds of up to twelve cows the animals are usually tied in a single row (see Fig. 3.6), but for larger herds a double row with the cows facing outwards is usually favoured. The length of the standing can be varied according to breed; e.g., 1·60 m (5 ft 3 in) for Friesians and 1·37 m (4 ft 6 in) for Jerseys.

When estimating the length of a building, double stalls, to take two cows lying side by side, are usually provided. These are generally 2·13 m (7 ft) wide for Friesians and 1·82 m (6 ft 6 in) for other breeds. The most satisfactory stall divisions are of an open tubular steel type of construction. Individual automatic water bowls are nearly always fitted. These may be of two types; in one the bowls are regulated individually and a bowl is only filled when the cow pushes a nose plate, and in the other a line of bowls is gravity fed from a regulating tank so that water is always present in the bowls at a constant level. A single neck-chain tying method is usually employed. Wheat straw is the most satisfactory bedding material.

The advantages of the cowshed system are that the cows can be

Fig. 3.7. Automatic water bowl.

given individual attention, food rationing can be controlled, and less bedding is required. The disadvantages are that the cows have less freedom and considerable manual effort is required in removing dung and carrying milking machine units to the cows.

Yards

In the yard and milking parlour system the cows are loose-housed in yards and taken to the parlour for milking. A floor area of about $7{\cdot}4\ m^2$ (80 sq ft) is allowed per head, and the cows are nearly always dehorned to avoid bullying and injuries. The yards may be entirely roofed over, but such buildings are expensive to erect and so most are only partially roofed. A bedded area under cover occupies from half to two-thirds of the floor space. Straw, woodshavings, or sawdust are suitable bedding materials. The remainder of the floor is used as an uncovered feeding space. Silage may be self-fed from a clamp adjoining this area, or hay and silage may be fed in communal mangers extending the full length of the yard. The feeding area should have a concrete floor, sloping away from the silage clamp, and be kept free from bedding so that the dung voided during feeding can be easily removed by a scraper fitted to the front of a tractor. A large communal water trough should be provided.

In covered yards, slatted floors may be used instead of bedding. Concrete or wooden slats have been tried. The slats are about 76·1 mm (3 in) wide at the top and are spaced at about 38 mm (1·5 in) intervals. A floor area of about 3·7 m^2 (40 sq ft) per cow should be provided, as, if the cows are given more room, the dung will not be pushed through the spaces between the slats. The dung and urine collect under the slats, and the slurry which forms is removed periodically. The cows should not be allowed access to a solid-floored area, for example a self-feed silage face, as they tend to lie on the dung-covered non-slatted area. Slatted floor housing has not proved very successful for dairy cows as injuries may be caused to udders, teats, hocks, and hoofs.

The advantages of yard housing are the greater freedom allowed the cows, the simplification of cleaning out by using mechanical aids, and, where straw is provided for bedding, the production of good quality manure. The chief disadvantages are additional requirements for bedding, unless slats are used, and the difficulty in rationing the food individually.

Cubicles

Cubicles may be fitted in yards under a roofed area to reduce the quantity of bedding required and allow the cows to lie in greater comfort and cleanliness. Each cubicle is about 2·08 m (6 ft 10 in) long and 1·06 m (3 ft 6 in) wide, and they are arranged in rows on either side of a 3·04 m (10 ft) passage. There are no troughs or water bowls in the cubicles and the cows can enter and leave them at will. The cubicle floors are raised 28 cm (11 in) above the passage and the faeces are voided into the passage, from which they can be removed with a scraper attached to a tractor. Alternatively, slatted floors may be fitted in the passage. Sawdust is commonly used for bedding and, as it is not fouled by dung, only about 6·3 kg (14 lb) is required per head per week to replace any which may have been scraped out.

Fig. 3.8. Cubicles.

Cow kennels
Cow kennels are similar in design but are not fitted in covered or partially covered yards. Each kennel consists of a roof and back wall, usually made of corrugated iron, and rows are erected on a concrete base. The rows of kennels are placed round an open yard.

Milking accommodation

Milking parlours
Milking parlours are designed in various ways but it is now usual for all types to be fitted with troughs so that concentrate foods can be fed during milking. In abreast parlours the cows stand side by side in a single row facing an exit passage. The floor may be on one level or the stalls may be 38–46 cm (15–18 in) higher than the floor on which the operator works. Tandem parlours are designed so that the cows stand nose to tail, usually at a level 61–76 cm (2 ft–2 ft 6 in) higher than the operator and enter and leave at the side of the stalls. The stalls may be in one line, in a U formation or occasionally in the form of an L. Herring-bone parlours are becoming popular. Under this design the cows stand in echelon in two rows, one on each side of the operator's pit. The edges of the pit are sawtoothed to facilitate

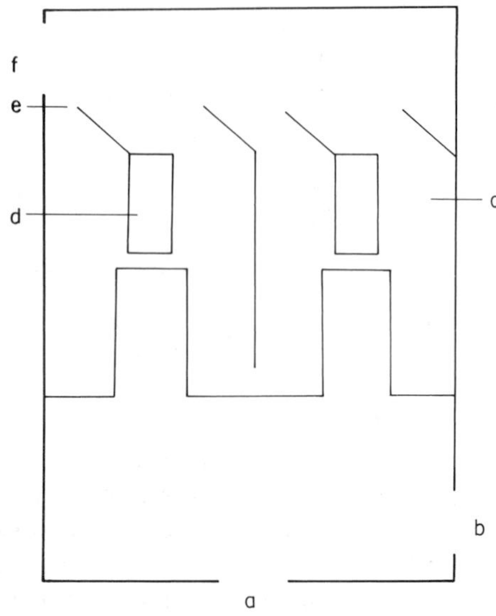

Fig. 3.9. Floor plan of a two-level abreast parlour. (a) Cows in, (b) door to dairy, (c) cow stalls, (d) milking unit, (e) stall exit gates, (f) cows out.

milking. The cows on each side have to leave as a group, and so a slow milker will delay the release of the other cows on that side of the parlour.

On some very large farms rotary milking parlours are used in order

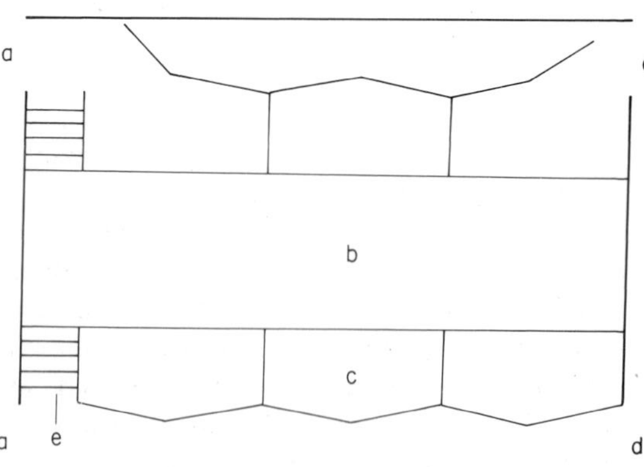

Fig. 3.10. Floor plan of a double-row tandem parlour. (a) Cows in, (b) operator's pit, (c) cow stalls, (d) cows out, (e) steps.

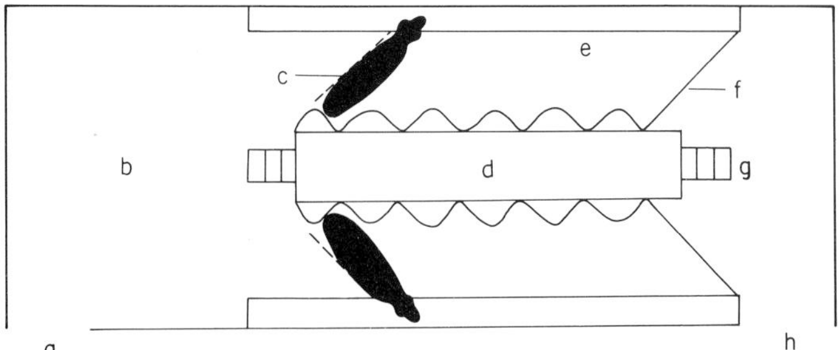

Fig. 3.11. Floor plan of a herring-bone parlour. (a) Cows in, (b) collecting area, (c) chain, (d) operator's pit, (e) cow standing area, (f) gate, (g) steps down, (h) cows out.

to increase the number of cows milked per man per hour. The cows stand on a moving platform, which rotates on a continuous or a stop/go principle. In the former type the entry and exit doors are fitted with pressure pads so that if a cow, on entering or leaving, applies pressure to the pad the plant will automatically stop, while in the latter type the movement is controlled with a foot switch. The three basic designs described under static milking parlours are found. In the abreast type the cows face inwards with the operator working on the outside of the platform. In the other two types the operator stands on the inside of the platform with the cows standing nose to tail in the tandem design and facing at an angle towards the outside in the herring-bone type. In most rotary parlours the food is delivered into the troughs from a hopper attached to a bulk food store.

Milking bail

A milking bail consists of a portable milking shed mounted on skids to facilitate moving by a tractor. It is divided into six or eight stalls of a walk-through type. There is no floor and no bedding is supplied, so the cows stand on the grass of the field. A simple portable hut is attached at the back. This is divided into two compartments, one to hold the motor driving the milking machine pump, and the other for the milk cans. The milk is taken back to a dairy on the farm for cooling. The cows are kept out in the field all the year round and the bail is moved to a new patch of grass every 2 or 3 days. The method can only be used with success in mild climates, and on light land which is not churned up by the cows in wet weather.

Loose boxes

Whatever system of housing is used for the cows a range of loose boxes is essential so that calving or sick cows can be kept on their own. Suitable floor sizes are 2·44 m × 2·44 m (8 ft × 8 ft) or 3 m × 3 m (10 ft × 10 ft).

Dairies

Farm dairies must be at least 18·3 m (60 ft) away from a cesspit or dungstead, but for easy working should be attached to, or placed near, the building in which the cows are milked. A milking parlour, in which cows are milked but not kept, is generally connected to the dairy by a door, but a connecting door between a cowshed, in which cows are housed, and a dairy is not advisable for hygienic reasons. Most farm dairies consist of a single room which is used for milk storage and equipment cleansing, but on some large farms, particularly those from which milk is retailed in bottles, two rooms are provided, the first for the cooling and handling of the milk and the second for the cleaning and sterilising of the milking equipment and bottles.

Bulls

Bulls should be housed individually in loose boxes about 4·5 m (15 ft) wide by 3·7 m (12 ft) long with a run 4·5 m (15 ft) wide by 7·3 m (24 ft) long attached. The wall of the run should be designed so that the bull can see out, as it has been observed that bulls which can watch movement in the farmyard are generally more tractable. However, bulls cannot be trusted and safety devices are usually incorporated in bull pens. In the box a corner manger with a catching yoke is a standard fitting. The door between the box and the run should be controllable from outside, so that the bull can be confined in either the box or the run as required. An escape gap about 0·3 m (1 ft) wide, through which a stockman can make an emergency exit, can be made in a wall of the run, but such a device is not advisable as it allows children or dogs to enter. A better alternative is to fix two upright steel posts in front of a corner of the run to provide a refuge for a stockman, from which he can climb over the wall. A dangerous bull can also be attached by a light chain which passes round the horns, through the nose ring to an overhead cable, one end of which is attached to the back wall of the box and the other to the end wall of the run. The cable should be at least 1·83 m (6 ft) from the floor so that the bull cannot reach it.

It is also advantageous to tether bulls out at grass during the summer months so that they can have additional exercise and are able to graze. There are several systems in use, but one suitable method is to have a centre spike or spikes which are driven or pressed into the ground with a pole about 2·12 m (7 ft) long, which can move round in a circle, fixed to the centre piece. The bull is attached to a chain at the free end of the pole and so can move round, but cannot dislodge the centre spikes with its head.

Feeding

Calves

Dairy calves are almost always reared by hand so that the milk which the dams produce can be sold. The colostrum produced by cows is commonly called beastings and is thick and yellow in colour. It is vitally important that each calf should receive colostrum from its dam, or from another freshly-calved cow, as soon as possible after it is born. From the first to the fourth day the calf needs 3·4 l (6 pt) of colostrum daily, in three equal feeds. After the fourth day normal milk is produced by the cow.

Calves can be reared on a liquid diet, supplemented by dry feeding, for about 10 weeks, but the labour requirements are high. Proprietary milk replacers are used instead of cows' milk because the cost per

Table 3.1. Examples of concentrate rations suitable for young cattle (parts by weight).

Food	Calves to be weaned from liquid food at 10 weeks of age	Calves on the early weaning system	Heifers
Crushed oats	20	20	30
Coarse weatings	10	—	—
Flaked maize	20	50	20
Linseed cake meal	40	10	—
Decorticated groundnut meal	—	10	20
Fish meal	10	10	—
Dried skim milk	—	10	—
Vitamin A and D preparation	1	1	—

unit is only about a third of the value of the whole milk it replaces. For the first 10 days 4·5 l (8 pt) of a mixture of equal parts of whole milk and milk replacer are given per day, and then 4·5 l (8 pt) of replacer only from the eleventh day, increasing to 5·1 l (9 pt) per day at the sixth week. This amount is held steady, and then gradually decreased until the liquid food is stopped at about 10 weeks of age. Good quality hay is given at 2 weeks of age and concentrate foods are fed from the third week onwards. An example of a suitable concentrate mixture is given in Table 3.1.

Fig. 3.12. Weighing concentrate food in a cowshed.

An early weaning system is becoming increasingly popular in which liquid feeding is discontinued at 5 weeks of age, from which time the calf is reared on concentrate foods, hay and water only. The concentrate diet used in this system must contain highly digestible ingredients with a good nutritive value. At 5 days of age milk and a milk replacer mixed in equal parts are fed at the rate of 3·4 l (6 pt) per head per day, and concentrate foods, hay and water are supplied. From the eleventh day onwards a milk replacer is fed according to the size of the calves, Jerseys being allowed 2·8 l (5·5 pt), Ayrshires and Guernseys, 3·4 l (6 pt) and Friesians 4 l (7 pt) per head per day given in two feeds. From the twenty-ninth day the amount is reduced by about 0·5 l (1 pt) daily so that liquid feeding ends on the thirty-fifth day. The consumption of the concentrate food is increased as rapidly as possible to 1·8 kg (4 lb) for Jerseys, 2 kg (4·5 lb) for Ayrshires and Guernseys and 2·3 kg (5 lb) for Friesians per head per day, but these levels are not exceeded. A concentrate ration formulation is given in Table 3.1.

Heifers

After about 3 months of age a moderate plane of nutrition is the aim. A low level of feeding retards growth and delays conception, while over-generous feeding can lead to early conception, which may give rise to dystocia, or to excessive fatness, which may reduce fertility. A general target is 0·5 kg (1 lb) liveweight gain per day, although 0·7 kg (1·38 lb) a day is the aim of those who wish to mate Friesian heifers at an early age for this breed so that they calve at about 24 or 27 months. Autumn-born calves can be turned out to grass in the spring and spring-born calves can be weaned on to grass at 8 weeks of age if the weather is warm. Concentrate foods need not be given provided the grass is plentiful and of a good nutritive quality. The heifers usually grow well during the grazing season and this has an economic significance since the use of grass, which is relatively cheap to produce, is maximised and the demand for expensive winter foods is minimised. Roughages and succulents will form the basis of the winter diet, good quality hay and silage being favoured. As a rough guide 1 kg of hay can be provided for each 45 kg of liveweight (2·5 lb/cwt) when hay is used alone. Growth rates on silage are likely to be lower because of the smaller amounts eaten. Roots, kale, and good oat or barley straw can be fed in moderate quantities, although straw alone will do little more than maintain the animal at a constant weight level. The young cattle are also given from 1·8 to 2·7 kg (4–6 lb) of concentrate food per head per day according to breed and the type of roughage being fed. During late pregnancy the heifers should be 'steamed-up' as described later under cow feeding. An example of a suitable concentrate ration is given in Table 3.1.

Bull calves being reared for stud

For the first 4 months of life the feeding of bull calves is similar to that advised for heifer calves being reared on a liquid diet. After that young bulls grow faster than heifers and so need more nutrients, particularly those providing energy, which are best supplied in a concentrate ration. The daily amount varies with individual animals but adequate quantities should be supplied to encourage rapid growth. Underfeeding retards the onset of puberty as well as reducing the growth rate. Good quality hay should be given *ad lib.* in addition to the concentrate allowance.

Cows

The rations for dairy cows are usually considered separately as the requirements for maintenance and production. The maintenance ration supplies the nutrients needed by the animal to maintain itself at rest without either gaining or losing weight and is related to body size. The production ration provides the nutrients needed to produce milk and can be calculated accurately by weighing the milk yielded. The bulk of the whole ration must be determined to check that the required nutrients are not contained in too great a bulk of food for the cow to consume. The most convenient method of judging the bulk of a ration is to calculate its dry matter content and then express this as a percentage of the cow's body weight. The approximate figures are, dry or low yielding cows 1·5–2 per cent, moderate sized, high yielding cows 2·5 per cent and large bodied, high yielding cows 3, or even 3·5, per cent.

Summer feeding

During the summer months grass forms the major part of the rations fed to dairy cows and should satisfy the maintenance requirements and a certain level of production. By careful management some farmers obtain sufficient nutrients from grass for the maintenance of a cow and the daily production of from 13·5 to 18·1 (3–4 gal) of milk throughout the grazing season, while others obtain this yield only during the early part of this period. To obtain maximum grass utilisation some farmers favour a carefully controlled strip grazing system by employing movable electric fences to regulate the area of grass grazed at a time. This system forces the cows to eat the whole grass plants, leaves and stems, and not select the leaves only. It also limits the amounts of grass damaged by treading and fouling. When grazed down the area is rested for a period of from 3 to 5 weeks, depending on the weather, and then grazed again.

Other farmers prefer a system of zero grazing, which means that the grass is cut and carted to stock housed indoors all the year round.

The grass should be cut daily or every 2 days and should be fed twice a day. Feeding may be from a trough or a self-feed trailer, and the quantity of fresh grass eaten will vary from 70 to 90 kg (150–200 lb) per head per day. Zero grazing is a demanding task which involves high machinery and labour costs for grass cutting and increased labour requirements for cleaning yards. Housing and bedding costs are high, and the disposal of the manure and slurry may present a problem. The advantages claimed are improved utilisation of grass as tramping and fouling are avoided, and capital expenditure is reduced because fences and piped water are not required in the fields. Labour is saved because no time is spent moving cows to and from the fields.

Winter feeding

For winter feeding the general aim is to use home-grown foods such as silage, hay, oat or barley straw, kale, or roots for the maintenance requirements at least. In most herds a maintenance ration suitable for all the cows is prepared, examples which satisfy the daily needs of Friesian cows being good silage 27·2 kg (60 lb); good hay, 9 kg (20 lb); or hay, 6·3 kg (14 lb) with roots, 18 kg (40 lb). During recent years silage has gained in popularity as a food for dairy cows but silage is very variable in composition. The intake of dry matter is reduced if silage has a high moisture content (over 80 per cent) and thus very wet silage is not suitable for high yielding cows. When silage is being self-fed for maintenance only, the cows may be rationed by time as it takes a cow about 2·5 hours to eat the required amount. The condition of heifers in their first lactation must be watched as they may be changing their incisor teeth which makes them unable to pull the silage out satisfactorily. Hay is an excellent conserved herbage for feeding to dairy cows for maintenance but its quality also varies widely. The barn drying of hay improves the quality during wet haymaking seasons.

Roots are not as popular as they were for feeding to dairy cows as the cost of production is high in relation to their nutritive value. However, they are relished by cows and often give better results in feeding practice than their chemical analyses suggest. Kale is an excellent feed for stimulating milk production and young, leafy crops have a high feeding value. Mangolds should be stored until after Christmas of the year in which they were grown before being fed since freshly-lifted roots may cause scouring. Turnips should be fed after milking for, if fed before, they may taint the milk. Not more than 25·4 kg (56 lbs) of any of the above should be fed per head per day. Dried sugar beet pulp has a higher feeding value than the roots mentioned and one unit will replace about two units of roots. Up to about 18·1 kg (40 lb) can be fed per head daily. However, dried sugar beet is more commonly used in the concentrate ration for milk production than in the roughage ration for maintenance. If more than 1·36 kg (3 lb) of dried sugar beet pulp is to be given in a single feed it should be soaked, using 4·55 l (1 gal) of water to 2·2 kg (5 lb) of pulp, otherwise it is liable to swell in the cow's alimentary canal and cause a digestive upset. On some farms there is a surplus of barley straw and this can be fed *ad lib.* because its bulk will limit the daily intake to about 0·45 kg (1 lb) of straw per 50·8 kg (1 cwt) liveweight per day. Thus Friesian cows eat about 5·9 kg (13 lb) per head per day. This only provides a small part of the energy required and so a supplement in the form of a mixture of equal parts of rolled barley and a proprietary mixture containing protein, minerals and vitamins must be given as well to form a satisfactory maintenance diet. The daily allowance per head is about 4·1 kg (9 lb) for Friesian cows.

Many farmers now prepare winter rations, composed of home-grown foods only, which, like summer grass, will provide for some milk production as well as maintenance. By adding any of the following quantities of foods to the maintenance rations previously suggested a ration for maintenance and 4·5 l (1 gal) production can be prepared; good hay, 3·2 kg (7 lb); kale, 12·7 kg (28 lb); good silage, 9·5 kg (21 lb). By adding double the amounts of good hay, kale, and good silage given above to the maintenance ration a home-grown ration to cover maintenance and 9 l (2 gal) production can be obtained, although the protein content may be low. If allowed to eat their fill, most cows on self-feed silage consume about 4·5 kg (10 lb) per 50·8 kg (1 cwt) liveweight per day, which is slightly over the theoretical allowance of good silage required for maintenance and the production of 9·1 l (2 gal) of milk. It is difficult to prepare a ration to provide for maintenance plus 3·6 l (3 gal) of milk from home-grown foods, but the following is an example: meadow hay, 4·5 kg (10 lb); clover hay, 6·3 kg (14 lb); marrow stem kale, 19·0 kg (42 lb); and crushed oats, 1·8 kg (4 lb).

For milk production in excess of that provided for by grass in summer or home-grown roughages in winter a balanced concentrate

Table 3.2. Guide to compiling balanced milk production rations.

Group	Feeding stuff	How to mix for balance			
Group II	*High protein foods*	Group II 1 part	e.g.	Decorticated cotton cake	1
	Cottonseed meal or cake	Group V 3 parts		Oats	3
	Soya bean cake	Group II 1 part	e.g.	Decorticated groundnut cake	1
	Decorated ground-nut cake	Group VI 2 parts		Maize meal	2
Group III	*Medium protein foods*	Group III 1 part	e.g.	Beans	1
	Linseed cake	Group V 1 part		Oats	1
	Beans	Group III 2 parts	e.g.	Linseed cake	2
		Group VI 1 part		Beet pulp	1
Group IV	*Balanced foods*				
	Brewers' grains—dried	All these are well balanced and can be mixed in any desired proportions or can be added to another balanced mixture.			
	Palm kernel meal				
	Coconut cake				
	Millers' offals, fine bran				
	Dried grass (best quality)				
Group V	*Starchy Foods*				
	Oats	Group II 1 part	e.g.	Soya bean cake	1
	Wheat	Group V 3 parts		Oats	3
	Rye	Group III 1 part	e.g.	Linseed cake	1
		Group V 1 part		Oats	1
Group VI	*Very Starchy Foods*				
	Flaked maize	Group II 1 part	e.g.	Decorticated cotton cake	1
	Maize	Group VI 2 parts		Maize meal	1
	Barley			Barley	1
	Sugar beet pulp—dried	Group III 2 parts	e.g.	Beans	2
		Group VI 1 part		Barley	1

ration is fed so that between 0·35 and 0·45 kg per l (3·5–4·5 lb per gal) contain the required nutrients.

Some farmers feed at this rate according to milk yield throughout the lactation, but it is best to vary the amount according to the stage of lactation as well as in proportion to yield. Early in a lactation, before the peak milk yield is reached between 5 and 10 weeks after calving, the cow should be fed generously so that she is not prevented from reaching her maximum potential. In herds where milk yields are recorded regularly this may be achieved by what is known as lead feeding, that is giving 1–2 kg (2–4·5 lb) per day of concentrate food above the theoretical requirement for milk yield. An alternative is to feed at the rate of 0·45–0·5 kg per l (4·5–5 lb per gal) during this period. Soon after calving a cow's appetite is poor, and although it rises over the following 2 months, even with lead feeding, high yielding cows are unable to eat enough food to meet their requirements in early lactation. They continue to produce extra milk by mobilising body reserves of fat to meet the deficit of energy. Thus they lose body weight during this period. However if the cow is to milk well and conceive this loss of weight must not be excessive. For a Friesian cow a loss of 25 kg (56 lbs) would be acceptable, but 50 kg (1 cwt) should be regarded as too great.

During mid lactation a cow's appetite is at a maximum and bulky food can be consumed to make up the body weight lost in early lactation. Thus the concentrate level can be reduced somewhat as extra concentrate food would tend to produce body fat rather than extra milk. In late lactation the level of concentrate food given should

Table 3.3. Balanced concentrate ration for the production of 4·5 l (1 gal) of milk.

Food	Amount	
	kg	lb
{ Decorticated groundnut cake	0·22	0·5
{ Flaked maize	0·45	1·0
Coconut cake	0·45	1·0
{ Bean meal	0·22	0·5
{ Crushed oats	0·22	0·5
	1·56	3·5

depend upon the condition of the cow and the quality of the roughage available.

The simplest way to compile a balanced ration for milk production is to mix foods according to their compositional quality. One classification of the common foods used for dairy cow feeding is given in Table 3.2 (Hunter-Smith and Gardner, 1953).

An example of a balanced mixture, made up according to the system given in Table 3.2 above is set out in Table 3.3, with the pairs bracketed together for clarification.

Other combinations of foods, balanced together according to the compositional grouping, will give similar results. In formulating a ration the cost of each of the foods included must obviously be carefully considered.

Pregnancy requirements

Cows and heifers must be well fed during pregnancy, particularly during the last 2 months, for they have to provide for the developing fetus and build up their body reserves in preparation for the coming lactation. Feeding in preparation for calving is generally known as 'steaming-up', and entails the giving of concentrates in addition to the maintenance ration. It is usual to start with 1·4 kg (3 lb) per day 8 weeks before calving and increase to 4·1–5·45 kg (9–12 lb) 2–3 days before parturition. The type of concentrate fed should be that normally given during lactation. The practical advantages of 'steaming-up' are that it increases the daily milk yield, lengthens the lactations of cows which milk for a short period, and increases the butter-fat percentage slightly.

Bulls

The fertility of bulls can be influenced by the diet, for bulls which are overfed become fat and lethargic while animals severely undernourished have not the necessary energy for the production of good quality semen. During the summer months quiet bulls can, with advantage, be tethered on grass, and during the spring and early summer the herbage will usually provide all the nutrients required. During the winter months most bulls are housed and are given about 1 kg of good hay and 1 kg of silage or kale per 70 kg liveweight per day (1·5 lb hay and 1·5 lb silage or kale per cwt). For semen production bulls are allowed from 1·8 to 2·7 kg (4–6 lb) of concentrates per day. Most farmers do not prepare a special concentrate ration for bulls but use the diet prepared for the milking cows. However, some bull owners mix a special ration and a suitable formula is crushed oats 2 parts, flaked maize 1 part, and decorticated ground-nut cake 1·5 parts.

Mineral requirements

Calves

When liberal amounts of milk or milk substitute, concentrates, and leguminous hays are fed mineral supplements are seldom necessary. A shortage of magnesium may cause hypomagnesaemia, the clinical signs being similar to those seen in cows and described later in the chapter, and a lack of iron may lead to anaemia. Both conditions occur in calves during the first 3 months of life, being fed on large

Table 3.4. The requirements of the important mineral elements.

Calcium	20 g (0·71 oz) for maintenance per day plus 2·8 g (0·1 oz) per 4·5 l (1 gal) of milk. A deficiency is unlikely to occur in a normal type ration. During the last 2 months of pregnancy add 17 g (0·6 oz) to the maintenance requirements.
Phosphorus	25 g (0·88 oz) for maintenance per day plus 1·7 g (0·06 oz) per 4·5 l (1 gal) of milk. In winter, rations with hay as the main roughage source may be deficient, and, in summer, cows grazing grass may not receive enough. During the last 2 months of pregnancy add 9 g (0·35 oz) to the maintenance requirements. The calcium/phosphorus balance is also important.
Magnesium	8 g (0·28 oz) per day for maintenance plus 0·75 g (0·03 oz) per 4·5 l (1 gal) of milk. Compounded dairy feeds for winter use are usually supplemented, but grazing cows receiving no supplementary foods need 50 g (2 oz) per day of calcined magnesite (magnesium oxide), particularly in herds where cases of hypomagnesaemia have occurred.
Potassium	Cows yielding 27 l (6 gal) of milk per day need under 50 g (2 oz) daily. Deficiencies are unlikely.
Sodium	9 g (0·35 oz) per day for maintenance plus 0·65 g (0·02 oz) per 4·5 l (1 gal) of milk. A winter diet of cereals and roughage may be deficient and grass varies widely in its sodium content.
Chlorine	13 g (0·45 oz) per day for maintenance plus 1·2 g (0·04 oz) per 4·5 l (1 gal) of milk. The chlorine content is generally increased when salt is added for the sake of sodium.

quantities of milk with no hay or concentrate food, which are growing rapidly. Feeding a concentrate ration to ensure a daily intake of from 2 to 4 g (0·08–0·15 oz) of magnesium and from 150 to 300 mg of iron are effective preventive measures.

Cows

The mineral content of the ration fed to dairy cows requires special consideration. Details of the quantities of the important minerals needed are set out in Table 3.4.

Vitamin requirements

Calves

A lack of vitamin A in the ration fed to a cow before parturition can lead to the birth of a weakly, blind, or dead calf. After birth the diets normally fed under practical management systems provide all the vitamins needed. Colostrum contains adequate vitamin A and the milk from well-fed cows contains the fat soluble vitamins A, D, and E. Proprietary milk replacers and concentrate weaning mixtures should be fortified with vitamins A and D by the manufacturers. Although the vitamin E requirements are met by the common diets the unsaturated fatty acids contained in cod-liver oil will destroy this vitamin and cases of vitamin E deficiency have been recorded when large amounts of cod-liver oil, for example 28·4 g (1 oz) per day, are given to correct a vitamin A and D deficiency. The deficiency produces a weakening of certain muscles leading to an unsteady gait, followed by an inability to stand, and death. The condition can be prevented or cured by supplying about 10 mg of vitamin E per day in the absence of unsaturated fats. The B complex vitamins required are provided by the rations fed in early life and are synthesised in the rumen as soon as this organ starts to function.

Cows

Vitamins A and D are the two which are essential for the well-being of dairy cows. A store of vitamin A is built up in the body during the summer if the cows are on grass, and this supply can be utilised during the winter months. Good quality silage and kale contain carotene and so provide an additional source of this vitamin if these foods are included in the winter ration. Vitamin D is mainly obtained by skin irradiation if the cows are out of doors during the summer and from well-cured hay during the winter. Concentrate rations are normally supplemented with synthetic vitamins A and D. The vitamin E requirements of adult stock appear to be small and should be adequately met if natural foods are given. Adequate amounts of the vitamins of the B complex are synthesised in the rumen, although the cow must receive an adequate supply of dietary cobalt for the formation of vitamin B_{12}.

Water requirements

Clean water should be supplied to calves even when they are being fed liquid food so that they can drink if they wish. If calves are being reared on the early weaning system it is essential that water should be provided *ad lib.* at the time when concentrate feeding is started. The quantity of water drunk by calves on a dry ration varies with the environmental temperature, but is roughly about 4·6 l (1 gal) per 45·4 kg (100 lb) body weight.

Older cattle can obtain a considerable part of the water they need from the food consumed if they are fed on succulent foods such as grass, kale, roots, or silage. The quantity of water which they drink will, therefore, vary, but dry cows of the larger breeds require between 36·5 and 45·0 l (8–10 gal) per day and during the last 4 months of pregnancy may need up to 70 l (15 gal) daily. If lactating, cows require in addition about 5 units of water for each unit of milk produced. It is important that dairy cows in milk should always have a plentiful supply of water available, as a shortage of water can seriously reduce the volume of milk yielded.

Breeding

Bulls are not used for service before they are 10 months of age, and a young bull should not serve more than 20 heifers in its first season. The maximum number of cows which an adult bull can serve in a year is normally 60. Heifers are generally not mated before 15 months of age in the case of Jerseys and from 18 to 21 months of age in other breeds. However some Friesian breeders are now serving their heifers at 15 months of age so that they calve when 2 years old. Even in well-grown animals of this breed at this age pelvic size may not be large enough to permit the easy passage of the fetus, and assistance at calving may be required in a number of animals.

Cattle will mate and breed at any time of the year, and cows come into oestrus at intervals of about 20 days, between the end of one period and the commencement of the next. The oestral period normally lasts for about 18 hours, but can vary from 6 to 30 hours depending, particularly, on the season of the year, the shorter periods generally occurring in the winter. During pregnancy cows rarely show signs of oestrus, but after calving oestral periods normally recur within 3–8 weeks.

The average time of ovulation is from 10 to 11 hours after the start of oestrus, and a suboptimal conception rate is obtained if cows are served early in oestrus. The usual advice is to mate or inseminate cows which come in oestrus during the morning on the afternoon of the same day, and those which are first observed in oestrus during the afternoon, on the morning of the next day.

The detection of oestrus is not always easy, particularly in winter and when animals are on their own. The important sign of oestrus is that the animal will stand when mounted by another cow or a bull. Other manifestations are attempts to mount other cows, restlessness, bellowing, swollen lips of the vulva, and a discharge of clear mucus from the vulva.

Cows are usually served somewhere about 80 days after calving, the aim being to obtain one calf per cow per year. The average lactation lasts for about 305 days which allows the cow to be dry for about 60 days before calving again. During this period the body reserves can be built up in preparation for the new lactation.

Occasionally bulls are allowed to run with the females and serve them as they come into oestrus. For example, a young bull may be turned out with a bunch of maiden heifers. In most dairy herds, however, the bulls are housed and service is controlled by the stockman. In this way a record of service dates can be kept.

It is important for the stockman to know as soon as possible whether the cows have conceived. The oestral period which occurs about 20 days after an unsuccessful service should be specially looked for, as its non-occurrence provides the stockman with the first sign of pregnancy. However, it is not an entirely reliable guide, for pregnant animals occasionally show signs of oestrus, while non-pregnant animals may not come into oestrus. It is most satisfactory to ask a veterinary surgeon to examine, by rectal palpation, heifers about 6 weeks and cows about 9 weeks after service to detect pregnancy. When carried out by an experienced veterinarian at the times specified this test is rapid and accurate.

The gestation period lasts for about 283 days. The mammary gland begins to develop within a few months of conception, but marked changes are not observed until about the sixth month of pregnancy. During the last month the udder increases greatly in size and the vulva also becomes swollen. Three to four days before calving the pelvic ligaments at each side of the tail slacken and the udder becomes distended. About 8 hours before the birth of the calf, drops of honey-coloured colostrum may appear at the ends of the teats.

The first stage of parturition, which comprises the dilation of the cervix and contractions of the muscles of the wall of the uterus, may last from 0·5 to 24 hours. The second stage, which terminates with the birth of the calf, varies from 0·5 to 4 hours in duration. The third stage, consisting of the expulsion of the afterbirth, is usually completed within 0·5 to 8 hours.

Artificial insemination

Artificial insemination has been employed very successfully in the breeding of dairy cattle, particularly by small farmers who find the process convenient and economical.

An artificial vagina is almost universally used today for semen collection, and the bulls at cattle breeding centres are trained to its regular use. An artificial vagina comprises a hard case containing a rubber lining. The lining is lubricated with vaseline and the space between the lining and the case is filled with hot water to give an internal temperature from 42 to 45°C (107·5–113°F). At the end away from that applied to the bull's penis there is a rubber cone and glass tube to collect the semen.

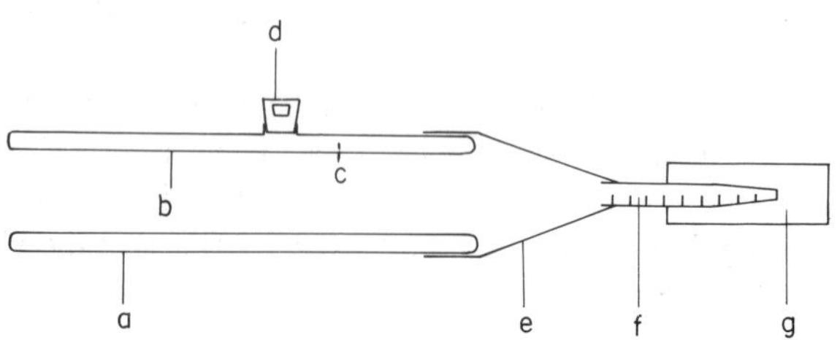

Fig. 3.13. Cross section of a bovine artificial vagina. (a) Rigid rubber cylinder, (b) rubber lining folded over ends, (c) space filled with water, (d) opening for introduction of water, (e) rubber funnel, (f) graduated glass tube, (g) sponge for protection and insulation.

Semen is usually collected twice weekly from each bull. The bull is led up to a 'teaser' cow confined in a metal crate and allowed to mount. When the penis is extruded the artificial vagina is held over its end so that the bull ejaculates the semen into the artificial vagina. The semen runs down into the glass collecting tube.

After collection the semen is carefully examined and, if of good quality, is then mixed with a diluting agent which increases the volume and prolongs its life. The rate of dilution depends on a number of factors, but is usually in the range from 1 in 50 to 1 in 150. It is generally recommended that dilution should leave at least 5,000,000 sperms per unit of insemination. Semen can now be frozen and held at very low temperatures and thus kept in a viable condition for a number of years. Liquid nitrogen at −196°C (−320°F) is the favoured freezing agent.

On the farm there are two methods of insemination in common use. In the first case the inseminator draws semen into a sterile plastic pipette by means of an attached syringe, inserts the pipette into the vagina and, if possible, into the cervix, and then expels the semen. A fresh pipette is used for each insemination. In the second method the required dose of semen is frozen in a hollow plastic tube in the laboratory after collection, and on the farm the semen is first thawed in warm water and then the tube is inserted into a metal holder with a plunger attached which retains the tube and expels the semen when the tube is in position inside the cow.

Milk production

It is generally considered that 10 months (305 days) is the optimum lactation length so that a cow may produce a calf each year and have a rest of 2 months' duration between the end of one lactation and the start of the next. However the actual duration of the lactation period can differ widely between cows.

The volume of milk produced daily during a lactation period varies according to the stage of lactation. For the first 5–10 weeks after calving the daily yield rises to a peak, which is maintained for a period varying from a few days to 2 weeks. After this the yield falls until the end of the lactation.

Quantity of milk produced

There are several factors which affect the total quantity of milk produced by cows during a lactation period.

The cows. As indicated earlier in the chapter the breed of cow influences the yield, and there are also variations due to strains within breeds. In addition body size has an effect, for large cows usually produce more milk than do small cows of the same breed. The age is also a factor as the yield of a first-calf heifer is about 75 per cent of its maximum, a second-calf cow about 85 per cent, and a third-calf cow about 90 per cent. The maximum yield is normally attained between the fourth and sixth lactations, after which there is an annual reduction.

Management. The level of feeding and the efficiency of milking are both important. The season of the year when the animal calves also has an effect for a cow or heifer calving in the autumn gives about 230–450 l (50–100 gal) more milk per lactation than an equivalent animal calving in the spring or summer. The reason is that an autumn calver is usually well fed on summer grass when pregnant and has a stimulus to increased milk production when turned out to grass the following spring, while a spring calver is often only moderately fed when pregnant because winter feeding is expensive, and the change from grass to concentrate feeding in the autumn tends to depress the yield.

Pregnancy and the dry period. The dry or rest period between lactations is important to allow the cow to prepare its body tissues for the strain of another period of milk production. Two months is usually considered to be the optimum length, and 6 weeks to be the minimum desirable, as a shorter interval will mean that the following lactation yield will be smaller than expected. This is particularly noticeable in the case of immature animals between the first and second lactations. As a cow is normally mated somewhere about 85 days after calving it will be pregnant while lactating, and pregnancy reduces yield and shortens the lactation length. A cow which does not conceive can be expected to yield about 30 per cent more milk before going dry than a similar animal which became pregnant less than 3 months after calving.

Quality of milk produced

There are also a variety of factors which determine the compositional quality of the milk.

The cows. As mentioned under the breed descriptions the fat content of the milk differs according to the breed, and there are also strain and individual differences within breeds. Cows of high yielding fat breeds nearly always produce milk with a high S.N.F. content, but this does not always hold true for individual animals. There is, on average, a fall in fat of about 0·03 per cent and in S.N.F. of about 0·1 per cent from one lactation to the next.

Management. Diets which are markedly deficient in fibre reduce the fat content of milk, and rations with a low energy level can lead to an appreciable reduction in the S.N.F. content. As the first milk drawn from a cow at a milking is poor in fat and the strippings are rich it follows that if a cow is not fully milked out the fat content will be low. If milking is carried out at unequal intervals during a day, e.g. at 10 and 14 hours, the fat percentage is lower after the longer period owing to the increasing pressure of milk in the udder which inhibits the action of the fat secreting alveoli.

Stage of lactation. The fat and S.N.F. percentages are high immediately after calving and then fall for about 2 months as the yield rises. The fat levels then tend to increase until the end of the lactation. The S.N.F. content remains fairly constant for a time and then falls if the cow is barren, or rises if the cow conceived about 3 months after the start of the lactation.

Health

A healthy cow holds its head in the normal, and not in a drooping position. The nose should be cool and moist, and the food ration should be consumed eagerly. The amount of flesh carried will vary considerably with the breed, but cows should not be too thin and the skin should move easily over the body muscles. Healthy cows ruminate at intervals and produce dung which is softer than that of horses. Cows frequently lick their own coats as well as those of others in the group. When being run as a herd cows tend to keep together and any individual remaining on its own should be carefully examined. Milk should flow easily from all four teats, and the udder tissue should be free from hard lumps. The daily yield of milk is normally fairly consistent and sudden falls indicate illness, although a reduction may occur when a cow is in oestrus.

Sick calves may scour, have sunken eyes, discharges from the eyes and nostrils or swollen or wet navels. They may also have swollen joints and walk with a stiff action. Some diseased calves show a very obvious abdominal movement when breathing.

The body temperature of a cow in milk varies from 38 to 39°C (100·5–102·5°F), and is generally higher when a cow is at peak yield. The pulse rate is about 60 or 70 beats per minute and the pulse is most easily taken at the artery which runs down the underside of the tail. The upper part of the tail should be grasped when standing behind the cow so that the fingers can feel the pulsations.

Common diseases and their prevention

Milk fever (hypocalcaemia) usually occurs during the first 72 hours after parturition, high yielding cows in their third or subsequent lactations being the ones most likely to be affected. A cow with milk fever staggers, lies down on its breast with its head turned backwards, becomes comatose, and if untreated, probably dies. The cause is a temporary dysfunction of the physiological mechanism involved in controlling the calcium level of the blood at a time when there is a sudden demand for calcium because of the start of milk production. In an attempt to reduce the incidence of this condition some farmers feed their cows a ration low in calcium during late pregnancy so that they are forced to mobilise their own body supplies of calcium thus keeping the physiological mechanism liberating calcium from the tissues active. Most affected cows recover if treated by an injection of calcium borogluconate.

Hypomagnesaemia (grass staggers or lactation tetany) is usually seen in lactating cows grazing on rapidly growing grass. The clinical signs are increased nervous excitability followed by staggering and falling, which may result in death through tetanic spasms. The condition is not caused simply by a deficiency of magnesium in the diet, for pastures on which it occurs may show a magnesium content as high, and sometimes higher, than pastures on which it does not occur. Although the primary cause is still not known a deficiency of magnesium in the blood is a factor and an effective preventive measure is the feeding of about 60 g (2 oz) of calcined magnesite per day to susceptible cows. Such a supplement must be given each day for a period before the cows go out to grass and continued until the danger has passed. The duration of the danger period will vary according to the system of grassland management on the farm. The supplement can be incorporated in the concentrate food fed to cows receiving such rations, but in herds being kept on low cost methods concentrates are not fed to the cows while they are grazing lush grass. In such cases heavy magnesium pellets may be administered to each cow by mouth for retention in the second stomach, so that small amounts of magnesium are released each day. Alternatively the pastures can be topdressed with magnesium oxide, an application of about 630 kg/ha (5 cwt/acre) being said to give protection for several years.

Bloat, or hoven, is the distension of the rumen with excessive gas formed by the fermentation of food, which in the case of dairy cows is nearly always herbage. The left side of the affected animal becomes markedly enlarged because of the ruminal distension, and the resultant pressure on the lungs and heart restricts the functioning of these organs and causes the animal to stagger and fall. In extreme cases death follows within a short period of time. The cause is grazing the cows on rapidly growing herbage which is low in fibre, particularly if it is rich in clover. The condition is seen most frequently in the spring before the ruminal bacteria have become adapted to herbage digestion. The incidence can be reduced, or even eliminated, by feeding hay before turning the cows out to grass, so that they do not gorge on the lush grass, or providing hay or straw on the pasture so that the cows can obtain fibre when they feel a need for it. Alternatives are to graze the cows on pastures likely to induce bloating for short periods at a time only, or to limit the amount of grass available by the use of strip grazing. This latter method has the added advantage that it forces the cows to eat the more fibrous stalks and not simply skim off the leaves which are relatively low in fibre.

Lameness can be a major problem in dairy cows. Abscess formation can occur in the feet of housed cattle forced to stand on rough concrete or concrete covered with wet slurry for long periods. One claw of one foot is usually affected. Cows walked over stony ground or grazed on kale stumps injure their feet, particularly in the skin between the two claws. As damp conditions soften the foot tissues the trouble is worse during long periods of wet weather. Bacteria gain entry through the wounds, which turn septic. Such lesions are very painful and the cow becomes acutely lame, goes off its food and shows a rapid reduction in milk yield. This condition is termed 'foul of the foot'. In both types of lameness prompt treatment is required to prevent the joints becoming infected. The walking of the cows through a foot bath containing either a 10 per cent solution of formalin or a 5 per cent solution of copper sulphate at regular intervals hardens and disinfects the feet and is a useful preventive measure.

Inflammation of the udder, termed *mastitis*, is, unfortunately, a relatively common disease of dairy cows. Most cases occur in the first third of a lactation, and about three-quarters of them are in a hind quarter of the udder. The incidence increases with advancing age. There are two clinical forms, acute and chronic. The acute form is easily recognisable as an udder quarter suddenly becomes swollen, hot and painful. The chronic form may follow acute mastitis and is accompanied by damage to the udder tissue, such as atrophy and fibrosis. In addition there is a subclinical form which cannot be recognised clinically but is revealed by laboratory tests. In both clinical forms if a few streams of milk are drawn on to the hand or the black disc on a strip cup, clots or flakes can usually be seen in the milk. In some chronic and in subclinical cases the milk will appear to be normal.

Mastitis is an infectious disease caused by a variety of microorganisms which gain entrance to the quarter through the teat. Predisposing factors are faults in the working of a milking machine, such as using too high a vacuum pressure, or bad milking techniques, such as leaving the machine on when all the milk has been removed, or hard pulling during stripping or teat-cup cluster removal, which damage the udder tissues and render them more susceptible to infection. Various methods can be employed to prevent the spread of the disease. Infected cows should be milked last and animals which are chronically affected should be culled from the herd. The regular sterilisation of all the milking equipment after use should be carried out efficiently, as a high standard of milking hygiene lowers the numbers of potentially pathogenic bacteria. The disinfection of the teats after milking by dipping them in a specially prepared iodophor/hypochlorite solution is a valuable help. When cows are being dried off at the end of a lactation all four udder quarters should be treated with an intramammary injection of an antibiotic as this practice has been shown to reduce the incidence of mastitis in a herd.

Contagious abortion, or brucellosis, in cattle is caused by the bacterium, *Brucella abortus*. In man the disease is known as undulant fever, and can be contracted through contact with infected cattle or through drinking infected raw milk. In a cow the clinical sign is the abortion of a fetus during the later months of pregnancy. The disease is spread through aborted calves or their afterbirths or the vaginal discharges of infected cows. Any of these may be licked by cows or contaminate foods being consumed by them. To prevent the spread of the disease in these ways aborted calves and their afterbirths must be burned. If the abortion occurred in a loose-box this must be thoroughly disinfected, and if it took place on a field the contaminated turf should be dug out. Aborting cows must be isolated for 2 months, or until all vaginal discharges cease.

Under the Brucellosis (Accredited Herds) Scheme efforts are being made to eradicate the disease on a national scale. Blood tests are used to detect infected animals and these can be eliminated from a herd. A herd which contains no reactors to this test is known as an accredited herd. The calves in these, as well as in other herds, are vaccinated between 4 and 6 months of age to give them protection against the organism for life. The vaccine used is a weakened strain of *Brucella abortus* known as strain 19, or, more simply, as S.19. The importance of the age of vaccination is that calves vaccinated before they are 6 months of age do not react to the blood test when of breeding age, whereas those vaccinated when over 6 months may give positive reactions to the test. This means that a positive reaction to the diagnostic test cannot be assessed, as it may be due to an active infection or a harmless reaction to the vaccine. Vaccinated calves are marked with a distinctive ear tag, although registered pedigree calves may be identified by a certificate if the owner prefers.

White scour is the most important disease of calves and it occurs particularly during the first, and to a lesser extent during the second and third weeks of life. It is caused by coliform bacteria which are normal inhabitants of the alimentary canals of adult animals. Calves which have not received colostrum have no resistance, and overfeeding on milk and cold damp housing conditions increase their susceptibility. The obvious clinical sign is a yellowish-white, pasty diarrhoea with faeces adhering to the tail and hind quarters. The calves go off their food, become lethargic and may develop a high temperature, and then, because the diarrhoea causes a marked loss of body fluid, they become increasingly weak. Death is a common outcome. Obvious preventive measures are the feeding of colostrum, or if this is not available its replacement by 28·4 ml (1 fl oz) of castor-oil, as a laxative, and 14·2 ml (0·5 fl oz) of cod-liver oil, to provide the vitamin A. The antibodies can be supplied by giving an injection of 500 ml (16·9 fl oz) of blood taken from the dam or another cow on the farm. Over-feeding, particularly on milk with a high fat content, should be avoided and the calves should be comfortably housed. The feeding utensils and pens used by diseased calves should be thoroughly disinfected before use by healthy animals as the virulence of the *coli* organisms is increased by passage through diseased individuals.

Handling

Dairy cows must always be treated with kindness if a satisfactory milk yield is to be obtained. They should be handled gently when entering and leaving buildings and not hurried when travelling to or from pasture. As they are creatures of habit it is important to establish a daily routine of feeding, milking, and general management and adhere to it.

Individual animals can be haltered in a manner similar to that described under horses, using a halter of a suitable size for either a cow or a calf, but unless trained are difficult to hold or lead with a halter. To catch a wild animal running loose in a yard it may be necessary to place a running noose around the neck, possibly manoeuvring it on the end of a long pole. A knot, possibly enlarged by the insertion of a wisp of straw, should be sited at an appropriate place in the rope to prevent the noose tightening round the neck and choking the animal. The other end of the rope should be passed round a strong post so that the animal can be held when it jumps forward and then drawn up to the post for handling.

During an examination an individual animal can often be quietened if its back is scratched. If this is not sufficient the best way to restrain a cow is to stand on the left side of the neck and hold the nose by inserting the thumb of the left hand into the left nostril and the first and second fingers into the other side, at the same time bending the head slightly to the left side. The right hand can be used to grasp a horn in a horned beast, or be placed round the neck of a polled animal. 'Bull-holders' or 'bull-dogs', which are metal nose clips, can be applied to the nostrils to relieve the strain on the fingers if the examination is likely to be prolonged. An alternative method of restraint, which may be used during an udder examination, is to raise the tail and hold it in an upright position. Both hands should be kept close to the base of the tail while holding it aloft.

Bulls, especially when over 1 year of age, are often treacherous and should have rings inserted in their noses. Bulls over 10 months of age must be ringed if they are to be exhibited in a public place, e.g. for show or sale. A bull pole with a snap link at the end of a wooden handle can be clipped on to this ring when the bull has to be led.

At the start of their first lactation some heifers kick while being milked. This can be prevented by arranging for an assistant to pass the heifer's tail between its hind legs, twist the tail round the front of the hock and hold it. If a leg is lifted to kick the pain caused by the pull on the tail checks the movement. Should this not be sufficient a strap can be applied above the hocks in a figure-of-eight pattern and tightened so that kicking is prevented. An alternative is to apply an udder kinch which is effected by passing a rope well up behind the udder, bringing the ends up on either side of the body in front of the hind legs and passing one end through a loop at the extremity of the other. An assistant pulls on the free end of the rope, and, although the heifer may tend to crouch, it will not kick.

Fig. 3.14. Cows in a crush.

In order to examine a cow's foot properly it is necessary to lift the leg. To raise a foreleg a rope can be tied just above the fetlock and passed over the back, which can be protected from rubbing by the insertion of a sack, to an assistant on the other side of the cow who pulls the leg up. Cows do not balance on three legs as well as horses and it may be necessary to place a bale of straw under the raised leg to give support. A hind leg can be lifted by inserting a pole in front of the hock to be lifted and behind and above the one on the opposite leg. A man on each end of the pole lifts the hock, at the same time leaning against the cow's quarters to help to maintain its balance.

A crush in which the cows can be closely confined is a useful item of equipment when cows have to be handled individually. The cows should enter a crush through a funnel-shaped approach passage. When the cow is in the crush it is checked by a barred gate in front and held in position by a bar of wood or steel tubing pushed horizontally through the crush walls just behind its tail. The position of the bar can be adjusted according to the length of the individual beast. The length is usually about 2·13 m (7 ft) per cow, and the width about 69 cm (2 ft 3 in) in the clear. The height of the sides varies but is usually between 1·22 and 1·52 m (4–5 ft).

Casting

For some manipulations it may be necessary to cast an animal. In the case of a calf this can be dome relatively easily if an assistant stands on the left side of the calf, grasps the lower jaw with his left hand and turns the calf's head towards its right shoulder. With his right hand he grips the upper part of the right hind leg, lifts it, and the calf falls on its left side. The head and neck can be held down with the left hand and knee while the right hand pulls the calf's right hind leg forward, thus exposing the udder regions for the operator. If older calves are being cast a collar, with a short length of rope attached, may be placed round the calf's neck. The rope can be run round the right hind leg, threaded through the collar, and held. This makes it easier to keep the leg forward.

The common means of casting older animals is known as Reuff's method. A rope about 9·2 m (30 ft) long is used. One end is attached to the headpiece of a strong head collar, or possibly made into a fixed loop and passed round the cow's neck. Should the animal be horned this loop can be placed round the base of the horns. The free end is brought back along the top of the neck to the withers and then passed under the chest and up to form a half hitch. The rope is next passed along the centre of the back and a similar loop made round the body just in front of the udder. The free end is pulled straight back by two men and the cow sinks to the ground. It is necessary to have a man at the cow's head to hold the head down as the cow falls and when it is on the ground. Should the manipulation take some time the legs can be roped together by a rope applied in the fetlock region.

Milking

The regular periodic complete removal of milk from the udder is one of the most important single factors in sustaining lactation. It is customary to milk cows twice a day, but milking three times daily produces yields about 15–20 per cent more. The value of this

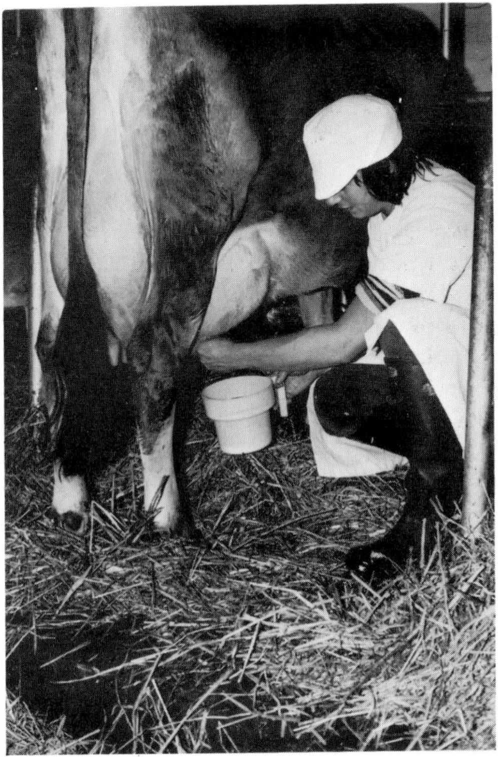

Fig. 3.15. A strip cup in use.

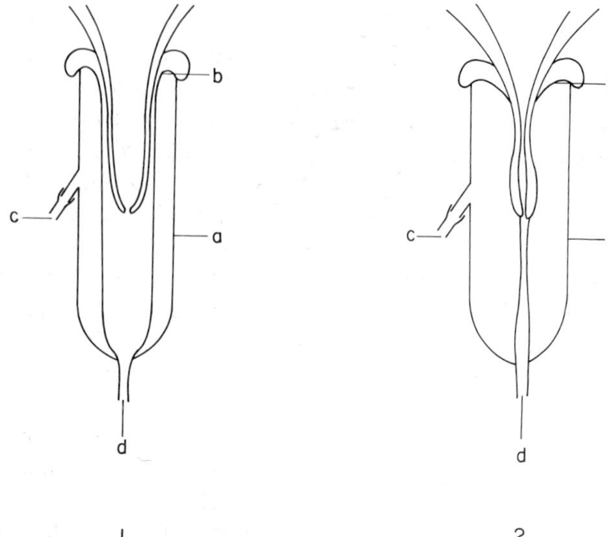

Fig. 3.16. Teat cup. (1) Milk flowing (vacuum at (c)), (2) teat squeezed (air pressure at (c)). (a) Metal shell, (b) liner, (c) air alternately withdrawn and admitted at this point, (d) constant vacuum.

additional milk is usually less than the extra cost of the labour required to carry out one more milking per day. On twice a day milking 12-hour intervals are best, but in practice intervals of 10 hours during the day and 14 hours overnight give satisfactory results. It is important that the routine, when established, should be followed regularly.

The person milking the cows must wash his hands before starting and keep them clean during milking. He must not smoke nor spit and must wear a clean overall and cap. The cows should be prepared for milking by having their udders washed, preferably with warm water, and then dried, disposable paper towels being very suitable for this purpose. A few squirts of milk should be drawn by hand into a strip cup before milking starts as this foremilk has a high bacterial content. Milk can be removed from the udder by hand, but a machine is almost always used.

A milking machine draws out milk by vacuum in much the same way as does a sucking calf. A teat cup is applied to each teat, the four teat cups required for a cow being known as a teat-cup cluster. Each teat cup consists of a metal outer shell and a synthetic rubber liner. A single direct vacuum cannot be applied continuously as this would trap blood in the lower parts of the teats, and so an additional component is introduced using an alternation of vacuum and atmospheric pressure outside the teat. This causes the teat to be squeezed and then released and so massage is applied. This is effected in practice by connecting the tube from the inside of the liner direct to the vacuum, and the tube to the metal shell, which operates outside the rubber liner, to a pulsator which applies vacuum and atmospheric pressure alternately. Thus, when there is vacuum pressure outside and inside the rubber liner the rubber is in its normal position and milk is withdrawn, but when air is admitted outside the liner the liner wall compresses the teat and the milk flow stops. The vacuum pressure should be about half atmospheric pressure, that is about 38 cm (15 in) of mercury.

When the cows are housed in a cowshed the milking machine pipeline extends along the line or lines of cows and the milker takes the milking units to the cows. In most cowsheds the milk from each cow is collected in a bucket so that individual cow yields can be recorded. The milk is then tipped into cans for removal to the dairy. Under the milking parlour system the milk goes from the teat cups into a pipe leading to the dairy. The milk may pass directly to the dairy or through a glass jar in which the yields from an individual cow can be measured. In some rotating parlours the operator only has to wash the udders, remove the foremilk and attach the milking machine teat cups as the teat cups can be automatically removed by a teat-cup removal device and the milk is automatically transferred to the dairy. It is possible, in such plants, for one man to milk 150 cows per hour.

Milk cooling

In the dairy the milk must be cooled to reduce the activity of the bacteria in the milk, as this improves its keeping quality. The cooling should be carried out as rapidly as possible, and the temperature should be reduced to, and maintained at, about 7·2°C (45°F). The method employed will depend on the type of milking machine being used. In bucket plants the milk flows over the surface of a cooler which has water passing through internal pipes. In systems where the milk is conveyed direct from the cow to a milk can, cooling is effected by means of a rotating in-can cooler. This is placed in the full milk can and water passing through the inside of the cooler

Fig. 3.17. Bulk milk tank.

causes it to rotate and stir the milk so that the cooling is uniform throughout the can. An alternative is to use sparge-ring cooling, when a perforated ring is placed over the neck of the can so that water runs over the outside of the can. Although in these three methods water from a mains supply is generally used it is possible to install chilled water units which provide iced water for circulation. This greatly increases the efficiency of the milk cooling.

On farms with large herds refrigerated bulk milk tanks are in general use. The milk passes through the pipeline into the tank without coming into contact with the outside air. The tank is thermostatically controlled so that the milk is cooled to, and kept at, a temperature of 4·4°C (40°F). The milk is continually stirred by an agitator so that this temperature is maintained throughout the bulk of the milk. A road tanker transport lorry calls daily at the farm and the milk is pumped from the farm tank into the lorry. As a tanker of this type can carry more milk than can a conventional lorry transporting milk in cans there is a considerable saving in transport costs.

Cleaning and sterilising milking equipment

Failure to keep milking and dairy equipment properly clean will lead to the production of milk which is of poor keeping quality because bacteria from dirty utensils infect the milk. All equipment must therefore be properly washed and sterilised after each milking.

The process starts with a cold water rinse which removes visible dirt and milk residues which might protect the bacteria from the sterilising agent used to kill them. This should be carried out immediately after milking as if the milk dries on the surfaces of the equipment it becomes very difficult to remove. After this rinse the equipment is sterilised by chemicals or heat, chemical sterilisation being the easiest and most common method. Sodium hypochlorite is frequently employed, its action depending on the chlorine content. It is used in conjunction with a detergent to remove any fat or grease. The detergent-steriliser wash should be used at a temperature of at least 46·1°C (115°F) and brushes should be employed for the teat cups and cleaning rods for the milk tubes. The chemical and detergent selected must be approved under the Milk and Dairies (General) Regulations 1959. After chemical sterilisation the equipment should be rinsed and allowed to drain and dry. Sterilisation by heat is more efficient, steam at 100°C (212°F) for 10 minutes being employed. Automatic electric steam raisers are generally used, as they are easy to work and produce steam rapidly. Small items of equipment and utensils are placed inverted on shelves in a chest to which steam is admitted, while pipeline parts can have steam forced through them in the reverse direction to the milk flow. On some farms where chemical sterilisation is normally practised steam sterilisation is used once a week as a safety measure.

Administration of medicines

Medicines are frequently administered to cows in the form of a drench using a narrow-necked bottle. The operator should stand on the right side of the cow in the region of the shoulder and pass his left hand over the cow's face so that its head can be held by inserting the fingers into its mouth behind the dental pad with the palm of the hand over the bridge of the nose. The cow should not be held by the nose. It is a help to have an assistant standing on the left side of the cow steadying the head by holding the upper part of a halter, or the horns if the cow is not polled. The neck of the bottle is eased into the right side of the mouth and the medicine poured in slowly in small quantities at a time. The fingers of the operator in the mouth cause the animal to move its tongue and facilitate swallowing. Care should be taken to see that the neck of the bottle is not bitten by the molar teeth and, should the cow cough, medicine pouring should be stopped and the head released.

Subcutaneous injections can conveniently be made behind the shoulder, intramuscular injections in the hind quarters and intravenous injections in the jugular vein in the neck. Intramammary injections are made into the teat canal, generally using specially prepared tubes which contain the medicament. Milk from quarters affected with mastitis treated with an antibiotic must not be sold until all the antibiotic has been excreted. Mastitis treatment packages have a label attached showing the maximum excretion time so that the period for which the milk must be withheld from sale is known.

General

Vices

Udder sucking

Some cows develop the habit of sucking milk from other cows or from their own udders. The application of a halter with blunted spikes protruding from the front of the noseband may deter the cow by causing pain either to itself when it sucks its own udder or to another cow it attempts to suckle which causes this animal to kick or move away. If this is not effective a bull-ring may be inserted through the nasal septum with a second ring dangling from it. This can be dangerous as the second ring may catch on tree branches or fence posts. An aluminium anti-sucking shield, which fits into the nostrils, and drops in front of the mouth, is the most satisfactory preventive.

Minor operations and manipulations

Disbudding

Cows kept in yards should have had their horns removed as they tend to be quieter and can be kept in smaller areas with less risk of injury. This ideal is best achieved by disbudding the calves at an early age. If the operation is carried out within the first week of life the horn bud may be destroyed by chemical cauterisation. A wetted caustic potash stick was the original agent used for chemical cauterisation, but the caustic tended to run and could injure the calf's eyes, and so a collodion solution of caustic chemicals is now used. As the collodion rapidly sets hard on exposure to air there is little risk of the chemical running. The hair round the horn bud is clipped away, the area is cleaned, and the collodion solution is rubbed well in with a brush. A second coat is then applied without rubbing so that a firm thick skin is formed.

After a calf reaches the age of 1 week an anaesthetic is required by law. The application of a hot iron is the disbudding method most frequently used, and the operation is best carried out when the calf is about 3 weeks of age. At one time irons heated in a fire were employed, but now electric disbudding irons are used. The tip of the iron is shaped to fit over the horn bud and is heated by electric elements incorporated in the iron. A ring of seared tissue about 2·54 mm (0·1 in) deep must be made all round the horn bud.

After the horns are partially or fully grown, animals can be dehorned with horn clippers or a saw, clippers being generally used on immature and a saw on mature animals. Whichever instrument is used, the horn should be cut off as near to the head as possible and it is advantageous to remove a ring of hair with the horn.

Hoof trimming

Cows tend to develop long and misshapen hoofs if kept in a stall in a cowshed or in a yard on straw for long periods, and some cows tend to wear their hoofs unevenly when on grass. Hoofs which are overgrown or unevenly worn should be trimmed by rasping or cutting the lower surface with a hoof knife, although overgrown toes may need to be cut back in front.

Minor trimmings may be carried out when the cow is in the standing position, but for major trimming operations it is usually necessary for the cow to be cast and experienced help should be sought.

Supplementary teats

Occasionally heifer calves are born with supplementary teats on their udders. These may be harmless, although unsightly, but may be harmful if situated at the base of a main teat. In either case, they should be removed before the calf is a week old. The calf should be cast and a local anaesthetic should be injected at the base of the teat and the unwanted teat cut off with a pair of curved scissors.

Ringing bulls

For ringing, a bull should be haltered and tied tightly to a beam or post. Using a special nose punch a hole is made in the septum dividing the nostrils internally. The hole should be sited behind the soft nasal tissue so that it is in the septum itself and neither in front of it nor too far back into the cartilage. A ring should be passed through the hole, closed and screwed up. These rings are usually made of copper, but may be of stainless steel or other metal alloys which do not rust. Because of constant rubbing on a manger or chain the rings sometimes wear thin and so should be examined periodically and replaced when necessary.

Milk marketing

In England and Wales the marketing of milk is controlled by the Milk Marketing Board. All farmers producing milk for sale must be registered with the Board, and the Board is obliged to purchase all milk of marketable quality which producers offer for sale. There is a hygienic quality scheme which provides for the payment of a reduced price for milk delivered in any month in which a test failure occurs within 6 months following an initial failure. Payment is also varied according to the compositional quality of the milk.

The Milk Marketing Board is also responsible for supervising the National Milk Records Scheme. The scheme is voluntary. Milk yields are noted and samples for fat analysis are taken monthly by a Recorder, who records this information and then forwards the samples to the regional laboratory. The documents are sent to Head Office for calculation by computer and presentation in the form of a monthly statement and this is sent to the member, accompanied by a certificate for each cow at the end of each 305-day lactation. Protein testing of the milk from individual cows is a supplementary service.

The keeping of cow milk-yield records enables the value of individual dairy cows to be assessed, this information being of particular value when breeding animals are being selected as stated

Fig. 3.18. Milk recording.

earlier. It also provides the only economical way of rationing concentrate food and the detection of a sudden drop in yield is the quickest way of noting the start of a disease.

In addition, business services of farm accounting backed by advisory work are run by the Board's staff. A farm's costs and physical inputs are analysed in detail and staff members discuss the results with the farmer to assist him to improve his production efficiency.

Welfare Code

Code No. 1 of the recommendations for the welfare of livestock made under the Agriculture (Miscellaneous Provisions) Act 1968 deals with cattle. The main provisions covering calves are that individual and group pens for unweaned calves should preferably have solid partitions from the floor of the pen so as to limit the spread of disease or vice, but that each calf should have an opportunity to see other calves. A dry lying area should be available and conditions which could produce chilling should be avoided, particularly for calves up to 4 weeks of age. Calves should receive colostrum as soon as possible after birth and liquid food daily during the first 3 weeks of life. They should normally have access to palatable unmilled roughage and fresh clean water by the end of the second week of life. A calf should not be removed from the farm of birth for at least 3 days, unless for suckling by another cow or direct to a place of slaughter. It should not be disposed of to a dealer or sent to a market before it is 7 days old, unless with its dam.

The requirements of dairy cows specially mentioned are that adequate facilities should be available for the proper restraint of animals when necessary, and that calving boxes should be provided. All cattle, whether tethered or in pens, should have sufficient freedom of sideways movement to be able to groom themselves without difficulty and sufficient room to lie down freely. Fractious or horned cattle should not be loose housed where there is danger of injury or bullying. Any injured or ailing animal should be segregated for treatment without delay, and veterinary advice sought if necessary.

References

ANDREWS A.H. (1973) *Vet. Rec.* **92,** 275–282.

HUNTER-SMITH J. & GARDNER H.W. (1957) *A short guide to the feeding of dairy cows.* Gardner, St. Albans.

CHAPTER 4
BEEF CATTLE

Introduction

Beef is a meat which is greatly favoured by consumers and its production constitutes an important sector of the agricultural industry. At one time the policy was to keep herds of cattle of the beef breeds to produce the animals required. The calves, after being suckled by their dams, were fattened on grass or in yards so as to give large carcases which provided the big family joints demanded. Now most of the cattle fattened are the progeny of dairy cows. These can either be pure-bred or result from crossing with beef bulls. During their early lives, they are reared artificially on milk substitutes and concentrate foods. The housewives' demands for smaller joints with less fat has meant that animals are now slaughtered at an earlier age.

The main reason for the change in beef production methods is that, in most areas, the older systems have become too expensive in their requirements for land and labour to be economically viable. This has led to intensification coupled with an increase in the scale of production, or, alternatively, to the keeping of the original number of animals on a smaller area, which allows more land to be used for other farming enterprises. The wider use of polling and dehorning assists this change by enabling more cattle to be kept in a given space. However, there are parts of the country where the land is poor or unsuitable for growing arable crops and in such areas it is still economic to keep beef cattle at grass. These areas are particularly the more mountainous districts and are used to maintain breeding herds to produce calves to be fattened elsewhere.

The fact that beef cattle are being slaughtered at an earlier age means a more rapid financial turnover, which is an economic advantage. It also means that less manure is produced per animal, which suits a change in arable farming methods. In the past the farmyard manure from cattle yards was greatly prized and the keeping of cattle for long periods was thus advantageous, but now farmyard manure is expensive to apply to arable land and artificial fertilisers are being spread instead. However, this policy change has caused difficulties. Slaughter at lower weights means that more beasts are required to produce a given weight of meat, and, partly for this reason and partly because of a reduction in the number of dairy cows, beef producers have been short of calves to fatten. This has led to a rise in calf prices which, along with other causes of inflation, has forced up the price of beef and this is meeting with consumer resistance.

In the future, high production costs appear to be inevitable with no comparable rise in the price of beef and so more efficient management will be required to maintain the existing levels of profitability. It also seems likely that an increased use of certain meat substitutes, chiefly made from soya beans, may pose a real challenge to the beef industry.

Definitions of common terms

Store cattle. Young animals of either sex kept growing slowly in preparation for fattening later.

Bullock or steer. A castrated male animal over about 1 year.

Veal. The meat from calves slaughtered when under 15 weeks of age.

Slink veal. The meat from calves slaughtered at a very early age which is mainly used for products like meat paste.

Beef. The meat from cattle killed when 10 months of age or over.

Bull beef. Beef produced from uncastrated male animals.

Cow beef. Beef from unwanted dairy cows.

Most beef is produced from bullocks, although heifers are also fattened. The daily liveweight gains of heifers are less than those of bullocks and fatter carcases are produced unless they are slaughtered at lower liveweights or fed on lower energy diets. Entire bulls are also used for beef production. They make greater weight gains than bullocks in a set time and utilise their food more efficiently. In addition the carcases are less fat. A considerable amount of beef is also obtained from culled dairy cows.

Principal breeds

Beef breeds

Cattle of all the beef breeds should have compact, wide, deep bodies with legs which are short below the knee and hock and set well apart. Viewed from the side a rectangular appearance should be presented, with the lower body line straight and parallel with the back. The hind quarters should be well developed with the back of the thighs descending almost to the hocks. The aim is to produce a carcase with the maximum amount of muscle and the minimum quantity of offal. There should be an even distribution of subcutaneous fat in the finished animal without any patchiness. Early maturity and rapidity of fattening are other desirable economic qualities.

Aberdeen Angus
Black polled early-maturing cattle producing high quality beef with an even deposition of fat in the muscle. The bone is small and fine and there is a high proportion of meat in the carcase. The high quality carcase characteristic is passed on to the progeny resulting from cross-breeding. Cattle of this breed are particularly adapted to intensive feeding in yards and produce carcases with high killing out percentages if slaughtered at a relatively early age.

Hereford
A red breed with a characteristic white face and other white markings.

Fig. 4.1. Hereford bull.

Although larger and maturing more slowly than the previous breed Hereford cattle are still relatively early-maturing and are particularly suitable for fattening on grass pastures as they are excellent grazers. They are hardy and the breeding cows are normally kept out at grass all the year round. Most strains are horned, but a polled type has been evolved.

Beef Shorthorn

A horned breed which is red, white, roan, or red and white in colour. Similar in size to the Aberdeen Angus, but heavier boned. There is a tendency for the muscle to be covered by uneven layers of fat.

Devon

A red-coloured, horned, early-maturing breed with fully grown animals attaining a large body size. Devon cattle are docile, hardy and good grazers able to utilise rough as well as good grass. The cows are probably the heaviest milkers of the beef breeds, and fattened animals produce high quality beef.

Sussex

Another red horned breed which produces high quality beef, but is slower maturing than the Aberdeen Angus. A hardy, good grazing breed, which can be kept under unfavourable circumstances as regards food and climate.

Galloway

A black or dun-coloured polled breed with a thick rough coat. They are hardy and can be kept on high hill land on which cattle of the breeds mentioned above will not thrive. Although slow-maturing, the carcases are of good quality without an excess of fat.

Highland

Very hardy cattle with a long thick coat and very prominent horns. Found in various colours, most being fawn, brown, or brindle. Have a low growth rate, but can be kept on poor quality moor land.

Luing

A breed which has been recently developed by crossing Beef Shorthorn and Highland cattle with the object of combining the early maturity of the former with the hardiness of the latter.

Fig. 4.2. Highland cattle.

Inter beef breed crosses

White Shorthorn bulls crossed with black Galloway cows produce Scottish Blue Grey Cattle. The steers can easily be fattened at from 18 months to 2 years of age and yield carcases with a high percentage of good quality meat. Cows of this cross are polled, hardy, and good mothers, and when mated with Aberdeen Angus, Beef Shorthorn, or Hereford bulls produce good quality beef cattle. Beef Shorthorn bulls mated to Highland cows produce hardy progeny which mature more rapidly than pure Highland animals. Mated to Aberdeen Angus or Shorthorn bulls females of this cross give birth to good beef type calves.

When used for crossing Aberdeen Angus and Hereford bulls 'colour mark' their calves. Aberdeen Angus cross calves are always polled and are either black or blue roan in colour, while Hereford cross calves always have a white face and other white markings. Beef Shorthorn bulls do not colour mark their progeny.

Recently imported continental breeds

Charolais

A French breed, which excells in rate of weight gain and in the high

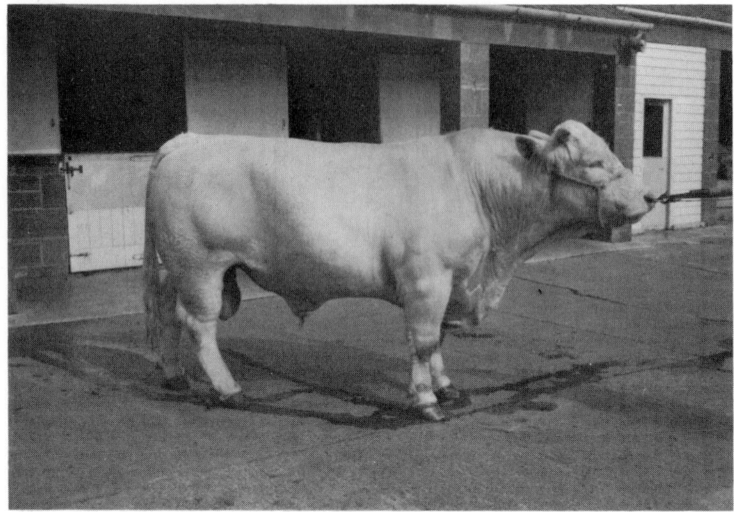

Fig. 4.3. Charolais bull.

proportion of lean meat in the carcase. The cattle are usually white in colour, although some are cream.

Limousin

Another French breed with a red body coat colour, shading to a lighter tan on the legs. Has a high carcase meat yield with a low percentage of bone.

Chianina

An Italian breed, which is the tallest in Europe. Adult cattle are white, with a black nose, black round the eyes, and a black tail switch. The calves are born tan in colour, but turn white when about 2 months old. Despite the large size of adult animals the calves are small at birth and so calving difficulties are encountered less frequently than in some other breeds.

Simmental

Originally from Switzerland, this breed is now popular in other parts of Europe. It is a dual purpose breed, red in colour with white markings on the head, belly, legs, and tail switch.

In general these cattle have a large mature size, grow rapidly and lay down fat at a later age than the smaller breeds. Bulls of these breeds are useful for crossing with cows of the dairy breeds, particularly Ayrshires, Guernseys, and Jerseys. Their early-maturing qualities improve the growth rate of their progeny out of slow maturing cows, and the carcases tend to have a higher muscle-to-bone ratio. However the calves of some of these crosses are large at birth, and calving difficulties and still-born calves are encountered more frequently than when the cows are mated to bulls of their own breed. Calves sired by a Charolais bull can be recognised by a general cream appearance, although white markings may be present.

Dairy breeds and their crosses

Friesian

Pure Friesian male calves are eminently suitable for intensive methods of beef production and can also be used for less intensive systems. However the progeny of crosses with beef breeds, such as Charolais and Hereford, are more suitable for the extensive systems.

Ayrshire

Male calves are not suitable for beef production, but crosses with Friesian, Charolais, Hereford, or Beef Shorthorn bulls produce beef type calves.

Jersey and Guernsey

The male calves are unsuitable for meat production but it has been found that calves out of Channel Island cows by Charolais bulls can be used for intensive meat production systems.

Breeding systems

Attention is now being paid to the testing of sires used for the breeding of both purebred and cross-bred cattle. Progeny testing is the most efficient method, but it is a long and expensive operation. It involves mating a number of females, certainly not less than twenty but preferably more, to the bull and keeping records of the growth rates, food conversion efficiencies, and carcase qualities of the progeny. A quicker method is performance testing in which bulls to be used for breeding are themselves recorded during their growing period. As growth rate has a high heritability a bull which has shown a good growth potential is likely to produce offspring with better than average growth rates. The best period for recording is from 90 days of age up to 300 days. There is much to be gained from measuring food conversion efficiency, as well as weight for age, in performance tests conducted under conditions of controlled feeding. In most cases there is a correlation between weight gain and food consumed per unit of gain because the shorter the time it takes an animal to reach a given weight the lower will be the requirements for purely maintenance purposes. However a fast-growing bull could have an unduly large appetite.

Identification

Beef cattle can be permanently identified by a number tattooed in the ear, but an animal has to be caught and held while this mark is read. For this reason this is not a practicable method for the easy identification of individual animals in large groups. Metal or plastic ear tags, which can be read without catching the animal, are probably the best method of identifying fattening cattle.

Housing

Calves

The actual type of housing provided for beef calves varies according to the method by which the animals are being fed. Early weaned calves can be kept in the type of accommodation described in the previous chapter.

Calves being multiple suckled. Under this system each nurse cow rears three or four calves at a time. The calves are housed, there being two common methods. In the first the calves are kept in groups in loose boxes and the cows are brought to the calves twice a day for suckling. In the second the nurse cows are tied and fed in a cow shed or covered yard twice a day, being allowed a width of 1·22 m (4 ft) per cow, and the calves are kept in calf pens and brought to the cows twice a day for feeding.

Calves being fed for veal. The calves are usually confined individually in wooden pens with slatted floors. The front should be made of bars so that the calves can see the animals in the opposite pens as calves have a herd instinct and are more content when they can see others. Veal calves are sometimes reared in multiple pens. These also have slatted floors and the calves may be fed in communal troughs or in individual pails. Veal calves should be kept in well-ventilated buildings maintained at between 18·3 and 21·1°C (65–70°F). The relative humidity should be below 80 per cent if possible. Subdued light is generally provided because it keeps the calves quiet and assists in the control of flies, but total darkness is not required for the production of white veal and should not be allowed.

Fattening cattle

The accommodation need not be elaborate, but there should be good ventilation without draughts at stock level. The design should allow for easy feeding and manure removal. When bulls are kept for beef they must be securely housed under conditions which permit their safe keeping.

Yards. After weaning beef calves are normally kept in small groups in yards. It is best to keep groups of cattle together from the early rearing stage through to slaughter, because they settle to a social order which is disturbed when groups are mixed. Animals within groups should be matched for size and it is best to have no more than twenty beasts per pen.

The yards are usually fully or partially covered with the fronts left open to ensure adequate ventilation. However, in the milder parts of Britain roofless yards are used and provided the cattle have good horizontal shelter from cold winds, often supplied by the retaining walls of self feed silage pits, yearling and 2-year old animals do as well as in covered yards.

When 6 months old each animal is usually allowed an area of 3·54 m^2 (40 sq ft) rising to double this figure at 2 years of age. The floor is generally solid with a bedded area covered with a thick layer of straw or other bedding material. However, slatted floors without any litter are used, a smaller area of about 2·31 m^2 (25 sq ft) being provided for animals at 6 months of age. On large units it is advantageous to have pens of differing sizes so that groups can be moved to larger pens as the need arises. Some feeders have found it beneficial to fit cubicles into the yards, as described under dairy cow housing.

The feeding troughs are generally placed along the open front of a yard so that they can be easily filled, probably from a self-unloading trailer. Adequate trough space must be provided so that all the cattle can feed at the same time. A length of about 0·69 m (2 ft 3 in) per beast is generally allowed. Self filling water troughs must also be provided in each yard.

Breeding cows

Spring calving cows of the pure beef breeds are generally not housed, but natural shelter in the form of stone walls or woods is required, particularly to afford the young calves some protection from the weather. If this is not available then cheap structures such as straw bale enclosures, usually three-sided, covered with a corrugated iron roof, may be erected for this purpose. Autumn-calving cows are often housed in yards during the winter months. Alternatively cheaply constructed cow kennels, similar to those used for dairy cow housing, may be provided.

Feeding

Calves

Early weaning system. The method described in the chapter on dairy cows can be used for calves of the dairy breeds or their crosses.

Single-suckling. The calves are allowed to suckle their own dams. To be successful with this rearing method costs in the form of food must be kept as low as possible because this is a low output system, one well-grown weaned calf per cow in the autumn of each year being the only product.

Some formers mate their cows so that the calves are born in the spring. The advantages are that the cows can be left out during the winter and cheaply fed on silage or straw and roots, that the spring grass stimulates the production of ample milk for the calf, and that the calf feeds off good quality grass. The calves to be sold do not suffer a weaning check as they are not taken off their mothers until the morning of the sale. Disadvantages are that the spring flush of grass may lead to the production of an excessive quantity of milk, causing the calf to scour, and that, because a calf is about 3 months of age before it becomes a really effective grazer, it will not take full advantage of the herbage until this is past its best.

Other farmers calve their suckler cows in the autumn. Advantages are that the calves have at least 6 months in which to make good use of the grass, and weigh heavier when sold in the autumn following the year of their birth. Disadvantages are that although the cows can calve at grass about October some in-wintering accommodation is required during bad weather, and the cows and calves have to be given conserved food. The calves have to be weaned about July in order to allow the cows to have a rest period before calving again and this means that the calves have to be given some concentrate food to compensate for the loss of their dams' milk. Although weaning before sale may cause a lack of bloom some buyers prefer to acquire animals that have recovered from the post-weaning check.

Double-suckling. Shortly after a cow has calved she is tied up in a loose box and a second calf is introduced and allowed to suck alongside the cow's own calf. The new calf may be rubbed with a portion of the cow's placenta to encourage acceptance. In a short time the cow usually suckles both calves, and can be turned out with them. The system can only be used with cows which yield sufficient milk to feed two calves and in areas where extra calves of a suitable type are readily available when required. It is anticipated that sufficient nutrients to produce the milk needed to feed the two suckled calves will be obtained from the grass. The bodily condition of the cows must be kept under observation for undernutrition will lead to a low milk yield and poor development of the calves. Concentrate food supplementation, if required, should be provided on the lines suggested previously for dairy cows.

Multiple-suckling. Nurse cows of a dairy breed are used to suckle four calves for 6 or 7 months, or four calves for 10 weeks, then another three for 10 weeks, followed by a further three for a similar period, and then two more for a final 10 weeks. The cows are run out at grass and brought in twice a day to suckle the calves which are housed. The attendant must supervise each feeding to ensure that all the calves receive an adequate milk ration. A balanced concentrate food, prepared as suggested under dairy cows, must be fed to the cows according to the estimated milk yield.

Veal production

Calves are reared on a liquid diet for from 12 to 15 weeks until they reach a body weight of 115–135 kg (250–300 lb). Bull calves are normally used, and Friesians give the best performance. However, because of the high prices which Friesian calves now fetch, cheaper Dairy Shorthorn, Hereford × Ayrshire, and pure Ayrshire, Jersey, and Guernsey calves have been tried. These calves are started on milk substitutes as described under dairy cattle. They then receive a specially prepared high fat milk replacer.

Beef production

There are three main methods of feeding beef cattle designed to produce animals ready for slaughter at 10–15 months of age (baby beef), at about 18 months old (young beef) and at 2½–3 years of age (mature beef).

Baby beef. This system is commonly known as intensive cereal or barley beef production. Although mainly used for the fattening of steers it provides the most satisfactory method of fattening uncastrated bulls. When 12 weeks old and weighing about 90 kg (200 lb) liveweight calves, usually pure Friesian or Friesian × Hereford crosses, are started on the high cereal ration. This consists of 85 per cent rolled barley, having a moisture content of not less than 16 per cent. The barley is rolled, and not ground, because grinding breaks down the long fibre of the husk and makes a dusty feed which is less palatable and may aggravate respiratory troubles. Grain with less than 16 per cent of moisture tends to produce an appreciable proportion of fine dusty material even when rolled. Cereals other than barley are sometimes used. The remaining 15 per cent of the ration is composed of a protein supplement with added minerals and vitomins.

The concentrate ration is fed *ad lib.* and sometimes no roughage, or only a little, is allowed. To reduce the incidence of bloat, which may occur in a high percentage of calves if the concentrate ration is given alone, about 0·9 kg (2 lb) of hay per head per day should be allowed. The cattle are slaughtered at between 10 and 12 months of age when weighing 360–410 kg (800–900 lb) liveweight.

Young beef. Autumn-born calves are particularly suitable for this fattening method. They are reared on an early weaning system and in the spring of the year after their birth are turned out to grass. They grow well on the grass during the summer, being allowed a small concentrate ration in the spring immediately after being turned out to help them adjust to the grass diet, and again in the autumn when the feeding quality of the grass falls. During their second winter they are fattened in yards. The concentrate ration may be the barley 85 per cent, protein/mineral/vitamin supplement 15 per cent, previously mentioned, but it is fed in regulated amounts, not *ad lib.* Initially, about 2·7 kg (6 lb) per head per day may be allowed rising gradually to about 4·5 kg (10 lb). This ration is supplemented with bulky foods such as hay, about 3·2 kg (7 lb) per head per day, or silage or barley straw and roots *ad lib.* The cattle are slaughtered at about 18 months of age when their weight is in the region of 450–508 kg (9–10 cwt). Compared with baby beef production this method slows down the turnover of capital and gives less efficient food utilisation, but it may show a better financial return because of the heavier slaughter weight and the use of a cheaper food in the form of grass.

Mature beef. Systems for the production of mature beef include a period, known as a store period, when the cattle are kept at a constant weight or growing slowly. The object is to utilise cheap rations while waiting for a suitable fattening period. Cattle of certain breeds, e.g. Friesians, develop large bony frames during a long store period, and so are better suited to one of the fattening systems previously described.

Grass fattening

The calves are housed in their first winter and fed on silage, or hay and roots, with a small allowance of cereals such as oats or barley. They are then grazed on moderate grass during the following summer and either left out at grass in their second winter, being given moderate quality hay or silage in bad weather, or yarded and fed on silage or straw and roots. They are then turned out on good pasture in the spring. The pasture must be capable of growing good leafy grass. Short-term leys or permanent pastures are both suitable. Some form of controlled grazing should be adopted so that the grass can be grazed and rested in rotation.

The aim is to attain an average liveweight gain of 0·9 kg (2 lb) per day so that they can be sold fat at between 610 and 760 kg (12–15 cwt) in the autumn. Animals which are not ready for slaughter then are housed in yards during the late autumn and finished on silage *ad lib.* and about 2·7 kg (6 lb) of barley per head per day. In some cases a small allowance of hay may be fed in addition.

Yard fattening

This system was at one time very popular. After a store period on grass during the summer, bullocks, at about 18 months of age, are run in groups of from twelve to twenty in open-fronted covered or partially-roofed yards. They are bedded on straw and the tramping of their faeces into the bedding to produce high quality farmyard manure is an advantage of the system. The ration fed depends mainly on the home-grown foods which are available. Silage is a valuable roughage, but oat and barley straw are used in the early stages for economic reasons. A limited amount of hay may be used in the later stages of fattening when it may be desirable to reduce the intake of bulky food.

The quantity of concentrate food fed will depend on the rate of liveweight gain expected, the quality of the roughage and the degree of fattening required. A simple mixture of barley, 3 parts and palm kernel cake, 2 parts is satisfactory, and if fed at the rate of 2·27 kg (5 lb) per head per day with good quality roughage can produce a daily liveweight gain of 0·91 kg (2 lb). The daily allowance can be increased to from 4 to 5 kg (9–12 lb) if rapid finishing is required. At this level the cattle should have a good covering of fat when they reach a slaughter weight of from 610 to 760 kg (12–15 cwt).

Cow beef

Cow beef is obtained from animals which have been discarded through age and from young cows culled because of infertility, udder disease, or poor milk yield. It is a question whether it is worth giving such cows, particularly the younger ones, more feed than is justified by milk yield during the later stages of lactation in order to improve carcase quality. The answer depends on the cost of the food, and the anticipated price per hundredweight for the meat. Normally the price levels do not justify additional feeding, and cow beef usually comes from cows slaughtered in the state of finish in which they happen to be when culled.

Mineral and vitamin requirements

The requirements of beef cattle are obviously similar to those of dairy cattle already described. A calcium deficiency in the diet is uncommon but a phosphorus deficiency may occur in animals grazed in areas where the soil is deficient in this mineral. The clinical signs are a lowered fertility rate in breeding cows and a poor growth rate in young animals. The soil may be treated by the application of a superphosphate fertiliser or bone meal may be supplied in troughs in the fields.

Vitamin A is the vitamin most likely to be deficient in beef cattle rations. Breeding cows fed high levels of barley straw are unlikely to receive enough carotene from which to synthesise the vitamin, and this can lead to infertility and abortion or the production of weak calves which are particularly susceptible to white scour and respiratory troubles. In fattening cattle the vitamin A content of the supplement fed with barley for the production of baby beef is particularly important because cattle given a high cereal diet have a greater requirement for this vitamin than do animals on a more conventional ration. The first sign of a vitamin A deficiency in cattle in this age group is usually blindness. A vitamin D deficiency in beef animals is rare because they are usually exposed to direct sunlight or fed on sun-cured roughages.

Water requirements

Beef cattle should have an abundant supply of water before them at all times. Mature beasts will drink an average of about 54·6 l (12 gal) per head daily, with younger animals requiring proportionately less. In yards the most satisfactory method is to have self-filling water troughs to which the beasts can go when they feel inclined.

Breeding

Cows in dairy herds of breeds other than Friesians being used for the production of calves for fattening are served by, or artificially inseminated with semen from, beef bulls so as to calve at the time required for milk production. The gestation lengths of dairy cows carrying calves sired by bulls of the imported breeds are from 3 to 14 days longer than is normal when the calves are the result of mating with British breeds. This is one factor responsible for the larger birth weights.

Cows of the beef breeds are usually not housed, but are run out with the bulls on hills or low quality pastures. Bulls of the beef breeds are quieter than dairy type bulls and so they can more safely be allowed to run loose. The bulls are not normally run with the herd throughout the year but are turned out about 2 months after the birth of the first calf, the aim being to produce one calf per cow per year. The bulls are left out for about 3 months, and any cows which do not conceive during this period are culled. The reason for this is that it is desirable for all the calves to be born within a relatively short period of time.

Although semen from beef bulls is widely used for the insemination of dairy cows, artificial insemination is not used to any extent in the breeding of pure beef cattle. In herds kept out it is difficult to determine when individual cows are in oestrus and to arrange for them to be confined for insemination.

Health

The signs of health resemble those exhibited by dairy cows but healthy beef cattle normally carry more flesh. Cattle wintered out grow long hairy coats as a protection against the cold. The body temperatures of beef cattle are normally in the same range as dairy cows, but the pulse rate is usually slower, from 45 to 50 beats per minute.

Common diseases and their prevention

Cattle being reared for beef are susceptible to most of the usual bovine diseases, but there are two conditions of particular importance, especially in barley beef units.

Pneumonia is probably the most important. Calves begin to cough when about 3 months of age, and some animals develop acute pneumonia. The mortality rate may be high. There are a number of viral and bacteriological agents involved in causing the condition, but the predisposing cause is bad housing with faulty ventilation, humidity and temperature. The main method of prevention is obviously to improve the housing conditions but vaccines have also been employed.

Bloat is usually associated with introducing calves too quickly to *ad lib.* cereal feeding, sudden changes in management or an inadequate provision of roughage. Such errors should be avoided.

Handling

Beef cattle should be gently handled, especially just before slaughter. If an animal is hit or injured the probability is that blood will escape from damaged blood vessels into the surrounding tissues and cause bruising, which results in discoloration of the meat and the production of a gelatinous infiltration. Affected areas have to be removed from the carcase before sale.

It is difficult to restrain individual bovine animals in a yard or field and so some provision for handling beef cattle should be made in the fields or near the yards according to the system being adopted. It is often possible and desirable to incorporate part or all of the handling facilities into existing buildings. A collecting yard should form the start of the handling system and should lead into a passage at the end of which a handling crate should be situated. This should be designed with a gate at each end, so that the cattle can be passed through, and should be fitted with a neck yolk on the outlet door so that an individual beast can be held while being examined or treated. A difficulty experienced in designing a crate for beef cattle is the variation in width required for handling adult and young animals. A width of 63·5 cm (25 in) is satisfactory for breeding cows and fat beasts, but is too wide for weaned calves, some of which may try to turn round and may fall or become jammed in the process. A width of 50·8 cm (20 in) or less is desirable for animals of this type. It is possible to design a crate with one movable side to accommodate animals of different sizes, but it is probably better to make an insert which can be fitted to one side of a wide crate to make it narrow enough to control calves. The crate should open into a holding yard where the cattle can be confined after being passed through. As it is advantageous to weigh fattening cattle at intervals to note their development a combined handling crate/weighing machine is a desirable refinement.

General

Minor operations

Fattening cattle which are not naturally polled should be disbudded as calves or dehorned as described under dairy cows. Cattle with horns require more yard space and more room at the food trough. In addition damaged hides and bruised carcases are more frequent in bunches of horned cattle.

Castration

Although male calves being reared for veal, and some of those being fattened for killing as baby beef, are left uncastrated, the majority of the male calves kept for meat production are castrated. There are three main methods of castration in general use, the rubber-ring method, the Burdizzo method, and a cutting method using a knife. For the efficient operation of any of the methods older calves should be cast on their sides, and young animals held in a sitting position. An anaesthetic must be employed for the castration of a bull calf over 3 months of age.

If the rubber ring method is employed it must be carried out within the first week of life. A special rubber ring is expanded on an instrument designed for the method and most of the scrotum and the two testicles are passed through the rubber ring, which is then released to encircle them tightly. This stops the blood circulation and in from 10 days to 3 weeks the part of the scrotum below the ring and the testicles shrivel and fall off.

In the second method a Burdizzo castrator, which is a pincer-like instrument, is pinched separately across each of the cords above the testicles, through the wall of the scrotum. This crushes the cords while the skin of the scrotum remains unbroken. In the course of a few weeks the testicles shrivel.

Older calves are usually operated upon by the use of a knife, although the method can be used for animals from 2 months of age. The scrotum is washed with a disinfectant solution and two incisions made at its base, one beneath each testicle. The testicles are then pulled out and the blood vessels twisted and broken in young calves, or tied with ligature in older calves, to prevent bleeding.

Meat marketing

A beast should have a good killing out percentage. The average is in the range of 54–57 per cent. Baby beef animals may reach 60 per cent, as may older well-finished beasts, although carcases from the latter may be rather too fat. Animals which have not been satisfactorily fattened will have killing out percentages in the 50–53 per cent range. The carcases must have a good thickness of muscle, particularly over the hind quarters and loins, from which areas come the favoured cuts. There should be a covering of fat over the carcase which is not excessively thick and there should also be fat distributed between the muscle bundles which is known as marbling fat. The meat should be bright red in colour and the fat should be white or creamy white, and not yellow.

Much of the cow beef is of poor quality as it comes from animals which have been discarded through age, but some of the beef from young cows culled because of infertility, udder disease, or poor milk yield is of quite a high quality.

Government subsidies

In the United Kingdom government subsidies are paid to beef producers to encourage home beef production and so reduce the quantity which has to be imported. A subsidy is paid for breeding cows of beef types kept in hill areas, and, at a lower rate, for beef breeding cows kept in other regions. Steer and heifer calves of recognised beef types, which have been reasonably well reared and are suitable for beef production, qualify for a calf subsidy. Calves which have received the subsidy have a circular hole, about 1·27 cm (0·5 in) in diameter, punched in the right ear.

Welfare code

One provision particularly applicable to veal calves is that the width of a pen for a singly-penned animal should not be less than the height of the animal at the withers. Beef type cows kept for breeding are covered by the requirements that consideration should be given to the provision of some form of artificial shelter if they are kept in exposed areas where natural shelter from cold winds is not available, and that, if out wintered, arrangements should be made in advance to ensure that adequate food can be made available to them in emergencies, e.g. heavy snow.

CHAPTER 5
GOATS

Introduction

Goats were one of the earliest species of animals to be domesticated and were valued because they produced meat, milk, and skins while living on vegetation found on hills which would not support cattle or sheep. Today goats are often termed 'the poor man's cows' because they will browse on shrubs and graze coarse vegetation on waste ground. Because the majority of the land is under cultivation the United Kingdom is not as well adapted to goat keeping as are countries abroad where goats are more widely kept. Their roving natures and browsing habits make it difficult to pasture them in herds like sheep or cattle without constant supervision, except in specially fenced paddocks, because of the damage they can inflict on trees and hedges.

In the United Kingdom most goats are kept in twos or threes for home milk production, although there are some large herds. Goats' milk is a valuable commodity and, in certain districts, the demand exceeds the supply. It has advantages as an article of diet for young children and older persons suffering from certain disease conditions. The goat is not valued as a meat producer in the British Isles, although kid meat is esteemed in many other countries and goat meat is consumed in some. Because they are inquisitive, amusing, and usually friendly creatures, goats also make good pets.

Definitions of common terms

Kid. Young goat up to 1 year of age.
Goatling. Female goat over 1 year but not exceeding 2 years.
Buckling. Male goat over 1 year but not exceeding 2 years.
First kidder. Milking goat after the birth of its first kid.
Nanny. Female after the birth of its first or second kid according to local custom.
Billy. Adult male.

Principal breeds

Goats are naturally horned, but polled types exist and are favoured. In all breeds the males are usually taller and heavier than the females. Good milking goats usually have a straight back, deep body, and well-developed hind quarters, giving a wedge-shaped appearance when viewed from the side like good quality dairy cows. The hocks should be sufficiently straight to avoid bruising the udder when the goat walks. The udder should be capacious, but carried well up under the body. When empty it should shrink and the tissue should feel soft and smooth and be pliable. The two teats should be hand-sized and set well forward. They should point straight downwards, as teats which point outwards tend to brush against an animal's hind legs when it moves and become bruised.

Anglo-Nubian

The head has a characteristic Roman nose and large pendulous ears. In size it is relatively large, with females weighing 63–91 kg (140–200 lb), and has a fine skin and glossy coat. Anglo-Nubian goats may

Fig. 5.1. An Anglo-Nubian kid.

Fig. 5.2. A Saanen kid.

Fig. 5.3. An Alpine kid.

be of any colour, and various whole colours are found as well as marbled and spotted colourings. The milk yield is only moderate, 1,360 l (300 gal) per lactation, but the quality is good with a butter-fat content of about 5 per cent and a solids-not-fat (S.N.F.) percentage of between 9 and 10.

The following three breeds have been developed in Britain from foundation stock imported from Switzerland.

Toggenburg
Moderate in size, the females weighing 45 kg (100 lb) with small, pricked ears. The body colour varies from deep chocolate to a drab colour resembling weak cocoa and milk, and there are white marks on each side of the face, on the legs, and round the tail and rump. They are reasonable milk producers, averaging about 1,450 l (320 gal) per lactation of good quality milk with a fat percentage over 4 and a S.N.F. percentage up to 8·7.

Saanen
Similar to the above but white in colour and slightly larger as the females weigh about 50 kg (110 lb). The yield is higher at about 2,180 l (480 gal) in a lactation. The fat percentage is about 4·3 and the S.N.F. percentage about 8·5.

Alpine
Black in colour with white markings in the same places as the Toggenburg. Long in the limb and benefits from more exercise than the other breeds. Some are heavy milkers with a lactation yield up to 2,230 l (490 gal). The butter fat content is average at about 4 per cent and the S.N.F. percentage approximates 8·3.

British
Can be of any colour. Many are good milkers, producing up to 2,275 l (500 gal) in a lactation of fairly high quality milk with a fat percentage between 4 and 5 and a S.N.F. percentage of about 8·3.

Identification

There is no generally recognised system for the permanent marking of goats for individual identification. Descriptions of the colours and markings can be used, particularly in Anglo-Nubian goats. Numbers can be tattooed, or tags can be clipped, in the ears.

Ageing

An estimate of the age of a goat may be obtained by examining the incisor teeth. Like cattle and sheep, goats have eight incisors in the lower jaw and none in the upper. By the first year the full set of temporary incisor teeth is in wear. At about 12 months of age the two central temporary incisors erupt and are replaced by the permanent teeth, and the other incisor teeth erupt in sequence, the last being the corners at about 2½ years. After 5 years of age the incisor teeth begin to show signs of wear.

An additional sign of age in the goat is the appearance of the beard. This begins to grow at 3 months and is fully developed by 3 years.

Housing

Goats only require housing of a simple type to provide shelter from the wind and rain or shade from the hot sun. In large herds they are usually kept in groups, but adult males and heavily pregnant females, and sometimes the milkers, may be housed individually. Buildings originally erected for other purposes can often be adapted to provide a dry, draught-free bed and room to feed and exercise. When goats are housed in buildings, either individually or in pens, adequate ventilation must be provided as goats are particularly susceptible to respiratory troubles if kept confined without ample supplies of fresh air. The method of ventilation employed must not produce draughts, but must prevent a build up of ammonia.

Kids and goatlings

Goats kept in groups are often left to give birth to their kids in the group, and this practice appears to work satisfactorily. However, it is better to house goats individually while kidding, and small loose boxes about 0·92 m (3 ft) wide by 1·52 m (5 ft) long are the most satisfactory form of accommodation. The walls should be at least 1·22 m (4 ft) in height. Whether the kids are to be suckled or hand reared they can best be housed in a loose box of the same size at first. It is advantageous to provide a large wooden box, placed on its side, so that the kids can sleep in it to obtain additional protection from draughts.

Goatlings are usually run in groups in a shed or covered yard. Ample trough space should be provided so that the smaller animals are not driven away from the food by bigger bullies. Some goat keepers fit a series of vertical slats to the upper part of the inside edge of the trough, about 0·152–0·228 m (6–9 in) apart, so that the goats have to push their heads between two adjacent slats in order to feed. This prevents the goats from pushing along the trough and so disturbing their neighbours.

Milking goats

Milking goats housed separately can be kept in stalls about 0·76 m (2·5 ft) wide and 1·37 m (4·5 ft) long. The stalls may need to be slightly narrower for small goats to prevent individuals turning round. The goats are tied by a neck collar and chain. Although shavings and straw are usually provided as bedding, goats are quite content to lie on wooden floors. If the house has been designed to accommodate a number of goats a feeding passage should be provided down the centre of the building to facilitate the distribution of food. Goats housed in either stalls or pens should be allowed out for exercise each day, and it is advantageous if the exercising areas adjoin the sheds so that the animals can run in and out easily.

Yards

Open concrete yards, with simple wooden huts, are excellent for group housing. An area of 42 m^2 (50 sq yd) is sufficient yard space for a group of up to ten adult animals, with about 0·5–1·1 m^2 (5–12 sq ft) of shed area per goat. The concrete keeps the goats' feet worn down and so prevents horn overgrowth. The yards must be well fenced as goats can jump or clamber over fences, and often stand with their forelegs resting on fences, which tends to distort wire structures so that they can be jumped more easily. Well supported chain-link fencing about 1·5 m (5 ft) high has been found to be quite suitable. Bolts should be used on the doors, as goats can open most other types of catch.

Whether kept individually or in groups hay racks should be provided to reduce the risk of hay wastage. Although most are made of wood, iron hay racks have a greater length of life. Clean water should always be available as goats are fussy drinkers and dislike stale, dirty water. Buckets are normally used as water containers, but goats readily drink out of automatic drinking bowls of the type provided in cowsheds. The water containers should be fixed at a height which prevents fouling with faeces or urine.

Tethering

Goats are often tethered when grazing. A light iron chain fitted with swivels, one close to the tethering pin and the other near the clip attached to the collar or head stall, should be used. The swivels prevent the chain from twisting into knots. The tethering pin can be moved frequently so that the goat is continually grazing fresh ground. Occasionally a running chain, as described under dogs, is used for tethering goats. The chief objection to tethering is that the animal has no protection from sudden storms. Light sheds on wheels, which can easily be moved, but can be fixed in position with a metal stake, have been tried but there is always a risk that the tethering chain may become caught in some part of the shed. Goats can be turned out in fields, but these must be strongly fenced to prevent escapes. If goats do escape they are liable to damage trees by eating the tender buds or bark.

Milking stand

Goats can be milked in loose boxes but not conveniently in stalls. In any case as goats are small animals it is not easy to milk them at ground level and so a milking bench is normally provided. This consists of a low table standing about 0·61 m (2 ft) above the ground with a solid front to which the goat can be tied, or a yoke, which can be closed round the goat's neck if necessary, but most goats do not require to be fastened during milking. Side rails prevent lateral movement and there is a hole, fitted with a detachable cover, in the floor for the bucket so that it cannot be kicked over and pendulous udders can be milked more easily. Goats soon learn to jump on to this bench, but some benches are fitted with a ramp which assists goats with pendulous udders to ascend. Most milking benches have feed boxes in the front so that the goats can eat while being milked.

Feeding

Kids

It is essential that the kids should have the colostrum and so they should be allowed to suck from their mother for 3 or 4 days. After this there are two methods of kid rearing, natural suckling and hand feeding. Valuable kids may be allowed to suckle their mothers for about 6 weeks, but as most goats give more milk than their kids require, they should be hand milked twice a day to remove the surplus milk. When the kids reach the age of 6 weeks they can be separated from their dams for part of the day, and some of the nanny-goats' milk can be withdrawn before the kids are reintroduced. This limits the volume of milk the kids receive, and so starts the weaning process. However, hand rearing is more common and it is usual to take the kid away from its dam when it has received the colostrum and rear it artificially so that the dam may be milked and its milk utilised. Most kids are taught to drink from a pail although some owners prefer to allow them to suck from an ordinary infant's feeding bottle. This latter method is time consuming, but the results are better.

Goats' milk should be given for the first week at least, after which cows' milk can be added in gradually increasing amounts until after a fortnight cows' milk can be given alone. Four feeds a day are required, and each feed should not exceed 0·57 l (1 pt) in volume for a female, although strong male kids may be given a total of 2·8 l (5 pt) a day. A week later, skim milk, reconstituted from powder, can be added in an increasing proportion from one-third to one-half. No feed should exceed 0·85 l (1·5 pt) or scouring may start, with a resultant check in growth. At about 6 weeks of age weaning can be commenced and three feeds a day should be sufficient. The kids start to nibble at solid food, such as grass or hay, when about 2 weeks of age and this should be encouraged so that when 6 weeks old they will be eating grass, hay and concentrate food. A suitable concentrate mixture comprises equal parts of crushed linseed cake, bran, and flaked maize. Weaning may be completed when the kid is about 2½–3 months of age, and the ration fed to the adult milking stock can be given from this time onwards.

In addition to the above-mentioned foods the kids will probably obtain some green food when turned out for exercise. Kids are safe on poor quality pastures and can also be allowed to browse on shrubs, but good grass pastures are too rich and the consumption of large amounts of such grass may cause indigestion. If kids have to be run on good grass and clover pastures they must be allowed out for 1 or 2 hours only and should, preferably, be given a feed of hay before being liberated to reduce further the grass intake.

The feeding of kids during their first winter, when they are from 6 months to 1 year of age, should be at an economical level, but good quality hay *ad lib.* and about 0·340 kg (0·75 lb) of concentrates per day are required, although some owners supply 0·454 kg (1 lb) or even 0·681 kg (1·5 lb) daily.

Goats

The feeding of adult goats is governed to some extent by their special feeding habits. Because of their mobile upper lips and prehensile tongues goats are able to graze very short grass and to browse on shrubs not normally eaten by other domesticated livestock. Goats do particularly well on the short grass found on downs and commons. In the latter areas they also browse on gorse bushes.

There is little scientific rationing of lactating goats, and most goat keepers have developed their own feeding systems. There is also a great variation in the type of diet provided. In the British Isles during the summer months from about mid May to the end of September, the exact dates varying according to the weather conditions, the rations will consist primarily of green food, either as grass from the grazing of paddocks or as leaves and small branches obtained by browsing. This diet will provide for maintenance and also some milk production, the amount depending on the nutritive quality of the food. During the winter hay can be provided for maintenance, about 1 kg (2·2lb) per day being sufficient for a goat weighing about 45·4 kg (100 lb) liveweight, while a goat of 63·5 kg (140 lb) requires about 1·59 kg (3·5 lb). About 1·36 kg (3 lb) of cabbage leaves, kale or roots or 0·34 kg (0·75 lb) of dry sugar beet pulp (soaked before feeding) can be used to replace 0·23 kg (0·5 lb) of the hay. Silage can be used provided goats are trained to eat it. Up to 3·56 kg (8 lb) per day can be allowed to replace 1·13 kg (2·5 lb) of hay. For production a dairy type cubed concentrate ration or a simple concentrate mixture such as decorticated groundnut cake, one part; flaked maize, two parts; and crushed oats, one part can be fed. Most goats need at least 0·454 kg (1 lb) of concentrate food for each 1·36 kg (3 lb) of milk produced. Some goat keepers ration their goats according to yield by allowing them to eat freely during milking. Thus, the more milk a goat gives, the longer milking takes and the more concentrate food the goat can eat. Others give an estimated amount with additions until it is clear that no increase in milk yield can be obtained by higher levels of feeding. The daily ration is then standardised until it is reduced as the yield declines towards the end of the lactation.

The change from winter feeding on hay and concentrates to fresh grass in the spring should be made gradually to avoid causing digestive upsets and scouring. This is best effected by putting the goat on grass for 30 minutes the first day, and then adding 30 minutes to the grazing period each following day until it becomes accustomed to its new diet.

Pregnancy requirements

During the last 2 months of pregnancy additional food should be given to provide for the growth of the developing embryo kids, and some goat keepers calculate the anticipated maximum level to be fed during lactation and give increasing amounts daily during the gestation period until this amount is reached 1 week before parturition.

Mineral and vitamin requirements

Little is known about the mineral and vitamin requirements of goats. Calcium and phosphorus are needed for bone development and it is generally considered that a goatling requires about 0·45g/kg (1·6 oz/2·2 lb) of calcium per day, and that the calcium/phosphorus ratio should be narrower than 1: 2. Bone diseases in goats can be caused by a calcium deficiency or imbalance, but they are usually only seen in animals kept housed and fed on a diet of hay, oats and bran with little green food. The conditions are aggravated by pregnancy. It is difficult to make most goats eat mineral supplements in powder form when mixed with their food, and so some owners administer tablets containing calcium and phosphorus to growing, pregnant and high yielding goats which are kept confined. The use of compounded foods in which minerals have been incorporated are useful under management systems in which goats cannot be allowed out at grass. Other goat keepers supply mineral licks containing the two minerals in question along with other trace elements.

Vitamins A and D are required, and the diet should be supplemented when goats have to be confined indoors for long periods. This can be achieved by a vitamin supplement incorporated in a proprietary concentrate diet or by the addition of about 7 ml (0·25 fl oz) of good quality cod liver oil to the diet daily.

Water requirements

It is very important to provide goats with an ample supply of fresh drinking water, as they often refuse to drink stale or tainted water. It is, therefore, a bad plan to leave water standing in a goat house, and clean fresh water should be offered morning and night. Dry goats living on grass or green vegetables may need about 0·57 l (1 pt) per day, whereas lactating animals will require from 3·5 to 6 l (6·2–10·6 pt), the actual amount depending on their milk yield.

Breeding

Billy-goat kids can be used for service when about 6 months of age, but it is probably better to delay breeding from them until they are 12 months old. During its first season a billy-goat should be restricted to six services. The effect of stud work on the condition of a male goat should be carefully observed, because excessive use during the autumn and winter can lead to body weakness and reduce fertility.

The odour of billy-goats is universally unpopular, and its seasonal production is a sexual characteristic. As the clothing of the attendant can become contaminated special overalls should be kept for use when tending male goats. If this is not done the offensive odour can be passed to the milk obtained from the nanny-goats.

Horns are undesirable in male goats as injuries to other goats, particularly during mating, may be caused. Unless polled, male goats should, therefore, be disbudded or dehorned. In polled strains many hornless males are sexually abnormal, and it has been shown experimentally that the genetic factor which prevents horn growth is closely associated with abnormal sexual development. Some of these intersexes are easily recognisable, but others, on superficial examination, appear to be normal males. All intersexes are infertile.

Female kids are sexually mature at about 6 months of age, or even earlier, but, unless well grown, should not be mated until they are from 15 to 18 months old to avoid retarding their development. If not pregnant, goats come into oestrus from September until February, or possibly a little later. During the summer the oestral periods normally cease. The duration of oestrus is about 24–36 hours, although there are variations, and oestrous periods recur at intervals of from 18 to 21 days. Oestrus is manifested by redness and swelling of the vulva, accompanied by a slight colourless discharge. The goat will also be restless and will bleat frequently and wag its tail. Under natural conditions nanny-goats are generally mated several times at each oestrus, but if mating is controlled, as it usually is in Great

Britain, it is best to arrange service, or insemination, at least 12 hours after oestrus is first observed. When first placed together male and female goats often start fighting, but, although they inflict blows on each other, do not engage in serious combat. Goats are normally mated to produce only one set of kids in a year, and March or April are the best months for kidding.

The average gestation period is 150 days, with a range of 143–157 days. It can be difficult to determine whether a goat is in kid, but as pregnancy advances the fetus can, possibly, be felt and seen to move on the right-hand side of the body. Most goats produce twins, but singles and triplets are not uncommon. Visible enlargement of the udder can be noted halfway through the pregnancy of maiden goats. Between 8 and 24 hours before birth the kids move towards the neck of the uterus and can no longer be felt on the right side, should this have been possible earlier. Other signs of approaching parturition are a rapid filling of the udder and a slackening of the tissues round the vagina. While giving birth goats bleat constantly and the first kid is usually born within 4 hours of straining being observed. The after-birth is usually passed about 30 minutes after kidding, although its expulsion may be delayed for up to 4 hours. There is one afterbirth for each kid. Most male kids are killed soon after birth, only those with high class pedigrees being retained for stud purposes.

Artificial insemination can be used in goats, but is not normally practised. Semen can be diluted and preserved at about 4°C (39·2°F) for 2 days, and recent work has shown that the freezing of goat semen, and therefore its preservation for long periods, is possible.

Milk production

Because of their restricted breeding season goats normally kid in the spring or early summer and most lactate for about 10 months, being dry for 2 months before kidding again. However, there is considerable variation as some only give milk in sufficient quantity to be worth drawing for about 7 or 8 months, while others will go on milking to within 1 month or 3 weeks of the day on which they are due to kid again if allowed to do so. Every effort should be made to ensure that an in-kid goat is dry about 6–8 weeks before kidding. It is, however, not always easy to dry off a high-yielding goat. The feeding of concentrate food can be stopped and milking can be carried out incompletely and at irregular intervals. It may be necessary to reduce water intake, but a goat should never be deprived of water. If not mated again some good milking goats will continue to lactate for 2 or 3 years without a break.

The maximum daily milk yield is attained between the eighth and twelfth week after kidding and then drops gradually. This means that the yield is more evenly spread throughout a lactation than is the case with dairy cows. The lactation yield after the first kidding is below that attained in the second, third and fourth lactations. After this the yield gradually falls, but many goats continue to give reasonable yields until their eighth or ninth lactation.

The compositional quality of goats' milk varies according to the breed, the system of feeding and management and the stage of lactation in similar ways to cows' milk. The butterfat is colourless and the fat globules are small and do not rise as rapidly as those in cows' milk. This is the reason why the fat in goats' milk is more easily digested by human beings than cows' milk fat. The proteins of goats' milk are also more easily digested, particularly by human infants, as they form a more friable coagulum. Another point is that goats' milk is alkaline, whereas cows' milk is acid in reaction. This occasionally is beneficial in the case of hyper-acidity in human milk drinkers. Legal standards controlling the composition of goats' milk do not exist.

Health

When in good health goats carry their heads erect, have bright and sparkling eyes and are lively in their movements, contrasting markedly with the mournful dejected appearance exhibited in illness. Healthy goats have a fine pliable skin with a smooth, lustrous coat. A constitutional disturbance is shown by a cessation of rumination, a dry nose, hot horns in horned individuals, a staring coat and a tight skin and an unnatural condition of the faeces. Goats do not carry much fat and as their basic areas of fat deposition are on the chest and belly their back bones tend to stand out in comparison with animals of other species. The legs should be strong and the hocks should be straight to avoid bruising the udder during movement.

The body temperature of the goat is similar to sheep, ranging from 38·9 to 39·4°C (102–103°F). The pulse rate is usually within the range of 70–90, and may be taken either at the artery in the lower jaw or at the artery which runs inside the hind limb at a position between the hock and stifle joints.

Common diseases and their prevention

Enterotoxaemia is a condition commonly known as pulpy kidney disease, which also occurs in sheep. It is caused by the absorption from the intestine of the toxin of a bacterium. This bacterium is normally found in harmless numbers in the intestine, but when there is a digestive upset the organism multiplies rapidly and produces large amounts of the toxin. The disease almost always attacks animals in good bodily condition and is usually associated with the grazing of lush grass or the feeding of green crops. In the acute form the animals exhibit abdominal pain and severe diarrhoea and die within 2 days, and in the chronic form they develop diarrhoea and become dull and listless, but usually recover. A vaccine is available for the protection of susceptible animals.

Coccidiosis. Most goats have a few of the internal parasites called coccidia in their intestines and immature forms are passed in their faeces. Under crowded unhygienic conditions the consumption of contaminated food or bedding can lead to a build up of numbers causing coccidiosis, which leads to diarrhoea, emaciation and death, particularly in young animals. Prevention is by the avoidance of crowded conditions in fields and unhygienic housing. The disinfection of infested premises is best carried out with a 10 per cent solution of ammonia.

Parasitic gastro-enteritis is caused by a massive infestation of the fourth stomach and intestine with small parasitic worms. The eggs which are laid in large numbers by the adult worms, are passed on to the ground in the faeces of infested animals. There they give rise to larval forms which migrate from the faeces on to the herbage where they await the grazing host. The main clinical sign is a progressive loss of condition, which may proceed to a state of emaciation. The appetite usually remains good until a late stage. Kids are particularly susceptible and healthy adults are relatively resistant. Prevention consists in avoiding restricted grazing, which can lead to a build up of infective larvae. The majority of the larvae die within 6 weeks of emerging from the egg, and so resting a pasture for this length of time will ensure a reduction in larval concentration to reasonably safe limits.

Lice. Infestation tends to increase rapidly during the autumn and winter. Clean animals usually become infested through contact with lousy goats, but lice can survive off their host for up to 18 days and so animals can become infested indirectly. Kids are most susceptible, heavy infestations of adult animals being restricted to those in poor condition. The clinical signs are constant rubbing and a scurfy, unhealthy appearance of the skin and hair.

Poisonous shrubs and trees. Although goats are often fastidious feeders they may browse on these and consume such a substantial quantity that toxic signs may result. Shrubbery plants which have caused poisoning in goats are yew, rhododendron and laurel. Yew causes sudden death, but most other poisonous shrubs give rise to indefinite signs such as salivation, colic or diarrhoea. Prevention is by fencing the shrubbery to keep the goats out or by digging up the poisonous plants.

Handling

Goats are less easy to control than the other domesticated species. It is difficult to drive goats, unless they are trained to conform, as they do not pack when approached but tend to scatter, relying on their individual agility to escape from suspected danger. Individual goats, properly trained, are seldom difficult to handle, and soon learn to come when called. Goats being handled frequently should be fitted with head collars or neck straps and controlled by holding these. Other goats are best held by the beard or by placing the hands firmly on the sides of the head. Although goats can be held by the ears, or, in horned animals, by the horns, they tend to resent being grasped by these appendages. It is not easy to sit goats on their hind quarters, but they can be thrown if necessary by grasping them by the beard and tail and drawing these two extremities together. They then stagger round in a circle and fall.

To examine the feet of a nanny-goat, or to trim overgrown horn, the legs can be raised in a manner similar to that described under horses, but in the case of large billy-goats it is advisable to rest the raised foot on a straw bale. If the hoofs are to be trimmed a solid wood block is preferable, as it gives greater stability.

When handling kids, e.g. for disbudding, it is best for the handler to sit down and hold the kid, which has been gently forced into a sitting position, firmly between his knees. He can then hold the kid's head with both hands, placing the fingers under its jaw and his thumbs on its cheeks.

Billy-goats, if handled in a kind but firm manner and exercised regularly, are not dangerous, but if kept chained and given little attention may become difficult to manage. Such billy-goats may charge children and adults who appear to be afraid. Billy-goats can best be restrained by grasping the beard and giving a tap on the nose if an attempt is made to get away.

Milking

Goats kept in large herds can be milked by machine, but most nanny-goats are milked by hand. Goats have only two teats, and these are large by comparison with cows and are connected to large collecting cisterns in the udder. This means that the pressure on the milk secreting tissue does not build up as it does in dairy cows and so milking intervals can be irregular without materially affecting yield. However, goats are normally milked twice a day.

A goat milking machine unit has only two teat cups and these are smaller than those designed for milking cows and are fitted with a claw piece that rests on the floor. There is a tendency for the teat cups to climb up the teats during the later stages of milking. The vacuum pressure recommended for goats is 25·4 cm (10 in) of mercury.

Hand milking is a skill which can be relatively easily acquired, goats being easier to hand milk than cows. The goats are best placed on a raised platform for milking. Most goat keepers milk from the right or left side, but some milk from behind. It is possible to milk while kneeling down, but it is more comfortable to sit on a low stool. Goats with large teats are best milked by grasping a teat in each hand and squeezing gently from the top downwards. This forces milk out of the teat. As soon as the stream of milk stops the hand is released and the process repeated. If the teats are too small to hold with the whole hand the top of the teat is grasped between the forefinger and thumb and these are slid down the length of the teat, so expressing the milk. Restless goats can be milked with one hand, the other being used to grip the back of the leg above the hock to prevent kicking.

The milking utensils should be cleaned and sterilised in a manner similar to that recommended under dairy cows.

Administration of medicines

Medicines are nearly always administered in a fluid form or by injection. For drenching it is best to straddle the goat's back and hold its body firmly between the knees. The mouth can then be opened and the neck of a small bottle inserted. The fluid should be allowed to trickle down the throat as slowly and gently as possible.

General

Vices

Nanny-goats may start to suck their own milk, and this habit is difficult to check. The noseband of the headcollar may be fitted with projecting spikes, as described for dairy cows, or a strap may be placed round the chest just behind the front legs, and a pole attached to it to run between the front legs to the noseband of the halter. This enables the goat to graze, but prevents it from getting its head round to its udder.

Minor operations and manipulations

Disbudding

Goats are much easier to keep if they are not horned, and when 3 or 4 days of age kids should be examined for the presence of horn buds. These appear under tight curls of hair on the head. If these curls are pressed, the points of the horn buds can be felt under the skin. They are movable as they are not yet firmly attached to the skull. Kids can be disbudded by the use of a collodion solution of caustic chemicals or a hot iron as recommended for calves. In the case of males the burning of a small area of the skin between and behind the horn buds destroys some of the scent glands which cause adult billy-goats to smell so offensively.

Hoof trimming

Overgrown hoofs are common in goats. Because goats are normally hill or mountain animals they have rapidly growing hoofs, and when kept on soft ground the amount of wear is not sufficient to control the horn growth. The walls of the hoofs become overgrown and turn under the sole causing discomfort and, in severe cases, deformity and lameness. The excess horn of the walls can be clearly seen and can relatively easily be pared down with a sharp knife. In the case of large billy-goats it may be easier to use a chisel.

Supplementary teats

Female kids are occasionally born with a supernumerary teat, which may be set apart or close to one of the normal teats. The surplus teat should be removed in the manner recommended for calves.

CHAPTER 6
SHEEP

Introduction

Sheep are kept in Great Britain for the production of both meat and wool, meat being the more valuable product. As a rough division British flocks can be separated into those kept on mountains and hills and those maintained in lowland areas. The methods of management vary, markedly in some aspects.

The upland regions consist of poor quality soil which in most cases cannot economically be improved. The sheep kept there must be adapted to the unfavourable environment conditions, and be capable of converting the mountain herbage into wool and mutton. This means that they are generally small in size and of low individual productivity. At one time wethers were kept till about 2 years of age on the hills, by which time they were ready for slaughter. Now the demand for mature wether mutton has ceased as housewives favour meat from young animals.

Sheep are usually kept on lowland farms as a sideline to some other main enterprise and are often regarded as scavengers which can utilise food which would otherwise be wasted. Because very little capital is invested in specialised buildings and equipment for sheep it is comparatively easy for lowland farmers to dispose of their flocks and change to some other farming system, such as cereal production.

Economically, sheep have the advantage that they produce two products, lambs and wool, and it often happens that if the price for one product is low the value of the other is high. The returns on meat sales are obtained more rapidly than in the case of beef, for lambs may be marketed 8 months after their dams were mated. The level of lamb production in the future will depend on a number of factors, including the possibility of sales to the European Economic Community. The demand for wool is likely to be affected by increasing competition from synthetic fibres.

Definitions of common terms

Lamb. A young animal under six months. Also the meat from such a sheep.

Hogg or hogget. A sheep of either sex between weaning and first shearing.

Shearling. A yearling between the first and second shearing.

Ram or tup. An uncastrated adult male.

Wether or wedder. A castrated adult male.

Gimmer A female between the first and second shearing.

Ewe. An adult female.

Draft ewe. A breeding ewe not considered hardy enough to continue breeding on a hill, sent to a lowland farm.

Broken mouth ewe. A ewe with one or more incisor teeth broken or missing, which prevents the animal grazing efficiently.

Mutton. The meat from a mature sheep.

Numerous other local terms are used in different districts.

Principal breeds

There are about fifty breeds of sheep which are kept in the British Isles, but a number of these are of little commercial importance. One method of classifying them is into Longwools, Downs, other Shortwools, and Mountain, and this system has been adopted.

Longwools

These tend to be large in size with white faces and legs and are hornless. The fleece is heavy but the meat is, comparatively, poor in quality. Sheep of this type are quiet in temperament and are kept on fertile land.

Border Leicester

A big, upstanding, active type of sheep, particularly famous for its rams, which are widely used for crossing purposes. The nose is markedly Roman in type and the ears are long and lie forward. The back is broad and long, and the legs are comparatively long. A disadvantage is that pure bred wether lambs may be too big and carry too much fat to produce a carcase which satisfies current consumer needs.

Romney Marsh

This breed was developed on the Romney Marshes in Kent. It is a hardy breed well suited for grazing lush wet pasturage as it is said to be resistant to liver fluke and foot rot, which can cause so much trouble in sheep kept on wet land.

Wensleydale
A distinctive breed with a bluish coloured head and a long fleece finishing in knots or curls. This breed is pre-eminently a crossing breed.

Downs
These tend to be smaller than the Longwools, although some Down breeds are relatively large. They all have coloured faces and legs and are hornless. The fleece is dense, the wool being the finest obtained in Britain, and very high quality meat is produced.

Hampshire Down
Like all the Down breeds its value is in the production of pre-potent, rapidly maturing, sires for crossing with other breeds of sheep. This breed is characterised by being able to gain weight more rapidly than most others, provided, of course, that the plane of nutrition is adequate to support the maximum growth rate. The extremities are

Fig. 6.1. A Hampshire Down ram.

Fig. 6.2. A Southdown ram.

Fig. 6.3. A Suffolk ram.

a rich dark brown in colour with a head which is large in proportion to the body.

Southdown
A small, thick, blocky, well-fleshed sheep with a mousey-grey face and small rounded ears. The fleece is of very high quality. Although an excellent meat-producing sheep the small body size is reducing its popularity as a crossing-ram breed.

Suffolk
Rams of this breed are very popular for crossing, as the lambs they sire grow rapidly and have good muscling on light bone. The extremities are jet black in colour. The face is acquiline, the ears are drooping and there is no wool on the poll.

Other Shortwools
These are similar in size and fleece and meat quality to the Downs, but may have white or coloured faces and can be horned or hornless.

Clun Forest
A very popular breed because the ewes are in demand for crossing with Down rams for fat lamb production. The face is brown in colour with a slight covering of wool on top of the head. The ears are upright and there are no horns.

Dorset Horn
This breed is gaining in popularity because of the fecundity of the females. The ewes can produce two crops of lambs a year, although three crops in 2 years is a more common average. Both sexes are normally horned although a polled strain has recently been produced.

Kerry Hill
The ewes reach a high level of productivity on good pasture, and, like the Clun Forest, are in demand for crossing. There are clearly defined black and white markings on the face and legs, which make them attractive in appearance. They are polled.

Mountain
These are small, active, hardy breeds, which are slow maturing. In general the fleece is lighter and the wool coarser. Most mountain

breed wools are used for making carpets or hard-wearing tweeds. Their great asset is their ability to live in exposed areas where other, more productive breeds, could not survive.

Cheviot
A white-faced breed with a small, but longish, body. Both sexes are hornless. Some of the lambs can be sold fat from the hills, but the majority are fattened on lowland farms. Ewes from the hills are in favour for crossing purposes in grassland areas.

Scottish Blackface
A very important breed. The body size is small, the face is black with white spots, and both sexes are horned. Although the breed is noted for its hardiness the lambs are earlier maturing than are those of most other mountain breeds. The lambs produced on the hills are usually fattened in the lowlands. Draft ewes from the hills are valued for crossing.

Swaledale
Similar to the Scottish Blackface but is longer in the body and stands

Fig. 6.4. A Welsh Mountain ewe.

higher on the legs. The upper portion of the face is dark with a greyish muzzle. The ewes are in demand for crossing purposes as they are good milkers capable of rearing strong lambs under poor conditions.

Welsh Mountain

A small, active, hardy, white-faced breed. The rams are horned but the ewes are hornless. The meat quality is excellent.

Cross-breds

There are numerous popular crosses which have been produced for many years, and have proved very successful.

Scottish Half-bred

Produced by crossing a Border Leicester ram with a Cheviot ewe sold from the hill. The ewes from this cross are in demand for mating with Down rams to produce fat lambs. These half-breds have white faces, Roman noses and longish ears, and a striking feature is their uniformity.

Welsh Half-bred

A Border Leicester ram on a Welsh Mountain ewe produces a small ewe which is relatively cheap to feed and is useful for mating to rams of other breeds.

Greyface

The ewes which result from the mating of a Border Leicester ram with a Scottish Blackface ewe are famed for hardiness, prolificacy, and milking ability. Mated to Down rams they produce lambs of excellent meat conformation and quality.

Masham

Masham ewes are usually bred by crossing Swaledale ewes with Wensleydale rams, although certain other crosses are sometimes called by this name. They vary in face colour from black to almost white but are hardy and can be kept on poorer land than any of the cross-breds previously mentioned.

Hybrids

As with other species hybrid types are being produced by using breeding methods similar to those which have proved successful in fowls. The aim is to develop a strain in which the ewes are highly fertile with a good milking capacity and which produce lambs which grow rapidly and have a good food conversion efficiency. The first of these was the Colbred which was developed by crossing four breeds, the Clun Forest, Dorset Horn, Suffolk, and an imported European breed, the East Friesland. By selection after hybridisation a breed which is fairly uniform in type has been developed. Most Colbred sheep are big with long wool and white faces. A feature is the ability of the ewes to produce twins or triplets which are vigorous at birth, and to provide ample milk for them. When the ewes are mated with Down rams lambs with a good growth rate and carcase quality are produced. Other hybrids have now been developed.

Breeding systems

The breeders of sheep have not paid as much attention to the keeping of performance records of the male and female animals kept for breeding as have the breeders of other types of livestock. Most sheep kept for breeding have been selected on the basis of handling and visual assessment. Now some shepherds are keeping simple records. The prolificacy of a ewe is of significance, and some breeders of low-land sheep try to breed from ewes which were themselves one of twins. The birth weight of a lamb is a useful guide as large lambs are more likely to survive and usually grow faster than small ones. Although the feeding and management of a ewe during pregnancy influence lamb birth weights, individual ewes vary in the size of lamb they produce, as some ewes use food for the formation of fetuses to a greater extent than others.

However, the weighing of lambs at birth is a time-consuming task at a time of the year when shepherds are busiest, and so in some flocks it is not considered feasible. The weight gain of lambs by 4 weeks of age gives an indication of the milking ability of a ewe, and, although milk yield is influenced by management, it is also controlled genetically. The grading results of the fat sheep sold and the weight and quality of the wool shorn also yield valuable information.

Breeding rams should come from prolific ewes capable of producing good lambs which grow rapidly and yield a good carcase. Obviously the only method of ensuring that rams are out of such ewes is by recording. The performance testing of rams, as described under beef cattle, has been used in sheep breeding to a limited extent.

Identification

As a temporary measure various types of proprietary marking materials, in different colours, can be applied to the fleece to indicate ownership. These marks are obviously removed whenever the sheep are shorn. The ears of sheep can be notched or punched, also to indicate ownership, but this method is not entirely satisfactory as sheep ears are easily damaged and injuries can destroy the markings. Most farmers do not attempt to identify individual sheep, but for improvement schemes to be effective it is necessary to be able to identify individual animals. Various methods are employed. In horned breeds individual numbers can be burnt on the horns. Ear tattooing is satisfactory in white-eared breeds, using black ink, but is not as useful in dark-eared animals, although green ink can be tried. Ear notching may be employed, using a numerical sequence similar to that described for pigs. Numbered metal or plastic ear tags are popular. On some farms small tags are used for lambs and large ones for adult sheep.

Ageing

Like cattle, sheep have four pairs of incisor teeth in the lower jaw and none in the upper. The pairs are named according to their position in the jaw, as in cattle. When born a lamb may show one or two temporary incisor teeth or none at all. All the temporary incisors will have erupted by the time the lamb is about 1 month old. The temporary incisor teeth are smaller, narrower and more conical than the permanent teeth. The permanent central incisors appear at about 1 year 3 months, the medials at about 1 year 9 months, the laterals at about 2 years 3 months, and the corners at about 2 years 9 months. A yearling sheep can be aged with certainty by its two broad teeth, but in older sheep the teeth are less reliable as an index of age. There are variations between breeds, and the figures given above apply to the lowland breeds. In general the teeth of the improved lowland types are shed earlier than those of the hardier mountain types, and some mountain sheep take 4 years to acquire a full permanent set of incisor teeth.

After all eight permanent incisors are present it is only possible to make an approximate estimate of a sheep's age. When over 6 years the teeth appear longer than normal because the gums have shrunk, displaying the roots. The teeth become discoloured and loose and begin to fall out, although some sheep retain a full set of incisor teeth until they are 8 years of age or older. Sheep fed on roots tend to lose their teeth earlier than those not given roots in their diet.

Housing

Sheep are not normally housed in permanent buildings in the British Isles, although the winter housing of sheep has been practised in certain other countries, particularly those with severe winter conditions. However there is now an interest in in-wintering ewe lambs during their first winter and in-lamb ewes on hill farms, and in-lamb ewes on some lowland farms with heavy land which is likely to be severely poached if sheep are run on it during wet winters. The advantages are that protection from winter weather improves the condition of the sheep, the shepherd can work in more comfortable conditions, and the stocking rate can be increased because the sheep do not damage the pastures during the winter and so increased nutrients are obtained from the grass during the spring and summer. The disadvantages are the capital cost of the buildings and the higher food costs, as no nutrients are obtained by winter grazing.

Elaborate buildings are not required, simple covered yards being quite satisfactory. It is very important to ensure that the buildings are well-ventilated or respiratory troubles will become prevalent. A very successful design is to leave the yards open on one side and to leave a space between the ground and the base of the walls on the other three sides so that air can enter all round. The floors can be solid, in which case straw is used for bedding, or slatted floors raised about 1·53 m (5 ft) above ground level can be fitted. Ewe lambs are allowed 0·45–0·55 m^2 (5–6 sq ft) of floor space per head and in-lamb ewes from 0·74 m^2 (8 sq ft) in early pregnancy to 1·21 m^2 (13 sq ft) near parturition. To prevent heavily in-lamb ewes crushing each other at feeding times it is essential that adquate trough space is allowed, 0·45 m (18 in) per ewe being the minimum length advised. Water troughs must be provided but these should be sited away from the food troughs to stop sheep carrying food in their mouths and fouling the water. Sheep yards of this type generally have storage space for food and bedding. It is an advantage to have the troughs and pen divisions made of a movable type of construction so that, as winter advances, it is possible to increase the sheep area and decrease the food and bedding storage area.

On most lowland farms some form of artificial shelter is provided for ewes at lambing. Simple lambing pens can be made out of straw bales or thatched hurdles, roofed with corrugated iron sheets, but permanent lambing pens in a building are preferable. The individual pens are about 1·67 m^2 (18 sq ft) in size, and are provided with infra-red heaters which can be used for weakly lambs. Ewes normally stay in these pens for a maximum of 48 hours before being moved outside to a sheltered field.

Fences

As most sheep spend a considerable part of their lives at grass good fences are important for fields being grazed by sheep. Stone walls and hedges are used in certain circumstances but posts and wire mesh are the most suitable materials when new fences have to be constructed. Sheep fences should be 1·07 m (3·5 ft) in height, but often the wire mesh rises to a height of 0·91 m (3 ft), and two strands of barbed wire are added to make up the required height. The best mesh size is 0·31 m (1 ft) as sheep will not get their heads caught if they attempt to reach through. Wood posts are in general use but metal posts last longer and are easier to fix. Concrete posts give good service over a number of years. Electric fences can be used for sheep of the quieter types, but two wires are needed, one at about 0·25 m (10 in) and the other at about 0·45 m (18 in) from the ground.

Feeding

General

A higher proportion of natural food is utilised in the diet of sheep than with other species and so concentrate foods play a smaller part. However, they must be fed in certain circumstances. The systems of feeding vary according to the type of farm, but, as the bulk of the foods fed are growing crops, accurate rationing and feeding standards are difficult to apply. Another reason why they are not used is that rations for individual sheep are not measured, the flock being the feeding unit.

There are certain practical points which must be observed in the feeding of sheep of all types. Sheep are competitive eaters and weakly individuals may be forced away from the food source. Ample space for feeding tends to prevent this happening, as does the keeping of sheep of a similar age and size together. Sheep are clean eaters and do not take kindly to soiled or musty food. As with other animals any sudden change of food upsets the digestion. This applies to the transfer of grazing sheep from grass to a crop such as rape, as well as to alterations in the composition of a concentrate ration. All changes should therefore be made slowly.

Hill and mountain

The economic returns from the traditional methods of hill sheep farming are low. The land is rough and most cannot be either cultivated or fertilised. On the highest hills heathers and bent grasses occur, while on lower hills bent and cotton grasses are found. The quality of the herbage varies throughout the year, rising from March to a peak in May or June and then falling, being particularly low from November until March. On most hills the feeding is essentially natural grazing throughout the year, except during periods of snow and hard frosts when hay may be given. Such a system is inflexible and the stocking rate is substantially determined by the winter-carrying capacity of the land, and is usually one sheep per from 1·6 to 4·1 ha (4–10 acres). This means that the summer stocking rates are low in view of the increased herbage production at that time.

The consequence of this annual pattern of food intake change is a cycle of body weight changes in the ewes. The ewes are mated in late November, when they are starting to lose weight, and continue to lose weight during pregnancy. The loss of condition at mating reduces fertility and so the number of lambs born is small. This is an advantage as twins are not wanted. The weight loss during pregnancy may affect fetal development adversely and, thus, birth weight and lamb survival. There is certainly not enough food to allow the ewe to build up twins. Hay is only given in very severe weather to prevent starvation and the ewes are normally expected to live on the plants growing on the hill. The reason is mainly economic, for the production of one lamb and one fleece per ewe per year does not allow for expensive food costs, but also one of management, for supplementary feeding makes hill ewes lazy as they wait for food to be fed instead of seeking out the available herbage.

The lambs are born during late April and early May when the herbage quality is improving and so the ewes can sustain the milk yield required for a single lamb, but on some hills undergrazed, dying, material dilutes the feeding value of the growing herbage. The growth of hill lambs is probably more dependent on the ewes' milk yields than is the growth of lowland lambs on a productive pasture.

Various methods of improving production have been suggested, but most involve capital costs and it is essential to ensure that the increased profits give a reasonable return on the additional money invested. The housing of the ewes during the winter is one method, so that the hill is only grazed, and so can be fully stocked, during the summer. To offset the costs of housing and winter feeding a substantial rise in the number of sheep kept will be required. An increase in herbage production through the sowing of new species and fertilisation is possible on some farms. On others limited areas are capable of improvement, and these can be utilised to the full by fencing and controlled grazing methods. It may be possible to graze cattle with the sheep during the summer months and thus improve the pasture quality by reducing the ratio of dead to green herbage.

The lambs born on the hills are generally weaned at about 4 months of age, and the wethers and the ewe lambs surplus to stock replacement needs are sold. A few may be sold fat but the majority are sent to lowland farms to be fattened. It is an advantage to have the choice of the two markets, the fat and the store. On some hill farms there is a small area of cultivated land which can be used to grow rape or a root crop on which the more forward lambs can be finished for killing. The lambs sold as stores usually respond quickly to the improved feeding on the lowland farm and fatten rapidly.

The replacement ewe lambs are usually sent to a lowland farm for their first winter from about the beginning of October to the end of March, although this return date is too early in most seasons because there is little growth of hill grasses until late April. The ewe lambs benefit from the improved keep on the lowland farm and the hill benefits from the lowered stocking. However, it is becoming increasingly difficult to obtain suitable lowland grazing, and some hill farmers are establishing reseeded pastures on the best area of their farm and wintering their ewe lambs there. Others are housing the lambs and feeding them mainly on hay. On a few farms the ewe lambs to be retained for stock are never weaned, but stay beside their dams on the hill all through the winter. Advantages claimed for this method are that there is no check due to weaning, and the ewe teaches the lamb by example to make the best use of the hill grazing at all seasons of the year.

Lowland

The ewes in some flocks are mated so that they lamb from late

December to early February in order to produce fat lambs for the Easter trade, while in others they are mated to lamb in late March or early April so that the lambs are born at the same time as the spring grass growth and feeding costs are minimised.

The early lambing ewes will need to be fed hay from November onwards at a rate which varies with the quality of the hay and the size of the sheep, but is about 0·91 kg (2 lb) per head per day. Good quality silage can be used as a substitute, allowing about 3·61 kg (8 lb) per sheep daily, or a mixture of hay and silage. In addition a high energy cereal ration must be given during the last 6 weeks of pregnancy, about 0·22 kg (0·5 lb) per head per day being allowed 6 weeks before lambing, rising to a maximum of 0·91 kg (2 lb) during the last fortnight. A suitable mixture consists of crushed oats, 2 parts; flaked maize, 2 parts; bran, 1 part; and decorticated cotton cake, 1 part. Roots, such as mangolds, swedes, or turnips, can be used to replace silage and part of the concentrate ration. They are palatable, but have lost popularity because of the cost of growing, harvesting, and storing them. Roots are generally fed sliced, although some are fed whole. The flock may run on the field to eat the roots where they were grown as this saves the expense of harvesting, but there is some waste as a part of most roots is left uneaten. About 2·23 or 2·7 kg (5 or 6 lb) is the usual daily allowance per head. In late lambing flocks hay or grass silage is supplied from January onwards plus supplementary feeding before lambing, as in the case of the early lambing ewes.

Following lambing the food allowance of the ewes needs to be increased. There is considerable individual and breed variation, but ewes will yield from 1·14 to 4·51 l (2–8 pt) of milk per head per day. In general it is considered good farming practice to feed lactating ewes fairly liberally, for lambs make economical weight gains when suckling. Good quality pasture should be provided as soon as possible, as this is the most economical feed, but hay or a combination of hay and silage and concentrate food should be given until the grazing supplies sufficient nutrients. On large farms it is a good policy to separate the ewes with twins from those with singles, giving the former group more liberal rations or the benefit of better grazing.

Flushing ewes

This is to provide a method of feeding which will ensure that the ewes are in rising condition at the time of mating. This stimulates multiple ovulation and so there is a higher proportion of twins and triplets born. It is important to see that the ewes are only just thriving in the summer, and then they are run on a fresh productive pasture or a field of rape or given concentrate food so that their bodily condition improves. Flushing is practised in flocks of all types, except hill flocks where the ewes are out all the year round because one lamb is all that a ewe can manage, and a few select ram breeding flocks, where large single lambs which grow rapidly are desired.

Fattening lambs

The lambs can be finished on ewes' milk, on grass, or on specially-grown forage crops. For the first method early-maturing lambs born in mid February to mid March are required which have been suckled on ewes provided with a generous concentrate ration initially and given free access to good quality grass as it becomes available in the spring to encourage high milk yields. The lambs should reach 31·7 kg (70 lb) liveweight in from 10 to 15 weeks and are not weaned before killing. Lambs to be finished off grass are usually born at the end of March and the ewes are only given small amounts of concentrate

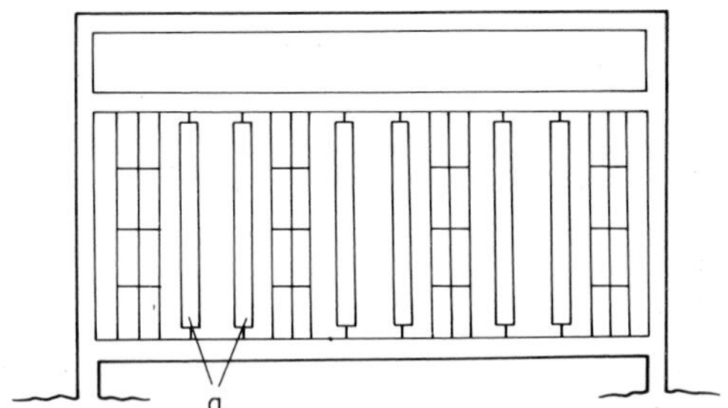

Fig. 6.5. A lamb creep to be fitted in a fence between fields. (a) Vertical rollers set to allow lambs, but not ewes, to pass through.

food. The lambs, when about 3 weeks of age, should have areas where they can graze away from the ewes so that they can make full use of the pasture. This is generally done by erecting a fence with a creep through which the lambs, but not the ewes, can pass. These lambs are usually weaned at about 12 weeks of age and should weigh about 40·8 kg (90 lb) liveweight by the middle of October. For finishing off forage crops a larger type of lamb is required. They should be born in April and their dams should be fed only small amounts of concentrate food. After weaning, when about 12 weeks old, these lambs are usually placed on stubble fields or rough grazings for a time and then finished on a forage crop grown specially. Suitable crops are rape, thousand-headed kale, mustard, turnips, and swedes. Surplus arable products and arable crop residues can be fed as well. Sugar beet tops are available from October onwards, but should not be fed until they are wilted which takes about 1 week after harvesting. Brussels sprouts stalks and leaves can be used when the sprouts have been picked for market, as can waste cabbage and cauliflower leaves. The lambs are usually marketed in the late winter when prices tend to be high.

Rams

Ram lambs being reared for service are kept on good pasture and generously fed on a ration of the type recommended for breeding ewes. During the breeding season adult rams should be fed the same foods as the ewes, but in slightly larger amounts. During the rest of the year pasture is usually adequate to provide their needs, but supplementary feeding may be required.

Mineral requirements

The mineral requirements of sheep are not very different from cattle, both being ruminants, and are normally satisfied by the grazing on both hill and lowland farms. However the mineral content of the herbage varies with the soil and the species and varieties of the clovers, grasses, herbs, and weeds and there are areas where the herbage is deficient in certain mineral elements. Neither calcium nor phosphorus deficiencies are likely to occur, but, as mentioned later, copper is concerned in the causation of the disease of lambs called swayback. Cobalt can be of special importance in sheep nutrition as in certain areas the soil is deficient. Sheep grazed in these regions develop a wasting disease known as nutritional pine which is cured or prevented by the feeding of cobalt in minute doses. On farms where the magnesium level is abnormally low deaths from hypomagnesaemia may occur in a manner similar to that recorded under dairy cattle. Particular mineral needs can be satisfied by the provision of suitable mineral licks or the incorporation of minerals in the ration being fed to sheep receiving concentrate food.

Vitamin requirements

Vitamin deficiencies are uncommon in sheep kept in the traditional ways. Vitamin A is synthesised very efficiently from the carotene in green food, and under outdoor conditions vitamin D is formed by sunshine activating certain sterols in the sheep's own skin. In the case of sheep confined in sheds during winter a supplementary source of these two vitamins in the diet may be required. In general, the foods consumed by sheep, especially well-cured hays and green pastures, contain ample quantities of vitamin E. Vitamins of the B group are synthesised by sheep, but cobalt is essential for the formation of vitamin B_{12}, and a deficiency of this vitamin is directly linked with the causation of nutritional pine.

Water requirements

Sheep require water in the same way as all other animals, but their needs are less than those of most domesticated species. Mature animals may drink about 4·5 l (1 gal) per day and fattening lambs about half this amount when kept on dry food. Adult sheep, other than lactating ewes, can often satisfy all their water requirements from the dew falling on a pasture and the moisture contained in the herbage. Lactating ewes do require water to drink, and their milk yield may be reduced if they are unable to drink freely. The water supplies for hill sheep need consideration. During a drought in summer hill pasture can become dried up and access to streams must be provided. During a severe winter when the water supplies are frozen sheep will eat snow to satisfy their water requirements, but in very severe conditions the snow may become too hard to be eaten and sheep may die of thirst.

Artificial rearing of lambs

Lambs in excess of a ewe's milking capacity which cannot be fostered on to another ewe can be removed when between 24 and 48 hours old after they have had the colostrum, and reared artificially. On most sheep farms, when only small numbers of lambs are involved, they should be housed in a well-bedded loose box and initially some artificial heat is required so that an environmental temperature of 18·3°C (65°F) is maintained. They must be fed on reconstituted milk of an appropriate composition. Cows' milk is not suitable as it contains less fat and protein than does ewes' milk, and milk substitutes used in calf feeding are also unsatisfactory for the same reason. There are now several proprietary milk substitutes formulated specially for lambs which can be used, or, in an emergency, full cream dried milk can be made up as specified by the makers, but using whole cows' milk instead of water. The resultant fluid approaches ewes' milk in its fat and protein contents.

Feeding should be at least three, and preferably four, times daily and a lamb will consume about 0·57 l (1 pt) of milk per day at first, rising to about 1·14 l (2 pt). The milk is allowed until the lamb is 5 weeks of age, after which the quantity is gradually reduced and the lamb is weaned when about 6 weeks old. The milk is usually given from a bottle fitted with a rubber teat, but lambs can be trained to drink from a bowl. After weaning a simple diet of 95 per cent barley and 5 per cent fish meal can be used or a more varied ration such as flaked maize, 2 parts; crushed oats, 2 parts; groundnut meal, 0·5 part; and fish meal 0·25 part can be prepared. Both diets should be fortified with a mineral/vitamin mixture and can be made more palatable by the addition of a little molassine meal. The lambs can be turned out to grass when about 5 weeks old, but should be provided with some form of shelter. The concentrate ration can be withdrawn as soon as the lambs eat grass freely, and at about 8 weeks of age they can be turned out with the main flock.

On some large farms where highly productive ewes are kept and a number of lambs have to be reared artificially each year a mechanised feeding system is required. Methods in which a number of lambs can be suckled on teats communicating with a central milk supply have been used successfully, provided the lambs have received colostrum from their dams. As the sooner lambs can be weaned off milk substitute and fed on dry food the easier artificial rearing becomes, the lambs are encouraged to eat some concentrate food and drink water when they are about 1 week old. Lambs reared in this way are usually kept confined and fed on dry foods until slaughtered at about 3 months of age.

Breeding

Ram lambs may be used for service provided they are well grown. A ram lamb is mated to about twenty ewes and an adult is allowed to run with forty or fifty. Rams are normally 'raddled' before being run with the ewes. This term means that their briskets are smeared with a coloured substance, or they are fitted with a harness to which a block of a coloured compound is fixed, so that they mark the fleeces of all the ewes they serve. The colours are changed every 2 weeks so that the shepherd can tell how many ewes have returned to service.

Ewe lambs of the lowland types can be bred from if born early in the year and well reared, but ewes of the mountain breeds are not mated until they are about 18 months of age. Ewes of most breeds have definite breeding seasons, coming into oestrus during the autumn and winter only. However, Dorset Horn ewes have an extended breeding period from July to May. Ewes of all breeds come into oestrus at intervals of about 16 or 17 days and the oestrous period lasts for from 1–3 days, with an average of 27 hours. The ovulation rate is influenced by the nutrition of the ewe at mating time, and, as previously stated, ewes are normally 'flushed' to ensure that they are in a rising condition at the time of mating.

For economic reasons it may be desirable to bring the ewes into oestrus earlier than is normal for the breed so that the resulting lambs can be sold before the majority are ready for slaughter and thus can command a higher price. The presence of a ram can stimulate earlier oestrus and, although this may only advance the breeding season by 2 or 3 weeks, in early fat lamb production this may be of economic importance. Vasectomised rams may be used for the purpose of stimulating the ewes, so that when the breeding ram is introduced there will be a number of ewes ready for service and a batch of lambs of approximately the same age will be produced. The injection of hormones outside the normal breeding season can also be used to induce oestrus at various times in the year, but is not a practical proposition in normal commercial sheep farming.

The gestation period is about 147 days. Normally a ewe about to lamb separates from the flock, lies down with the head raised slightly and starts to strain. An hour or two later the water bag appears and eventually this bursts and the first lamb will be born. Any more lambs are usually produced soon afterwards.

The birth weight of a lamb is considerably influenced by the number of lambs born. In any breed the average birth weight of singles is greater than that of twins, and of twins greater than that of triplets. Unless newborn lambs are adequately cared for they may die, cold being the principal cause. There is evidence to suggest that lambs which are small at birth are less viable than are larger sized individuals. This is one reason why single lambs are favoured in mountain breeds, which lamb in the open, and ewes of the lowland breeds, producing a proportion of twins and triplets, usually lamb in pens affording some protection from inclement weather. Lack of ewe's milk is another cause of death, and lambs should suck colostrum as soon as possible after birth. Ewes and lambs should be kept in small groups until the lambs are suckling well and, if possible, ewes should never be separated from their lambs in the first few weeks of life.

Artificial insemination

Artificial insemination can be used in sheep breeding, but is not a common practice in the British Isles. There are two methods by which semen can be collected from a ram. The first is by the use of an artificial vagina of appropriate size in a manner similar to that used for bulls. The second is by electrical stimulation of the lumbar region of the spinal cord. The semen from a ram is much more concentrated than that from a bull, and investigations are proceeding into the best conditions for dilution and storage, including deep freezing. The insemination of the ewe requires the use of a graduated syringe to deliver the small volume of semen required, and an instrument to dilate the vagina adequately.

One major complication in the use of artificial insemination in sheep is the difficulty in determining when the ewes are in oestrus, as there are no reliable observable signs of oestrus, except the ewes' behaviour towards a ram. Therefore vasectomised rams, or entire rams prevented from serving by having a sack tied under their bodies over their penes, must be used to detect the ewes which are in oestrus. These rams are raddled so that the ewes they mount are colour marked. The marked ewes are caught for insemination.

Health

Sheep in a lowland flock tend to keep together and an individual on its own may be diseased. This does not apply to the mountain breeds where individuals scatter in search of food. The head gives an indication of the state of health. Healthy animals have keen, bright eyes and pricked ears, while dull eyes and drooping ears are signs of disease. Unshorn sheep of all types should have unbroken fleeces carrying a natural bloom, and ragged wool may be indicative of rubbing caused by skin irritation. If the wool is divided with the fingers the healthy, supple, underlying skin can be seen. Sheep should have sound feet and should not show signs of lameness when walking.

The body temperature of sheep is between 39 and 40°C (102 and 104°F) and tends to be higher in animals with a heavy fleece, particularly in hot weather. The pulse rate is usually between 70 and 90 beats per minute, but may rise to 100 during the summer months. It may be taken at the lower jaw, as in cattle, but is more easily counted at the artery which runs inside the hind legs about half way between the stifle and hip joints. As sheep are not usually accustomed to being handled the pulse may rapidly accelerate from nervousness when it is taken. Time must therefore be given for a sheep to quieten if an accurate count is to be made

Common diseases and their prevention

The description of *enterotoxaemia* in goats covers the disease in lambs. Single lambs, being done well by their dams, and fattening lambs being over fed are the most susceptible as these conditions predispose to the digestive upsets which favour the multiplication of the causal organism.

Lamb dysentery is caused by another strain of the same organism as enterotoxaemia, but the disease only occurs in lambs during the first 3 weeks of life. Clinical signs are sudden collapse and death, and, in animals surviving sufficiently long, blood-stained diarrhoea leading finally to exhaustion and death. Further losses in the season in which deaths have occurred can be prevented by injecting serum into all the lambs soon after birth. To prevent losses in future years the ewes can be vaccinated at the time of mating and again about 1 month before lambing, and the vaccinated ewes pass on an immunity to their lambs in the colostrum.

Foot rot is a highly contagious disease caused by bacteria and is particularly prevalent on wet land. Affected horn degerates and the hoof becomes underrun and the tissue becomes necrotic, causing severe pain. This pain restricts movement and so limits feeding and a severe loss of condition follows. Every attempt should be made to eradicate the disease from a flock, and this is possible because the causal bacteria can only live away from a sheep's foot for about a fortnight. Eradication involves the removal of all infected tissue from all the hoofs in the flock, the passage of the flock through a shallow foot bath containing a 10 per cent formalin solution and the removal of the flock to a clean pasture. A corrugated bottom to the foot bath is desirable as the raised ridges tend to force the clefts of the feet open and give better exposure to the solution. It may be necessary to cull a few badly infected individuals, but after about four such treatments and moves the flock should be clear of infection.

Parasitic gastro-enteritis is described under goats and is applicable to sheep. Routine preventive dosing with an anthelmintic is commoner in flocks of sheep than in herds of goats, and the two most important times are the dosing of the ewes at about lambing time and of the lambs during June. At the time of dosing the sheep should be moved on to clean land, by which is meant pasture which has not carried sheep during the current grazing season.

Liver fluke is mainly a parasite of sheep kept in wet areas, but flukes can live in several other hosts such as cattle, goats and some wild animals. The adult flukes live in the bile ducts of the liver and their eggs pass out in the faeces of the sheep. These hatch in moist conditions and the larvae bore into mud snails and develop into a new type of larvae which emerge from the snail. These swim and can only survive in moisture. They then encyst on blades of grass and, if swallowed by a suitable host, develop into flukes which burrow from the alimentary canal into the liver where they feed on the liver tissue as they grow and eventually reach the bile ducts. The clinical signs shown while the larvae are growing in the liver are a rapid loss of condition, a disinclination to eat, listlessness, and death. The adult flukes in the bile ducts have a more chronic effect. There is anaemia, slow wasting and fluid accumulations in the tissues, particularly under the jaw. Preventive measures consist of keeping sheep off grazing areas where mud snails are plentiful and the draining of damp land, although the first is often not practicable and the second may be economically out of the question. The snails may be killed by applications of copper sulphate, but this is expensive in labour, as up to six dressings a year may be required if the treatment is to be really effective. Sheep must not be allowed to graze an area treated with copper sulphate until it has been washed into the soil because of the risk of poisoning.

Blowfly attack occurs during the summer months and is commonly known as 'fly strike'. Blowflies or greenbottles are attracted to areas of the wool soiled by faeces, e.g. below the root of the tail, or by plant juices, e.g. on the shoulder region in bracken areas. Sheep on lush pastures with loose faeces are most likely to be attacked. The flies deposit their eggs on the skin of the sheep and these hatch in about 24 hours to produce maggots which feed on the inflamed skin. The clinical signs are a withdrawal from the flock, a wagging of the tail or rubbing of the body due to the irritation, and a close examination reveals the presence of large numbers of maggots. If untreated death may follow in about 4 days. Prevention is by clipping the ewes' wool on both sides of the hind quarters below the tail, commonly known as dagging or crutching, to reduce the risk of faeces accumulation, and by dipping the flock in early summer with a dip, such as an organo-phosphorus preparation, likely to deter the flies from laying eggs and kill those maggots which hatch from any eggs which might be laid. Even if dipped a flock must be inspected frequently to note signs of scouring and fly strike.

Sheep scab is caused by mites which burrow into the skin and are spread from sheep to sheep by direct and indirect contact. The mites go through a period of dormancy during the summer, becoming active again during cold winter weather. The mites obviously cause intense irritation and the clinical signs are rubbing, leading to a loss of wool and a marked fall in body condition. A severe infestation can cause death. Prevention is by dipping. Every flock must be dipped between 16 August and 13 November each year in an approved dip, and each sheep must be immersed for 1 minute.

Handling

Butting is the only means of defence which sheep have, and their instinct is to run away from danger. Sheep in the British Isles are usually kept in flocks and rarely become tame. Hill sheep, particularly, have a wild, nervous temperament. Sheep do not appreciate being scratched, like pigs, or petted, like horses and dogs. When handling is necessary sheep should be driven as a flock into a pen and a well-trained sheep dog can be of great assistance in effecting this. In the pen individuals can be caught with a shepherd's crook, being hooked just above the hock of the hind limb. Alternatively the leg above the hock can be grasped in the hand, but a leg should not be seized below the hock as the sheep may injure a joint or, possibly, break a bone while struggling. Horned sheep can be caught by the horns, and some shepherds use a crooked stick to catch an animal round the neck. Sheep should not be seized by the wool as this may tear and blemish the fleece and bruise the body.

To hold a sheep it is best to stand at one side and place one hand under the jaw and the other on the opposite flank. Should the animal try to run back the second hand can be transferred to the animal's hindquarters to check this movement. Horned sheep can be held by the horns with the handler standing in front of the animal, at one side, or astride its shoulders.

For an examination of the feet a sheep should be held in a sitting position by placing the left hand under its neck and the right hand under the belly near the right hind leg. The animal can then be lifted into a sitting position, placed in front of the handler's legs and supported against them. This position is also suitable for injections into the inside of the thigh.

Pregnant ewes or fat sheep are liable to roll on to their backs while trying to rise from their normal lying positions and remain in this state unable to get up. A sheep in this state should be lifted gently to its feet and supported until it is steady on its feet, otherwise it may stagger and fall on its back again. Sheep left on their backs for any length of time will die and so in-lamb ewes and fat sheep should be inspected regularly.

When lambs have to be castrated or docked the handler should sit on a bench and hold the lamb's back between his thighs with its head resting against his body. The lamb's right fore and hind legs are held in the right hand and the two left legs in the left hand. Thus the scrotum and tail are presented to the operator.

It is almost impossible to husband even a small flock of sheep properly without the use of pens and on farms where large flocks are maintained a sheep handling unit is essential. Considerable thought is generally given to the design of such a unit so as to reduce the labour needed for shearing, dipping, dosing and lamb sorting to a minimum. This unit should be sited near a hard road so as to facilitate the loading of stock into wagons when necessary. All the pens should have concrete floors to make cleaning and sheep handling easier and to avoid muddy conditions which may harbour infectious agents. The unit should start with a collecting yard which should be large enough to hold a flock. Sheep require about 0·37 m^2 (4 sq ft) for small and 0·56 m^2 (6 sq ft) for large ewes per head of standing room. The sides are usually constructed of post and rail or sheep netting, and should be at least 1 m (3·3 ft) high. The collecting yard should lead to a drafting race, which is a passage about 3 m (9·8 ft) long and 450 mm (17·7 in) wide for mountain breeds and 500–550 mm (19·7–21·3 in) for larger breeds through which the sheep pass in single file. It can be used for dosing, injecting, checking ear marks or sorting sheep. It leads into one or two smaller holding pens. Two are preferable to one as they can be filled alternately, and the fact that one is full attracts other sheep to enter the second.

A foot bath is required, which may be built separately or included in the drafting race. It is better to have them separate. It is usually possible to position a portable spray race at the end of either the drafting race or the footbath. There must also be a dipping bath, three types being used. The commonest is the short swim bath, which is suitable for any flock up to 1,000 sheep. These may be made of galvanised iron or built of concrete and brick. They contain from 473 to 1,820 l (115–400 gal) according to the size required for a

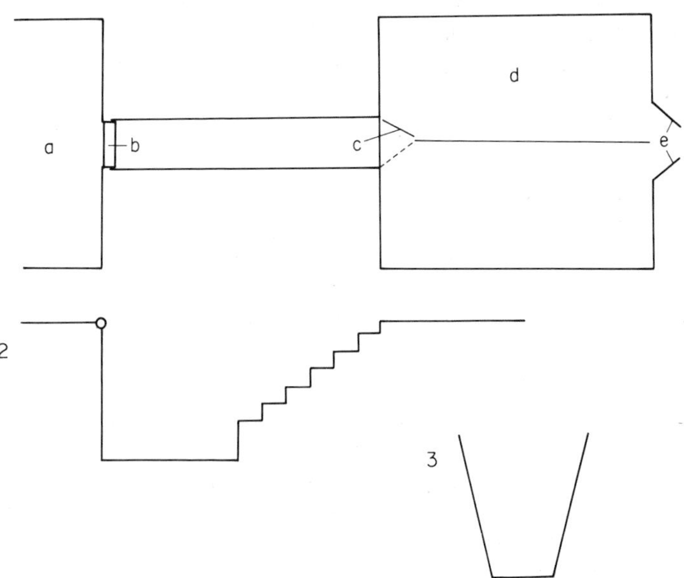

Fig. 6.6. Sheep dipping bath and pens. (1) Floor plan, (2) longitudinal section of bath, (3) cross section of bath. (a) Collecting pen, (b) roller, (c) single gate, (d) draining pens, (e) exit gates.

particular flock. The long swim bath is similar but is longer and deeper and the circular bath is designed so that the shepherd stands on an island in the centre and controls the sheep as they swim round. These last two types are only economic for flocks of at least 1,000 sheep.

Whatever type of dipping bath is used two draining pens should be erected side by side at its exit, so as to catch the surplus dip which drains from the fleeces shortly after dipping and return it via a sieving device to the sheep bath. When dipping is in progress both pens are used. When the second pen is full the first is emptied and refilled and so on. If each draining pen is designed to hold about twenty sheep each sheep is allowed about 10 minutes to drain before being let out on to a field. Portable yards with drafting races and foot and dipping baths can be used on farms on which it is not convenient to collect the sheep together in a central handling area.

Shearing

Sheep are shorn once a year in late May or early June in lowland flocks and in July in hill flocks. The warmer weather causes the grease to rise in the wool and when this is well up in the fleece shearing can begin. Also these are times between the worst of the cold weather and the onset of very hot weather, and the lambs are old enough not to be upset unduly by the disturbance caused by gathering the flock for shearing. Before a day's shearing the sheep are driven into yards and held there for at least 6 hours. This ensures that their rumens are comparatively empty which helps to prevent digestive upsets following the considerable amount of handling required during shearing. Whenever possible the yards should be under the cover of a Dutch barn so that the fleeces remain dry in wet weather. The shearing pens should be kept clean and free from hay, straw and dung which might foul the wool. The machine shearing of sheep is a skilled task usually carried out on large farms by skilled shepherds and on smaller units by contract shearers who tour a district. A helper should be available to wrap the fleeces so that they are wrapped soon after shearing to prevent contamination. The neck wool should be twisted into a cord and bound round the fleece. Fleeces must not be tied with string as fine strands of string can become entwined with the wool from which they cannot easily be removed. After shearing the sheep are left in the pens for a time to see that none have been harmed during the procedure and are then

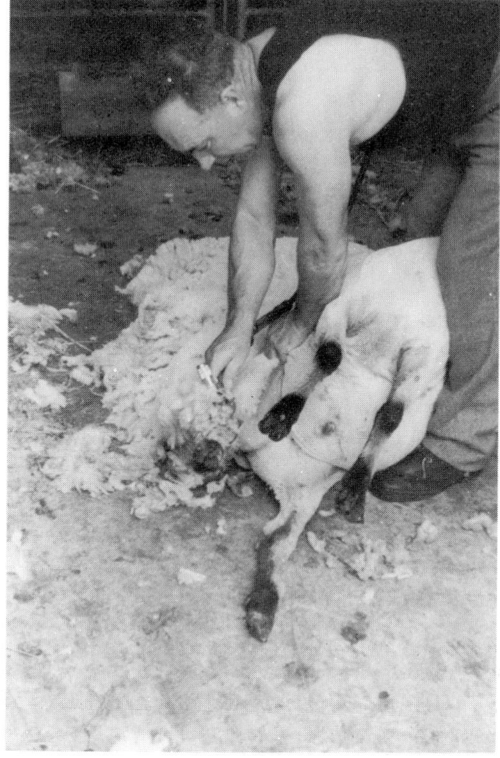

Fig. 6.7. Shearing a sheep.

returned to the fields. Should cold, windy, rainy weather occur soon after shearing the sheep must be given some protection by being placed in a yard or sheltered field if losses are to be avoided.

Washing

Sheep may be washed before shearing, usually from 4 to 10 days before. Washed sheep are easier to shear and washed wool commands a higher price than unwashed, but weighs less. Washing is a practice which has almost disappeared on lowland farms, but some upland farmers who have running streams on their farms continue to wash their sheep.

Dipping

Before dipping commences the sheep should be starved for at least

Fig. 6.8. Sheep in a dipping bath being guided by a crutch.

6 hours. The dip should be carefully prepared according to the maker's instructions and the bath should be topped up to replace the dip removed by each sheep in its fleece as recommended by the manufacturer. Dipping should not be carried out in the late afternoon when the temperature is falling nor while it is raining. Flocks should not be dipped during the last 2 weeks of the ewes' pregnancies nor during the first 4 weeks of the lambs' lives. When dipping precedes weaning it is best to dip the ewes before the lambs. The ewes are then dipped while the tank is full, and their udders will have dried before the lambs are released. Each sheep should be lifted by placing the left arm under its neck and the right hand on the loose skin of the right flank and then lowered tail first into the tank, at the same time pushing it forward. The head and neck can then be immersed in the dip by pressing on the shoulders with a crutch, which is a pole with a cross-bar at the end.

Spraying

In spraying the sheep are soaked by jets from coarse pump nozzles at high pressure. The powered spraying units used constantly agitate the liquid to ensure that it is always at the correct strength. The principal advantages are that there is less disturbance for the sheep and that there are no risks of drowning or dip ingestion. Although spraying must not be practised to control sheep scab it is effective for the eradication of other external parasites.

Fostering

As a ewe has two teats it can conveniently rear two lambs, and in lowland flocks it is desirable to ensure that each ewe rears two lambs if possible. To achieve this fostering lambs on to ewes which have lost their own lambs or only given birth to a single lamb is necessary. The lambs to be fostered may be orphans, or the shepherd may remove one lamb from a set of triplets. As ewes do not readily accept lambs which are not their own, care in fostering is necessary. Ewes recognise their newly-born lambs by smell and so a lamb to be fostered can be thoroughly rubbed with the afterbirth from the ewe on which it is to be fostered, or, alternatively, if the ewe already has a lamb both this and the foster lamb can be rubbed with a rag soaked in a substance with a distinct smell. If a ewe has a dead lamb this can be skinned and the skin can be fastened over the body of the foster lamb. When the ewe has accepted the foster lamb the skin can be removed. Whatever method is employed the ewe and the foster lamb or lambs must be penned together and not released until the shepherd is sure that the ewe is suckling the lambs satisfactorily.

Administration of medicines

As a prevention against a build up of internal parasites, sheep in most flocks are drenched several times a year. When only small numbers are involved a small, long-necked bottle can be used. The handler backs the sheep into a corner of the pen, stands astride the sheep's shoulders, holding its body firmly between his knees, and with one hand holds the sheep's head level while inserting the neck of the bottle into the mouth of the sheep in the space between the incisor and molar teeth with the other. The medicament should then be poured slowly into the mouth. It is quite easy to see that the sheep is swallowing properly while the drench is being given. Some people find a syringe convenient for dosing because the amount of the medicament required can easily be measured. In the case of large flocks a special drencher is used to expedite the dosing process. This type of syringe or gun is connected by a long rubber tube to a sack on the operator's back and the nozzle of the gun is inserted into the mouth just far enough to prevent the fluid from running out of the sides of the mouth. It should not be pushed down the throat as far as it can go. By pressing a control lever a measured dose can be delivered. If the sheep are driven down a drafting race each one can be restrained by grasping the head, without the need to hold the body.

Tablets on the whole are not as satisfactory as liquid drenches for sheep. If they have to be used they are generally administered by a specially designed gun. It is not easy to force tablets down the throat of a sheep with the fingers and can be dangerous to the sheep through choking and to the handler through bitten fingers.

General

Minor operations

Castration

In most flocks all the male lambs, except those intended for breeding, are castrated, although, as ram lambs put on weight more rapidly than wether lambs, under very intensive husbandry systems where all the lambs are sold for slaughter at an early age, castration may not be practised. The setback to the growth of lambs is least if castration is performed when the lambs are young, preferably under 2 weeks of age. The three methods described under beef cattle can all be used for lambs, but the rubber ring and Burdizzo methods are favoured. The Burdizzo castrator used for lambs is smaller than the instrument designed for calves. An anaesthetic must be administered if a lamb over 3 months is to be castrated.

Docking

In lowland flocks the tails of the lambs are docked to reduce the risk of fly strike, because long tails become soiled with faeces particularly when the sheep are fed on forage crops or luxuriant grass. Docking should be carried out so as to leave about 25·4 or 50·8 mm (1 or 2 in) of tail. The operation can be performed by the application of a rubber ring between two vertebrae, using the expanding instrument mentioned under calf castration. The lower part of the tail quickly drops off. This is best carried out when the lambs are a few days old. Alternatively the tail can be crushed with a large size Burdizzo instrument, or seared through with a hot iron. The best age for these methods is about 2 weeks. For convenience, docking and male lamb castration are usually performed at the same time. In hill flocks the tails are generally left undocked, or only cut to hock height. The reasons are that a tail protects the udder in bad weather, and that the scouring which occurs in lowland flocks on good pasture is less likely to be seen in sheep grazing on hills.

Hoof trimming

Overgrown horn needs to be cut back as in the case of goats. It is best to sit the sheep on its hind quarters with its back resting against the operator's legs. A sharp knife or a pair of foot-rot shears can be used.

Wool Marketing

Wool is an important product of sheep farming, and all wool in the British Isles is sold through the Wool Marketing Board. The weight of a fleece varies from up to 6·35 kg (14 lb) in some of the longwool breeds to about 0·91–1·36 kg (2–3 lb) for Welsh Mountain ewes kept under hill conditions. The price varies according to the quality, the finer wools suitable for cloth making fetching a higher price than the coarse wools from hill breeds which are used in the manufacture of carpets. In the hill breeds wool coarseness is thought to be linked with sheep hardiness because of the increased weather proofing, and so an improvement of wool quality is not usually considered desirable.

There are a number of properties which influence the value of a fleece to a wool buyer. Fine wool is wool in which the individual fibres are of narrow diameters, while coarse wool has fibres which are wider on cross section. Fine wool will spin out to a greater length than coarse wool and is suitable for cloth making. Coarse wool is used in the manufacture of carpets. The length of the staple is significant, the longer wools being of value for the production of worsteds, and the shorter wools for the manufacture of other woollen goods. The physical condition of a fleece is important. All fleeces contain a certain amount of dirt and grease and so when wool is cleaned there is always a loss in weight. Adhesive plant seeds such as burrs, tar, and some branding fluids are particularly difficult to remove from wool. Stained or discoloured wool is not favoured as it cannot be dyed a light colour.

Government subsidies

A subsidy is payable on ewes in flocks which are kept on hill land and which graze there for the greater part of the year. A supplementary grant is made for ewes of specified hardy breeds and crosses kept in flocks which are self-maintaining and are kept in accordance with the recognised practices of hill sheep farming in the district.

CHAPTER 7
PIGS

Introduction

Pigs are kept for the production of meat for human consumption. Unprocessed pig meat is consumed in Great Britain as pork, and, when minced, is the main ingredient of sausages and forms the basis of many tinned luncheon meats. Pig meat is more easily and effectively preserved than other meats and cured bacon and hams are popular items in the human diet. Pig fat, known as lard, was at one time widely used in cooking, but is now to some extent being replaced by vegetable oils.

Pig farming can be carried out extensively with the breeding sows being run at grass with cheap shelters for protection and the offspring being fattened in groups in yards, or it can be carried out intensively on a small area of land with the pigs kept in specialised buildings which protect them from changes in the weather. Pig keeping can be conducted on a small scale as a part-time occupation or in conjunction with other farm enterprises or large scale specialist operations can be developed.

To make a financial success of pig production farmers must continually study market prices as the values of pig meats tend to vary markedly. The changes are usually linked with production levels, for when pigs fetch good prices farmers increase numbers until a stage is reached when supply exceeds demand and prices fall. This leads to a reduction in pig numbers so that the supply is insufficient to meet the demand, leading to a rise in prices, and the cycle starts anew. It is possible to increase pig numbers fairly rapidly because sows are so prolific, producing two or more litters per year with ten or more pigs in a litter. Another financial consideration is that, because of the nature of the digestive tract, the growing pig must be fed a maximum of concentrate food and a minimum of roughage. Thus, when concentrate foods are high in price pig production costs will rise.

The future prospects for pig farming appear to be reasonably satisfactory. It is likely that people will continue to eat pork, bacon, and ham and over the last few years there has been a marked increase in the consumption of pork all the year round due to the increased use of refrigeration, the old idea that pork should not be eaten during the summer months having vanished. However, there are wide differences in the financial returns obtained by the best and worst farmers as pig farming is very susceptible to the efficiency with which it is organised. Most of the large-scale specialist enterprises now in the majority are well run, some being conducted along lines similar to the conveyer belt systems used in factories with young pigs and feeding stuffs being introduced at one end and pigs ready for slaughter being taken out at the other. The future prospects for such really efficient pig farmers should be good, but some of the less efficient are likely to experience difficulties in making a profit.

Definitions of common terms

Piglet. Young unweaned pig, usually under 8 weeks of age.
Runt. The smallest piglet of a litter.
Store pig. From 8 to about 12 weeks of age if there is a period between weaning and fattening.
Hog or barrow. A castrated male.
Boar. An uncastrated adult male.
Gilt. A young female.
Sow. An adult female.
Bacon. Pig meat sold after being cured by a process which usually involves immersion in a solution of brine
Pork. Pig meat sold and eaten in a fresh state.

Principal breeds

Bacon type

Bacon pigs should have relatively small heads and legs and well-developed hind quarters and long backs to provide the maximum amount of meat which can be marketed as bacon and ham. Most of the breeds mentioned under this heading will also produce good pork if appropriately managed and are now widely used for this purpose.

Large White

One of the most popular breeds, widely distributed throughout the

British Isles. The pigs are large with adult sows weighing between 204 and 272 kg (450 and 600 lb). The skin and hair are white. The head is long with medium-sized ears which incline forward but do not flop down. The body is long and fairly deep. The sows are prolific and good milkers and the young grow rapidly.

Fig. 7.1. A bacon pig of the Landrace breed. The tail has been docked.

Landrace

This breed originated in Denmark and breeding animals were imported into the British Isles from Sweden in 1949. Since then the breed has increased in popularity and numerically is second to the Large White. The pigs are white in colour, have small heads with floppy ears, light shoulders, long backs, and heavy hams. They are not quite as heavy as Large Whites. The sows are fairly good mothers and the piglets grow rapidly and utilise their food efficiently.

Welsh

The Welsh breed is a white pig with lop ears and is intermediate between the Landrace and Large White breeds, although in its modern form it is more like the Landrace.

Tamworth
A chestnut red-coloured pig with a long, relatively narrow body. The pigs are hardy but mature and fatten slowly, and the sows are not particularly reliable mothers. The breed is rarely seen on commercial farms.

Dual purpose type
Generally kept for crossing with either bacon or pork type boars.

British Saddleback
The pigs are black with a white saddle circling the body in the region of the shoulders. The forelegs are white, and white markings are often found on the hind legs and tail. The ears are of medium size and bend forwards without flopping. The sows are prolific and good milkers, and can be kept at pasture satisfactorily.

Large Black
An old breed which is not now seen frequently. The body weights are nearly as high as the Large White, but the body is shorter, deeper, and wider. The skin and hair are black and the ears long and hang down over the face. The pigs are very docile and thrive well on grass. The sows are reasonably prolific and make very good mothers. When crossed with a white boar the piglets are blue or blue and white in colour.

Pork type
The pork requirements are met by a thick compact pig of medium length having a thick loin, a short leg, and a minimal thickness of back fat. The following breeds produce good porkers, but will not produce good bacon pigs.

Middle White
This white-coloured breed has lost popularity of recent years. The head is short with a turned-up nose, and the body is short, wide, and compact. The young pigs mature early and fatten rapidly. The sows are not very prolific.

Berkshire
Another breed which is not widely kept now. The pigs are black in colour with white feet, a white tip to the tail and a white mark on

the nose. The head is short, but the nose is not turned up. The conformation is similar to the Middle White, but the body is slightly longer. The sows are not as prolific as the larger breeds. The merit of the breed is that the pork produced is of very high quality.

Cross-breds

The production of first crosses between breeds has long been popular in pig breeding and a very common cross is the Landrace × Large White. Dual purpose breed sows have been used because of their mothering ability and crossed with bacon type or pork type boars in order to breed pigs most suitable for either bacon or pork production.

Hybrids

In commercial pig production hybrids are produced from selected cross-bred sows which are usually mated to selected boars of a third breed. By testing it has been ascertained that the resultant pigs grow rapidly, utilise food efficiently, and have high quality carcases.

Breeding systems

Pig breeding improvement schemes yield results relatively rapidly because pigs are comparatively fast growing and have a multiple birth rate. Pig breeders have been quick to realise the value of both progeny and performance testing in improving the quality of their pigs. These methods of selection are carried out on individual farms and on central testing stations. On farms the breeding stock can be selected from among the fastest growing animals with the best body conformation and the parents of the bacon pigs with the highest grading returns from the bacon factory. Additional records of some value in selecting the breeding pigs are the regularity of farrowing in the sows and the numbers of piglets born and reared, although the heritabilities of these traits are low. On central testing stations selected animals from a number of breeders are reared under the same conditions from an early age. Usually each breeder sends two boars, one hog and one gilt. The liveweight gains and food conversion efficiencies of the boars are recorded and the thickness of the back fat is measured by ultrasonic waves. The hog and gilt are slaughtered and their carcase qualities assessed. Such observations can be regarded as performance testing, for the boars can be used for breeding if the findings are satisfactory, and as progeny testing, for the results of the carcase examinations of the two slaughtered pigs can be related to parentage. The advantage of the testing station system is that the performance of the pigs from one farm can be compared with those from stock from other premises, all having been kept under standard environmental and management conditions.

Identification

Several methods are available for the individual identification of pigs. The most satisfactory is ear tattooing, which can be performed when the piglets are 3 weeks of age or younger. Although very successful in white-coloured pigs it does not work very well with coloured breeds. Ear notching is an alternative method, generally carried out,

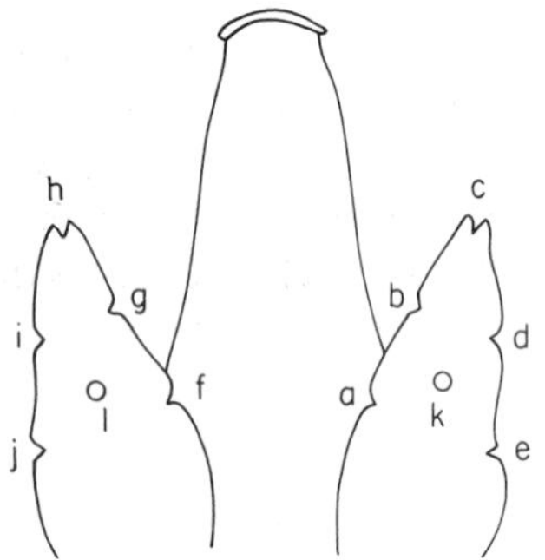

Fig. 7.2. Ear notches for pig identification. a=1, b=2, c=3, d=4, e=5, f=10, g=20, h=30, i=40, j=50, k=100, l=200.

as shown in the diagram, by the use of a special ear marker or a harness punch. A number of notches, in one or both ears, will be required to give a total number of points equal to the number of the pig. There is no universal system, each producer having his own plan. Plastic ear tags can be used, and these may be numbered, to identify individuals, or plain, but of different cclours, to indicate members of groups. Slap marking is an acceptable method when identification is required for a temporary period only and is in general use before pigs are sent to a slaughterhouse. This method leaves a tattoo on the skin over the shoulder or ham of the pig and is applied by a special marker which is slapped against the skin in the required position. Pig marking sticks can be used to leave temporary marks of different colours on the heads and backs of pigs. This method is of considerable value when groups of pigs are being treated; those which have been attended to being marked so as to be easily differentiated from those which have not been treated. Adult sows can have large tags inserted in their ears or collars placed round their necks. Both tags and collars are usually made of pastic in various colours and are numbered.

Ageing

Pigs have incisor and canine teeth in both jaws and can be aged by an examination of these teeth, but this is rarely attempted for the following reasons. Tooth eruption times vary more widely in pigs than in other species; it is difficult to examine the teeth of adult pigs and young pigs are generally sold by weight, and age is not important.

There are three pairs of incisor teeth, and the permanent teeth do not erupt from the centre outwards as in other species. The corners erupt first at about 7 months, then the centrals at about 1 year and finally the laterals at about $1\frac{1}{2}$ years. The canine teeth, or tusks, are well developed and erupt at about 9 months of age. These brief details enable the age of pigs over 7 months to be determined approximately.

Housing

Pregnant sows

Can be run in groups on strongly-fenced, well-drained paddocks, with huts of a simple type of construction to provide shelter. Alternatively they can be kept in partly or fully covered yards. Under both systems water should be provided *ad lib.* and ample trough space allowed. To reduce the risk of bullying at feeding time individual feeding stalls can be fitted in yards or introduced as movable units into the paddocks.

Pregnant sows can also be housed in individual stalls. These are usually 0·61 m (2 ft) wide and 1·60 m (5·25 ft) long for tethered and from 1·83 to 2·14 m (6–7 ft) long for untethered sows. The sows can stand up or lie down, but cannot turn round. A feeding trough and automatic water bowl are fitted at the front, and a dunging passage, which may or may not have a slatted floor, is provided at the rear.

Fig. 7.3. Sow in an individual feeding stall attached to a yard.

Farrowing sows

Well over 10 per cent of piglets born die in the early days of life, most through chilling or crushing, and so efforts must be made when designing buildings for farrowing to minimise such losses. About a week before the sows are due to farrow they should be introduced to their farrowing quarters so as to become accustomed to their new surroundings. Various types of accommodation are provided, but in all cases warmth and some method of preventing the sows from lying on their piglets must be supplied.

Indoor system

Farrowing crates. Under the indoor system the sows may first be placed in farrowing crates. A crate is a space about 2·14 m (7 ft) long by 0·76 m (2·5 ft) wide and 0·88 m (2 ft 10 in) high railed off in a pen about 1·68 m (5·5 ft) wide. Metal bars form the sides, a solid wall is in front, and there is a metal gate at the rear, on the inside of which is fitted a bar 228 mm (9 in) from the floor and 228 mm (9 in) from the gate to prevent the sow crushing the piglets against the gate. This area enables the sows to stand up and lie down, but not to turn round. On either side of the crate, occupying the remainder of the pen floor, are nests to enable the piglets to avoid being crushed when the sow lies down. These are provided with infra-red heaters and covered with a light roof of hardboard to retain the heat. The sows and their litters may be moved to some other type of accommodation when the piglets have become active at about 4 days of age, or may remain in the crates for up to 3 weeks after farrowing if the piglets are to be weaned at this age.

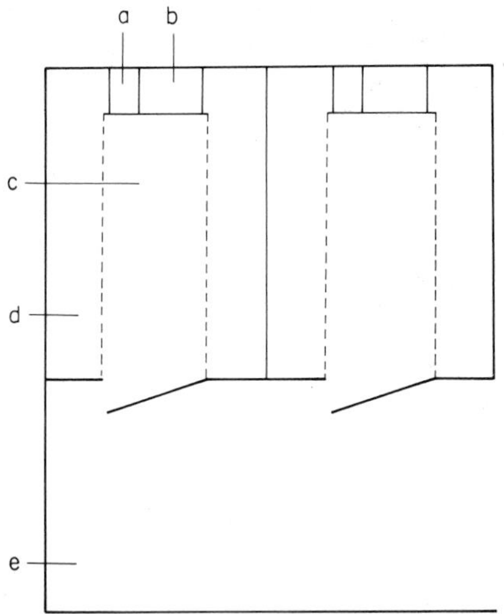

Fig. 7.4. Floor plan of two farrowing pens fitted with crates. (a) Water trough, (b) food trough, (c) sow crate, (d) piglet nest, (e) passage and sow exercise area.

Farrowing and rearing pens. On farms where farrowing crates are provided this type of accommodation is used from the time the piglets are about 4 days of age only, but on other farms the sows also farrow in these pens. The commonest design is to have a totally enclosed farrowing house divided into a series of pens each 3·05 m

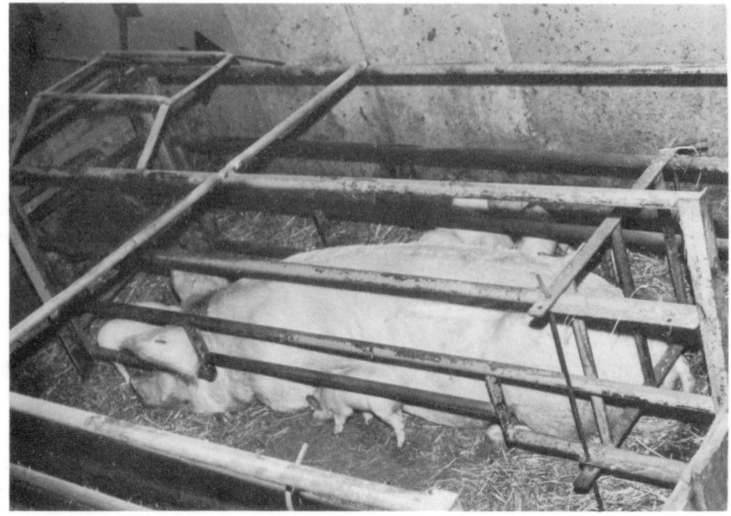

Fig. 7.5. Sow and litter in a farrowing crate.

Fig. 7.6. Piglets under an infra-red heater.

(10 ft) wide and 2·44 m (8 ft) deep. The pens are fitted internally with farrowing rails 228 mm (9 in) above the floors and a similar distance away from the walls. If a sow lies down when a piglet is behind her the rail prevents her back from crushing the piglet against the wall. Each pen has a combined nest and creep compartment 0·76 m (2·5 ft) wide down one side, the wall dividing this from the pen being so designed that the piglets can enter, although the sow cannot. Heating lamps suspended from the creep ceilings enable additional heat to be provided for the piglets. This arrangement also allows special food to be supplied to the piglets, which the sows cannot eat. At the opposite end of the pen is a dunging passage. Doors at the sides of the dunging passages attached to the pens can be shut back to enable the passages to be cleaned out over the full length of the building.

Multiple suckling pens. Under this system from three to five sows are mixed together in houses or yards when their litters are 3 weeks of age. After 3 weeks the sows are removed and the young are left together until ready for the fattening pens. At least 5·56 m^2 (60 sq ft) of space is allowed for each sow and litter, and a large creep area is provided giving each piglet about 0·185 m^2 (2 sq ft) of floor space.

Outdoor system

The sows are placed in movable wooden huts which are fitted internally with farrowing rails as previously described. The huts may have wooden walled runs attached, in which case the sow and litter are allowed free access to the area of grass enclosed in the run, or the sows can have a harness applied to the body. The harness consists of two straps, one round the neck and the other round the chest, joined by two short side straps on either side of the backbone and another between the forelegs. A chain attached to the chest strap is fixed to an iron peg driven into the ground. The peg is sited so that when the sow is in the hut the chain is almost fully extended to prevent the young piglets being caught in loose folds of chain. Under this system the piglets are able to run free in the field and there is some cross suckling between litters. Paddocks used for pigs must be strongly fenced. The fences need only be about 1 m (3 ft 3 in) high, but the posts must be firmly inserted in the ground and situated fairly close together. A convenient arrangement is to have major posts at about 3m (10 ft) intervals with about three minor uprights between them.

Pig netting is commonly used as the fencing material and the base should be set about 10 cm (4 in) below ground level because pigs' snouts impose severe strains on fence bottoms. A plain top wire, a plain wire halfway down and a strand of barbed wire at the bottom give additional strength to the fence. Electric fences can be used for controlling pigs other than piglets, but two strands of wire are needed, one set at 25 cm (10 in) and the other at 50 cm (20 in) above the ground.

Boars

If the pregnant sows are kept in paddocks or in yards a boar may conveniently be run with them. If the sows are kept in individual stalls separate boar accommodation must be provided. A boar kept on its own should be allowed a floor area of not less than 7 m^2 (75 sq ft). If the pen is used for living and the service of sows an area of about 9·3 m^2 (100 sq ft) will be needed. In the latter case bedding should be provided in a lying area, and some litter should be spread over the service area to prevent slipping. It is an advantage to exercise housed boars two or three times a week; alternatively a yard about 7 m^2 (75 sq ft) can be provided adjoining the sty to which the boar has continual access. Exercise helps to maintain fertility and keeps the animal's legs in good condition, leg troubles being found in closely confined boars. The boar accommodation provided in association with sow stalls should be close to the stalls so that the boar can be walked behind the sows to help detect those which are in oestrus.

Weaner pigs

To reduce the shock of weaning some farmers remove the sow and leave the piglets for a few days in their accustomed surroundings before mixing them with other piglets in a yard. Others mix several sows and their litters of approximately the same age together a few days before weaning so that the piglets become used to running as a group before the sows are removed. The advantage, which is similar to that obtained by the use of multiple suckling pens, is that evenly matched groups can be selected for transfer to the fattening pens. However, some farmers prefer to fatten litters as a whole and move them from the rearing to the fattening pens without mixing litters of pigs together.

Early-weaned pigs

Under this system the piglets are removed from the sow at between

7 and 10 days of age, according to size, and accommodated in batteries of cages kept in subdued light in a building at 26·7°C (80°F). The piglets are kept in groups, and the cage size is adapted to suit the increased space requirements of the pigs as they grow. When they reach the normal weaning weight of about 18·2 kg (40 lb) they are transferred to conventional fattening units. This system is not an easy one to operate, but some farmers use it efficiently. The term early weaning is also loosely used when piglets are weaned at 3 weeks of age.

Fattening pigs

Danish piggeries

This popular method of housing consists of a building with a central feeding passage 1·22 m (4 ft) wide, bordered on either side by a range

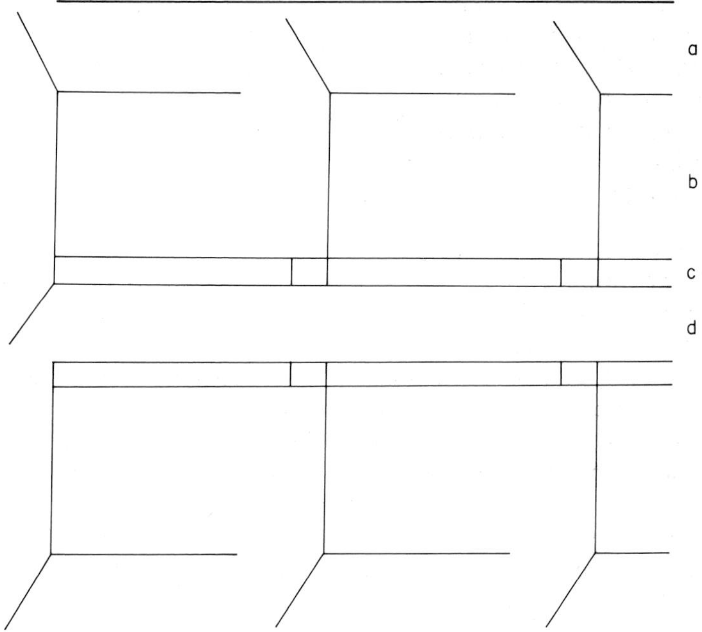

Fig. 7.7. Floor plan of a Danish piggery. (a) Dunging passage, (b) pens, (c) food (long) and water (short) troughs, (d) feeding passage.

of pens 3·05 × 1·83 m (10 ft × 6 ft) in size which in turn lead to dunging passages at the sides of the building 1·07 m (3·5 ft) wide. Eight pigs are usually placed in each pen, and the pen floor should be well insulated and may be raised about 76·2 mm (3 in) above the level of the dunging passage. The feeding troughs border the central passage, so feeding is simple, and doors between the pens and the dunging passages enable the pigs to be shut in the pens while the dunging passages are cleaned over their entire lengths.

Cottage type

The building consists of a line of low-roofed, straw-bedded pens each leading through a small door to an outer, uncovered yard. The food and water troughs are sited at the front of the yards. The pigs tend to dung and urinate in the yards. This simple design has been modified into the McGuckian and Harper Adams systems.

Solari type

An open-sided yard has fattening pens fitted under its roof on either side of a central feeding passage. The pens comprise roofed sleeping quarters, termed kennels, 3·05 m (10 ft) long by 2·14 m (7 ft) wide by 1·07 m (3·5 ft) high, each leading to an open-topped yard 3·05 m (10 ft) long by 1·86 m (6 ft) wide. The kennel roofs can be raised to regulate temperature and ventilation. The floors of the yards are set 0·46 m (18 in) below the kennel floors and are covered with straw. Gates at the side open for cleaning.

Sweat-box house

This consists of a row of totally-enclosed, well-insulated houses, each about 2·44 m wide and 4·88 m deep (8 ft × 16 ft), with half-heck doors, resembling loose boxes. About twenty pigs are placed in each house. An adjustable window allows about 2·58 cm^2 (4 sq in) of ventilation space per pig. Under these conditions a 'Turkish Bath' atmosphere develops with temperatures about 23·9–29·4°C (75–85°F) and relative humidities of nearly 100 per cent. The floors slope either to the front or the back of the house where there is a dunging channel covered with slats for drainage. There are no troughs, and the pigs are fed a pelleted food on the floor. Water is supplied in automatic drinking bowls.

Straw yard

On general farms, yards formerly used for fattening bullocks may be available, and groups of twenty or more pigs are run together, bedded on a thick layer of straw under which they can burrow in order to keep warm.

Feeding

General

The cost of feeding represents about 80 per cent of the total costs of pig production and so is a major factor in determining the financial success of a pig enterprise.

In the British Isles pigs are generally fed on concentrate foods. They can be self-fed and allowed to consume as much food as they wish, or controlled-fed and given a determined quantity each day. Self-feeding from hoppers can only be operated with dry food, but under controlled-feeding systems food may be given either wet or dry.

A simple feeding plan on a pig farm is to prepare four types of concentrate feed mixture: a creep feed for piglets from 10 days of age, a rearing feed for pigs from 23 kg (50 lb) to about 55 kg (120 lb), a finishing feed for bacon pigs from 55 kg (120 lb) to slaughter, and a breeding ration for sows and boars. Typical examples for these classes of stock are given in Table 7.1.

Table 7.1. Specimen meal mixtures (percentage by weight) from Davidson (1964).

Ingredient	Creep feed	Rearing feed	Finishing feed	Breeding ration
Sugar	7·00	—	—	—
Barley meal	23·00	35·00	48·00	57·50
Ground wheat	15·00	20·00	20·00	15·00
Bran	—	—	—	15·00
Fine weatings	20·00	—	—	—
Maize meal	20·00	30·00	20·00	—
White fish meal	7·00	7·50	—	2·50
Soya bean meal	7·00	7·00	10·00	7·00
Salt	—	0·25	0·50	0·50
Dicalcium phosphate	1·00	0·25	1·00	2·50
Ground limestone	—	—	0·50	—
Vitamins A+D	+	+	optional	+
Riboflavine (g/ton)	2·00	2·00	1·00	2·00

Breeding stock

Pregnant sows

During pregnancy a sow or gilt must provide for the needs of the developing fetuses and build up body reserves which will be used during lactation. Some form of individual feeding is desirable to prevent some sows being overfed and others underfed. The amount of food required per day will vary according to the individual animal, but from 2·23 to 2·5 kg (5–5·5 lb) per day of meal is a fair average. When in-pig sows are kept on good quality grass it is usually assumed that the grass consumed will replace from 0·91 to 1·36 kg (2–3 lb) of meal per day. The practice of flushing sows just before service is favoured by some breeders. This is generally effected by increasing the meal allowance fed during the period before mating. The level of feeding during early pregnancy has little effect on the development of the embryos, but in late pregnancy the requirements of the fetuses rise rapidly and an increase in the feed allowance is necessary. On the day before the sow gives birth the quantity of food should be reduced, and a slightly laxative diet given to prevent constipation.

Lactating Sows

The nutrient requirements of a sow during lactation are considerably greater than during pregnancy. After farrowing a sow is given about 2·3–2·7 kg (5–6 lb) of food on the first day and the quantity is subsequently raised according to the size of the sow and the number of piglets. One method is to feed daily 0·9 kg (2 lb) for the sow plus 0·5 kg (1 lb) for each piglet. Some breeders consider that 5·5 kg (12 lb) should be the maximum amount fed daily to any sow, however many piglets are being nursed.

Boars

Boars are usually fed on the rations prepared for sows at a corresponding rate according to liveweight.

Young pigs

Suckled pigs

During the first weeks of life milk provides the sole source of nutrient intake. Unless stimulated to commence earlier piglets begin to eat solid food when about 10 days of age. They should be fed a creep feed *ad lib.* in a creep to which the sow cannot gain access.

Early-weaned pigs
Some farmers are now weaning at from 7 to 10 days of age, when the piglets should weigh a minimum of 1·5 kg (3·3 lb) body weight. One suitable ration consists of 50 parts full cream milk powder, 50 parts skim milk powder, 1 part citric acid and 0·34 part vitamins and minerals. The citric acid is added to lower the acidity of the milk replacer (*Farmers' Weekly* 1970). After a week the ration is switched to a mixture of equal parts of this milk-replacing diet and a creep feed type food. The milk can gradually be withdrawn after a further week.

Fattening pigs

Meal feeding
The treatment of pigs reared for pork or for bacon does not differ materially during the early stages of growth. Both types may be allowed unrestricted food intake up to a body weight of about 45 kg (100 lb), when some of the pigs being reared for pork will be ready for slaughter. To obtain a suitable carcase bacon pigs are then fed a restricted diet. This lengthens the period of keep, but reduces the amount of fat laid down in the body. The degree of restriction varies, but the quantity of food fed will probably not be allowed to exceed 2·72 kg (6 lb) per pig per day.

Whey feeding
Whey is the residue left after the extraction of most of the fat and protein during cheese manufacture. It is palatable, but may cause digestive upsets in young pigs, and so whey feeding is mainly confined to older fattening pigs. In most whey feeding units the material is stored in tanks and piped to drinking bowls and fed *ad lib*. A supplement of from 0·91 to 1·36 kg (2–3 lb) of a meal ration with a high protein content is allowed per head per day.

Lehmann System
The aim of this system is to utilise large quantities of roots or other bulky, but relatively cheap, foods. The design is to provide a basic meal allowance of 1 kg (2·20 lb) consisting of 7 parts of barley meal and 3 parts of fish meal shortly after weaning and to continue this ration until slaughter. As the pigs grow, cooked potatoes are introduced and fed in increasing amounts to appetite. The use of supplementary foods other than potatoes is seldom as effective.

Swill feeding

Swill is the waste and by-product material from the preparation and consumption of meals for human beings. Its value differs widely because of the variation in its constituents. Swill must be thoroughly cooked before feeding, as this softens fibrous material, separates meat from bone and sterilises its contents. This last point is important and is enforcible by law in order to prevent the transmission of infective diseases from meat residues in the swill to the pigs to which it is being fed. Concentrated swill is occasionally prepared by local authorities and other bodies. Raw swill is heated in cylinders for approximately 2 hours at about 100°C (212°F) and this gives a safe product which can be transported relatively easily.

Mineral requirements

Calcium, phosphorus, and sodium chloride need to be added to rations containing large proportions of cereals. As most grains contain reasonable amounts of phosphorus it is calcium and salt for which there is the greatest need. Both iron and copper are necessary for the manufacture of blood haemoglobin, but young piglets are the only ones likely to show an iron deficiency as described later. The addition of about 0·05 per cent of copper sulphate to the fattening diets normally fed leads to more rapid growth and greater efficiency in food utilisation. The mode of action is not yet fully understood. Pig rations should include about 230 g per 1,016 kg (0·5 lb per ton) of zinc carbonate to prevent a skin condition known as parakeratosis developing. The effect of a zinc deficiency is aggravated by an excess of calcium in the diet.

Vitamin requirements

Nearly all the known vitamins are essential for the maintenance of health in pigs, but under practical conditions of pig husbandry in the British Isles those most likely to be deficient in the diet are A and D. However deficiencies of members of the vitamin B complex are encountered. Pigs kept intensively and fed on home-mixed rations may be short of vitamin A, unless the diets are fortified. The clinical signs of a deficiency include unthriftiness, eye abnormalities, and a failure to breed. Animals kept outdoors on pasture and those allowed access to green food are not likely to be deficient. Housed pigs have little solar irradiation which would enable them to synthesise vitamin D, and so adequate supplementary sources of this vitamin must be supplied to pigs not exposed to sunlight. If this is not done rickets and painful joints may be observed in young pigs. Under conditions of normal feeding the usual diets containing whole cereals and milling offals supply sufficient of the B vitamins except B_{12}, but when unusual diets or foodstuffs are used clinical signs of lack of appetite, unthriftiness and loss of hair due to a deficiency of one or more of the vitamins of the B complex may occur.

Water requirements

Pigs must be provided with an ample supply of drinking water, except for animals being fed whey *ad lib.* Weaned pigs need about 3·4 l (0·75 gal) per head per day rising to about 5·7 l (1·25 gal) or 6·8 l (1·5 gal) by the time they reach bacon weight. Pregnant sows and boars require about 18 l (4 gal) per day, and nursing sows may drink 22·8 l (5 gal) per head per day, or even more. Water may be supplied in troughs filled by a hose-pipe or by hand, automatic water-bowls, or nipple-type waterers. Even the smallest piglets become accustomed to nipple drinkers provided they are within their reach, but they tend to play with them with a resultant risk of wet floors. The water supply should be sited away from the sleeping area so that the bedding is not wetted. There is always a risk that some pigs will defaecate in their water bowls and to prevent this some types of trough are fitted with a heavy lid which the pigs have to raise with their snouts before drinking.

Breeding

Boars are used for service from 9 months of age, but do not reach full sexual maturity until about 18 months old. Gilts are usually served when about 8 months old provided they weigh 127 kg (280 lb) or over. Pigs will breed at any time of the year and the interval between the end of one oestral period and the next is about 19 days. As oestrus lasts for about 2–2½ days sows usually come into oestrus at 3-weekly intervals. Sows come into oestrus from 3 to 5 days after the weaning of a litter at 4 weeks of age or over, but if earlier weaning systems are practised the time lag is longer, about 10 days, and not so precise.

When sows come into oestrus the main sign is a reddening and enlargement of the vulva, but the ova are not shed until between 24 and 36 hours after the visible signs. This is the period of maximum fertility when mating or insemination should be performed. During this period the sow will stand quietly for a boar to mount, and will also stand if a man sits on its back or rubs his hands on either side of the body behind the shoulders, where the forelegs of the boar would rest in normal service.

A boar may be run with a group of sows so that they can be mated as they come into oestrus, or the sexes can be housed separately and the sow taken to the boar to be served during the most fertile period. In the latter case sows are usually served twice during an oestral period, but in the former case a number of services may be given. Copulation in the pig usually lasts for about 15 minutes. It is usual to keep one boar per twenty sows. This should mean that a boar will give eighty services per year (two per oestral period, with each sow producing two litters per year) but as some sows will not conceive when first served it means that a boar will give about 100 services per annum. This is satisfactory provided the farrowing programme is fairly evenly spaced throughout the year, but a boar might be overworked if required to serve a number of sows at a short interval.

The gestation period usually lasts for about 115 days, with a range from 111 to 117 days. The signs of impending parturition are nervousness and uneasiness, an enlarged vulva and possibly a mucous discharge. Colostrum is present in the teats and some sows attempt to make a nest. About 4 hours is the average time taken by a sow to give birth to a litter, which can vary in size from two to twenty piglets, with an average of about eleven. Some pigmen remove piglets in excess of the number of usable teats on the sow, and, if possible, foster them on to other sows with small litters of piglets of a similar age.

Piglets are usually weaned at 6 weeks of age, but some breeders now wean at 5 weeks. This is quite successful provided the piglets are well grown. On some intensive systems weaning at 7 days or 3 weeks of age, or at some period between, is practised while in other units the piglets are weaned as soon as they reach 4·54 kg (10 lb) liveweight. Considerable care is needed if these early weaning systems are to be used satisfactorily.

Artificial insemination

Artificial insemination is practised to some extent in pig breeding. For semen collection a boar is trained to mount a dummy sow made from metal hoops and sacking. In the British Isles an artificial vagina is generally employed, the outer casing of which is similar to that used for bulls, but it is only 30·5 cm (12 in) long. Within this, starting 2·5 cm (1 in) back from the receptive end, is inserted a 1·25 cm (0·5 in) thick piece of corrugated foam rubber extending for about 12·7 cm (5 in). This serves as an artificial cervix. The whole is lined by a latex rubber sleeve. A 500 ml (17·6 fl oz) capacity polythene bottle is attached to the end by a latex rubber cone. Before use the space between the outer casing and the inner latex rubber lining is filled with water at about 80°C and the inside of the receptive end is greased with vaseline. Just before use the water is emptied out and a rubber bulb pump is attached by means of rubber tubing to the water vent. The boar mounts the dummy and protrudes its penis into the artificial vagina, which is then inflated to make contact with the penis. The temperature of the artificial vagina is no longer important, but a fluctuating pulsation must be maintained by means of the rubber bulb to reproduce the vaginal and cervical contractions of normal service.

There is a Japanese method in which no special equipment is used except that the operator wears rough faced rubber gloves. As the boar mounts a sow the penis is held in an encircling grip by the thumb and index finger of the right hand. The penis is directed away from the sow into a funnel and the semen is collected in a flask.

Undiluted boar semen will only remain fertile for about 6 hours, but diluted semen can be used satisfactorily for up to 2 days after collection, although the fertilising capacity is reduced after 24 hours. If stored under high carbon dioxide tension in sealed bottles the semen will remain viable for 3 or 4 days. Boar semen must not be refrigerated as the spermatozoa are destroyed by this process.

The greatest difficulty in the insemination of sows or gilts is to determine the optimum time at which to perform the operation. As previously stated insemination should be carried out on the second day of oestrus. Sows are normally inseminated twice during an oestrus period, and conception rates of over 70 per cent can be obtained.

Inseminations can be carried out by trained inseminators visiting the farm when called, or by the farmers themselves using diluted semen sent to them by post or rail through a semen delivery service. A development of the former method is a planned weaning inseminator service used on large units in which a number of litters are weaned together so that the dams come into oestrus at about the same time and the inseminator can inseminate a number of sows per visit.

The normal method of insemination in the British Isles is to use a catheter made of a piece of rubber tubing 1·25 cm (0·5 in) in diameter and 45·7 cm (18 in) long tapered at one end with raised spiral ridges to resemble the end of a boar's penis. This is greased before use and twisted anti-clockwise when inserted through the vagina until it becomes locked in the cervix. The bottle containing the semen is attached to the outer end of the catheter and the semen allowed to flow into the cervix by gravity. The whole operation may take 15 minutes. After use the catheter is washed and sterilised by boiling, but disinfectants should not be used because residues remaining might damage spermatozoa in the next semen sample for which the catheter was used.

Health

Healthy pigs move freely without lameness or stiffness or an arching of the back. They do not shiver or pant neither do they sneeze nor cough persistently. The body is well fleshed and the skin is free from scurfiness and spotting. In the so-called white breeds the skin colour is a light pink, not dead white. The tails are normally carried curled. Healthy pigs show a keen interest in food and have normal bowel movements without showing either constipation or diarrhoea.

The normal body temperature of a pig is in the range 39–40°C (102–103°F). The pulse rate is from 70 to 80 beats per minute and is best taken at the artery situated inside the hind leg midway between the stifle and hip joints.

Common diseases and their prevention

Piglet anaemia occurs mainly in housed suckling piglets, but is occasionally seen in such animals reared outside. The skin is pale, growth is retarded, and the piglets may appear to be thin and hairy. The cause is a shortage of iron in the bodies of the rapidly growing piglets. Piglets produced by sows kept confined have low iron reserves and sows' milk is deficient in iron. As soon as the piglets start eating creep feed, which contains iron, they make a rapid recovery without any treatment, but weak, anaemic piglets may be slow to start eating concentrate food. The administration of iron to the suckling piglets, usually by intramuscular injections of a proprietary iron preparation at an early age, is in common use as a preventive measure on intensive pig farms where the condition is common.

Diarrhoea is most likely to occur when piglets are about 3 or 4 weeks old and may be due to an upset in digestion following the eating of solid food for the first time. Sudden changes in the food fed to pigs at weaning may also cause diarrhoea. Both types are usually only temporary. An infectious form of gastro-enteritis of young pigs, starting at any age from a few days old, is also recognised which may cause death or lead to unthriftiness in survivors. This disease may attack successive litters on rearing premises over long periods of time. Prevention of this form of diarrhoea is by thorough disinfection of the premises.

Virus pneumonia can be contracted by pigs of all ages and adult animals can remain carriers for 2 or more years. The disease may take an acute or chronic form, the latter being the commoner. A cough is the most obvious clinical sign, which can usually be stimulated by exercise, and the pigs do not thrive as well as unaffected animals. In acute cases rapid breathing movements may be seen, the loss of condition may be marked and the pigs may die quite suddenly of respiratory failure. In chronic cases the cough may persist almost indefinitely and the failure of the affected animals to thrive as well as pneumonia-free pigs has important economic implications as they take several weeks longer to become marketable. There is no method of preventive vaccination. It is possible to develop herds which are free from the disease, but great care must be taken to avoid the introduction of the virus into such herds, as the animals will have no resistance and a high mortality rate may follow.

Round worms. The most important of these parasitising pigs is a large worm about 15–40 cm (6–16 in) in length. The adult worms live in the intestines of pigs and the eggs produced by the females pass out in the faeces. The larvae develop in the egg and can remain dormant for long periods. When swallowed by a pig the larvae hatch in the stomach, and then bore through the intestinal wall and reach the lungs via the blood stream. The larvae are swallowed for a second time and develop into adult worms in the intestines. A heavy infestation of developing larvae in the lungs will cause coughing and unthriftiness, and large numbers of adult worms in the intestines will also lead to a marked loss of condition and may produce a partial blockage of the intestines. These clinical signs are most obvious in young pigs, as an immunity appears to develop at about 5 months of age. As older pigs can be carriers of worms which produce eggs the routine dosing of pregnant sows is a method of preventing young susceptible piglets from becoming infested. Pigs should not be exercised in large numbers in small paddocks as the soil and pasture become highly infective.

Handling

Pigs can be difficult to handle as they are often stubborn, and their smooth bodies are not easy to grasp and hold. They cannot be led by the head as can other farm animals. Their eyes are deeply set and their instinct is to look for a gap and charge at it. Adult pigs should be approached cautiously as some individuals can resent handling and be savage, but quiet animals can often be humoured by scratching their backs and will usually lie down if rubbed along the belly. Before suckling piglets are handled the sow should be removed from the pen, since the squealing of the young may enrage the dam.

It is difficult to drive small pigs as a group as often one or more individuals rush away on their own and try to find an escape route. Pig handling boards about 92 cm × 61 cm (3 ft × 2·5 ft) with holes cut in the top of one of the longer sides for use as handles are of value for pushing pigs in the required direction or preventing them from going the wrong way. In a building small pigs are best controlled by being pushed as a group into a corner and held in this position by means of a hurdle. While packed fairly tightly as a group they remain comparatively quiet, but a pig becomes upset if it is separated from its fellows. Individual animals can be caught by seizing one hind leg above the hock. This is preferable to the alternative method of grasping the base of an ear. For examination a small pig can be held with an arm around its body.

There are several ways in which pigs can be held for castration but the best method is for the assistant to grasp the hind legs above the hocks, with the back of the pig towards the operator. The forelegs and head can then be swung between the assistant's knees and held there. In the case of small pigs it is easier for the assistant to sit and hold a foreleg and hind leg of each side in each hand with the pig's back resting on his knees. This position can also be used for docking the tails of piglets.

For the execution of various manipulations on medium-sized pigs they can best be crowded in a group behind a hurdle and held there without being given room to move. An assistant may steady a pig by grasping the bases of its ears. This crowding method is not so satisfactory in the case of large pigs, but injections can sometimes be made while they are feeding. It is possible to make some observations, such as taking the rectal temperature, while talking to an adult pig and scratching its back. If this is not adequate a twitch, as described for use on horses, may be used and the loop placed round the upper jaw.

Various types of pig catcher are also available, which work on the same principle. They consist of a metal rod about 60·96 cm (2 ft) in length with a wire loop at one end, the size of which can be altered by means of a sliding handle. The section of the wire to be inserted in the mouth is usually covered with rubber-tubing to prevent damage to the gums and tongue. When tightened the wire loop is fixed by means of a ratchet and the pig can be held by grasping the metal rod firmly.

Alternatively, a running noose can be made at the end of a piece of rope about 3·05 m (10 ft) in length and slipped over the upper jaw into the open mouth and pulled tight when it is behind the tusks. The free end can then be tied to a post and the pig will instinctively pull back and so be restrained. If the animal is very difficult to handle it may be necessary to drive it into a crate, such as a farrowing crate, before the twitch or other method of restraint is applied.

Administration of medicines

The administration of liquid medicines to pigs should be avoided as they invariably struggle and squeal during dosing and there is a danger that the fluid will pass into the lungs. Electuaries can be smeared on the teeth or the tongue. If the medicine is palatable it can be mixed with the food and given to the pig when it is hungry, if necessary after a period of starvation lasting about 24 hours. Most medicaments are now given by injection, into the inside of the thigh in small pigs or into the loose tissue behind the ear in older pigs. Young pigs can be held by a seated assistant who holds a foreleg and a hind leg of each side in each hand with the pig's back resting on his knees, while older animals can be crowded into a small space by using hurdles. An assistant holds the pig's ears while the operator makes the injection, although it is often possible to make an injection without the ears being held.

General

Vices

Tail biting

This vice occurs in intensively housed pigs, but is not normally seen in pigs on pasture. It can start quite suddenly in a pen while neighbouring pens, in which pigs are kept under similar conditions, are unaffected. It is probable that an injury to the tail of one pig attracts the attention of another pig, which bites the tail off. This pig may then attack the tails of other pigs, and the culprit may finally be identified as the only pig with an intact tail. However, as the vice is imitative, other pigs may start until all the tails in the pen are bitten off. Some breeders consider that a shortage of bedding and cold environmental conditions predispose to this vice by causing the pigs to crowd in an endeavour to keep warm. Distasteful tail dressings can be tried as a deterrent, but if the vice is prevalent on a farm the only effective method of prevention is to dock the tails of all the pigs placed in the fattening house.

Savaging

This can be a problem on intensive units. The vice is frequently encountered when strange pigs are mixed. It also occurs in established groups when pigs are overcrowded and kept in cold buildings without bedding. To avoid savaging in the former case, if pigs have to be mixed, strangers should be placed together after dark so that they can acquire a similar odour before they can see each other, or they can all be sprayed with a substance having a distinctive smell to mask individual differences. In the latter case a reduction in the number of pigs in a pen or the provision of ample supplies of straw may stop the savaging. Occasionally the savaging may be the work of one bully, and the removal of this pig is the only solution.

Minor operations and manipulations

Castration

Male pigs not required for breeding are normally castrated, and this operation should be performed during the first 4 weeks of life. Piglets can be castrated at 1 week old if a skilled castrater is available. The operation should not be performed within a week of weaning. A knife is used to make incisions in the scrotum over each testicle and the testicles are withdrawn and the spermatic cords are pulled gently until the blood vessels rupture. This rupturing reduces the risk of haemorrhage. It may be necessary to sever the spermatic cord. By law an anaesthetic must be administered to pigs over 2 months of age which are being castrated.

Docking

To prevent the vice of tail biting developing in intensively housed pigs docking (de-tailing) may be practised. The operation is usually performed when the piglets are 3 days of age, and, except in special cases, must be carried out before the seventh day of life. Pincers or large curved scissors may be used so that the part of the tail to be removed is quickly and completely severed. Generally a short stub of tail, about 1·25 cm (0·5 in) in length, is left.

Tooth cutting

Soon after birth some pig men cut the tips of the piglets' eight sharp teeth with a small pair of pliers to avoid undue tearing of the sows' teats and to stop the piglets biting one another on the face and ears when fighting for a place on the teats. Care must be taken to avoid injuring the gums. Other pig men do not consider this operation to be necessary as a routine and only cut the teeth if an excessive amount of damage is caused.

The tusks, or canine teeth, grow long in old boars and project from the sides of the mouth in both jaws. To avoid injuries to other pigs or persons handling a boar the tusks should be cut back. A running noose at the end of a thin, but strong, rope is inserted in the mouth and pulled tight behind the tusks, and the free end tied securely to a strong post. As the animal pulls back its mouth opens and the tusks can be cut with a strong pair of cutters.

Ringing pigs

By instinct most pigs do some rooting if kept out of doors, and adult animals can do considerable damage to pastures. To reduce the undesirable effects of rooting, breeding pigs kept outdoors are often ringed. Older animals are restrained by a rope through the mouth over the upper jaw while young pigs are held. Normally one or two metal rings, usually of the copper bull-ring type, are placed in the snout away from the edge of the cartilage but free from the bone, although some pigmen use a ring that is placed through the nasal septum. Rings should be set well back so that they do not catch in objects such as wire netting.

Pig meat marketing

The meat of some pigs is sold fresh as pork, the less valuable parts of the carcase being converted into sausage meat. The required liveweights are from 45 to 82 kg (100–180 lb), with the most popular weight being 68 kg (150 lb). The carcase must be lean to satisfy consumer demand, and be capable of division into chops of good quality meat. Uncastrated boars killed at a weight of up to 63·5 kg (140 lb) liveweight provide excellent quality pork.

Bacon pigs are sold at about 90 kg (200 lb) liveweight and must be lengthy animals without excessive quantities of fat. Carcase grading schemes take note of the carcase length and the thickness of fat over the shoulder and loin. The production of high quality bacon also demands that the flesh be cured in a manner which will give a flavour which is attractive to the consumer.

In recent years there has been a demand for heavy pigs, generally known as heavy hogs. These are slaughtered at about 113 kg (250 lb) liveweight and the carcases are cut so that all the parts can be used to the best advantage and manufactured into a number of products sold, particularly, in self-service stores. The manufacturers require a lean animal, although excess back fat can be trimmed off if necessary.

Welfare Code

Code No. 2 of the recommendations for the welfare of livestock deals with pigs. The internal surfaces of buildings which are accessible to pigs should not have any sharp edges or projections likely to cause injury. When pigs are fed on a system which does not allow continuous and unrestricted access to food all the pigs in the group should be able to feed at the same time. Groups of pigs should be carefully watched to ensure that persistent fighting leading to the causation of severe injuries does not take place. When farrowing, sows should be accommodated in pens which have a farrowing crate or farrowing rails so that the piglets are protected from injury.

References

DAVIDSON H.R. (1964) *The Production and Marketing of Pigs*, 3rd edn., pp 1–516. London, Longman, Green & Co. Ltd.

FARMERS' WEEKLY (1970) The Belgian Revolution, pp 10–22. *Pigs*, Summer Number, 24 July.

CHAPTER 8
RABBITS

Introduction

Before the last war rabbit meat was a common article of diet, but the outbreaks of myxomatosis in 1953 made housewives suspicious of rabbit meat. This prejudice has now been largely overcome and the production of rabbit meat is an expanding industry. Skinned and dressed carcases weighing from 0·9 to 1·2 kg (2–2·5 lb) are most popular, and the flesh should be light coloured with white, not yellow, fat. About 70 per cent of a rabbit carcase consists of edible meat, as compared with about 50 per cent in the case of broiler chickens. As a by-product white pelts are more valuable than coloured, those from immature animals killed for meat production at the above mentioned weights being used for the manufacture of felt.

Rabbits are occasionally bred for fur production, but this is not a developing branch of rabbit keeping. High quality furs can only be produced if the rabbits are pelted when in full coat, usually during the winter period when they are from 7 to 9 months of age. At one time there was a market for the wool from Angora rabbits, but this is not now a profitable industry.

Rabbit farming is an enterprise which can be conducted as a part-time occupation by persons who can organise some free time daily or as a full-time business. In general, profitable rabbit farms are either small units maintained by family labour or large mechanised units using labour-saving devices wherever possible.

The future outlook appears to be bright, provided production costs can be kept as low as possible so that the meat can be sold at a price which is attractive to the consumer. There is a rapidly growing market in the British Isles and there appears to be a considerable demand for rabbit meat in certain continental countries. As a member of the European Economic Community, Great Britain could play a large part in supplying this market.

Definitions of common terms

Buck. A male rabbit.
Doe. A female rabbit.
Grower. Young rabbit after weaning and before reaching slaughter weight.

Principal breeds

Over forty rabbit breeds are recognised in the British Isles, but only a few of these are used primarily for meat production. The requirements are for a strain of rabbit in which the does conceive readily, produce large litters and yield ample supplies of good quality milk, and the young rabbits grow rapidly, produce well-fleshed carcases and have a good conversion ratio of food into flesh.

New Zealand White

These are self-white in colour with pink eyes and adult animals weigh from 4 to 5·5 kg (9–12 lb). The bucks are lighter in weight than the does and are stockier and shorter in the body. The hairs in the coat are longer than are those of the Californian. The young have a low food conversion ratio and the flesh is of good quality.

Fig. 8.1. New Zealand White rabbit.

Fig. 8.2. Californian rabbit.

Californian

These have a white body with black markings on the nose, ears, feet, and tail and produce a white pelt. The eyes are pink in colour. Adult animals weigh from 3·6 to 4·5 kg (8–10 lb). The body type is blocky and the neck is short so that the head appears to be set on the shoulders. The legs are short, and the bone structure is fine. The fur is dense, but not long, and the hide is relatively light in weight. Most animals have a placid temperament and the does are usually good mothers. The young are well fleshed on the back and haunches, grow rapidly, and give carcases with a good meat-to-bone ratio. The flesh is excellent in quality.

Dutch

Distinctively coloured with markings on the cheeks and hind part of the body separated by an area of white. A small, compact variety producing a well-fleshed carcase. The adult weight is about 2·25 kg (5 lb).

Flemish Giant

A large, coloured breed with adults weighing between 5 and 6·3 kg (11 and 14 lb). The young tend to grow slowly, but produce carcases with well-fleshed backs.

Fig. 8.3. Dutch rabbit.

Beveren

Most are coloured, but some are white. The adult weight is from 3·2 to 4 kg (7–9 lb). The young are slow-maturing, but have well-developed backs and hind quarters.

Cross-breds

The first two breeds described, the New Zealand White and the Californian, are the most popular, for they are both white-fleshed, produce white pelts and have rapid growth rates. A first cross between these two breeds produces an excellent meat rabbit, combining the best qualities of each breed. The other breeds are also crossed to produce young rabbits giving carcases well covered in meat, but such crosses are losing popularity.

Hybrids

Hybrid rabbits have been bred by using three or more breeds or distinct strains to produce separate buck and doe lines. Individuals of these lines, when mated, produce excellent meat rabbits of uniform quality, but rabbits of this meat-producing generation cannot be used for further breeding as their progeny are not uniform in size and appearance. The production of really good hybrid rabbits is a skilled task involving the use of large numbers of animals and should only be undertaken by large-scale operators.

Fur breeds

The fur breeds can be subdivided into two classes, the normal coated and the Rex varieties. Coat colours imitating the rarer fur producing animals, such as the Silver Fox, Sable, and Chinchilla, have been developed in the normal coated breeds. The Rex coat is very fine and dense in texture as the long guard hairs are not present. Rex rabbits can be bred in any colour, including the varieties mentioned above. Angora rabbits have long hair which can be cut or plucked at intervals and made up into soft silky clothing.

Breeding systems

Successful rabbit breeders select their future breeding stock with care, although planned progeny and performance tests are rarely carried out. The bucks are selected because they have good conformation with a compact body and grew rapidly during the rearing period. The does are chosen from dams which have produced large litters regularly and reared them satisfactorily, so that all the young in a litter are approximately the same size at weaning. This necessitates the making of an adequate nest, covering the young after birth and producing enough milk to feed the young until they start to eat solid food. Young does from such dams selected for breeding should be of good conformation and should have eight or more functional teats.

Identification

To enable record keeping to be carried out efficiently rabbits must be identified individually, and ear tattooing is the most popular method with commercial breeders. Specially made tattooing numbers are held in forceps and small wounds made in the ears. These wounds are then rubbed with a special ink, so that the numbers are permanently marked. Numbered ear tags punched through the ears may also be used, but there is a risk that these may be pulled out. Occasionally wing bands made for marking chicks are used as ear tags. Rabbits of the fur breeds, which may be used for showing, are usually ringed at about 8 weeks of age with closed rings supplied by the British Rabbit Council. Different sizes are produced for use in different breeds and each ring is stamped with the year of issue and an individual number. The ring is slipped over one hind foot to a position just above the hock joint, and, as the bones grow, the ring is kept in position. Ear notching has been tried but is not very satisfactory.

Ageing

There are no definite signs which can be used to establish the age of a rabbit. The two lower and the two upper incisor teeth continue to grow throughout a rabbit's life in order to compensate for wear. The toe nails grow continuously, but are usually clipped if not worn down naturally, so their length cannot be used as an age guide.

Housing

Adult rabbits should be housed individually, while young rabbits are usually run in groups until the required killing weight is attained. Outdoor hutches can be used provided they are soundly constructed with a weather-proof, over-hanging roof, sloping to the rear. However it is far better to place the hutches in a building because a more even environmental temperature can be maintained, and the attendants are protected from inclement weather conditions. Rabbits can stand fairly wide temperature ranges, but must be kept in well-ventilated surroundings which are free from draughts. Breeding stock will produce young all the year round at air temperatures between 10 and 15°C (50 and 60°F), while fattening rabbits after weaning grow most rapidly at temperatures between 15·5 and 18°C (60 and 65°F). Good lighting is important to encourage regular breeding in adult animals, and a high food intake and rapid growth in young rabbits, and some rabbit keepers recommend a constant 14-hour day with a light intensity of about 10 lux.

In the past, wooden, solid-floored hutches, usually built in tiers three hutches tall, were in general use. Maintenance requirements were high, bedding materials had to be supplied, and the hutches had to be cleaned out once or twice a week. Nowadays wire-mesh cages are in general use, designed to reduce labour costs to a minimum. The wire-mesh floors allow the faeces and urine to drop through into a droppings pit in single-tier units or into droppings trays in multiple tiers. Rabbits housed in metal cages are obviously more susceptible to temperature fluctuations than those kept in hutches, so cages should always be fitted in well-insulated buildings. The majority of hutches or cages have a floor area of 0·75 m^2 (8 sq ft), as this is a satisfactory size for a doe and litter, or weaned young up to killing weight. The height is usually between 45 and 50 cm (18 and 20 in). Some cages with a floor area of 0·37 m^2 (4 sq ft) are also required for housing future breeding animals individually.

When wire-mesh cages are used for breeding nest boxes must be provided not later than 21 days after mating. These are usually made of wood, but disposable cardboard boxes are sometimes used. They are open at the top and measure about 53 × 25 cm (21 × 10 in) with sides 25 cm (10 in) high to protect the young from draughts. The pop-hole should be 15 cm (6 in) square and begin 10 cm (4 in) from the floor. This last height is important as the ledge prevents the young rabbits from leaving the nest too early, and also brushes back young rabbits holding on to the teats if the doe is disturbed and leaves the nest while suckling. Nest boxes need not be provided in wooden hutches as the doe will make a nest in a secluded part.

To reduce labour costs pelleted foods are usually fed in food hoppers and hay in wire-mesh racks, both of which can be filled from outside the cage without opening any doors. Water must be provided, particularly when rabbits are fed solely on concentrate food. Glazed earthenware or galvanised iron drinkers are used on small units, but in large-scale enterprises some form of automatic watering device, such as a nipple-drinker system, is fitted.

Colony system

A colony system may be used for rearing young rabbits from weaning at 4 weeks of age until killing weight is reached. About forty, and not more than fifty, young usually form a group and are run on solid floors covered with wood shavings about 8 cm (3 in) deep, allowing a floor area of 0·14 m^2 (1·5 sq ft) per rabbit. The unit is cleaned out at the end of the fattening period only. Alternatively, the rabbits may be run on 1·25 cm (0·5 in) square wire-mesh floors allowing 0·09 m^2 (1 sq ft) per animal. The faeces drop through the mesh on to the floor below, and are removed when the rabbits are marketed. Under either system water is piped to each pen and food is supplied from self-feed hoppers with large storage capacities. There is usually a certain amount of bullying and although some rabbits reach meat weight at 8 weeks of age, others take longer.

Outdoor system

The folding of rabbits on grassland in movable wooden arks can be carried out in spring and summer when sufficient pastureland is available. The arks need to be moved regularly to ensure an adequate supply of grass and prevent too heavy a concentration of faeces. Feeding costs are low, particularly in the spring when the grass is lush, but the carcases tend to be darker and therefore less attractive to the consumer and the high labour costs render the system uneconomic.

Feeding

The rabbit has a large caecum and colon in which there is some microbial digestion of starch and cellulose. The products of this microbial synthesis, e.g. certain water-soluble vitamins, are utilised by the practice of coprophagy. The rabbit passes hard faeces during the day, which are voided, and soft faeces during the night, which are eaten direct from the anus. Undigested components of the diet which have been broken down by the micro-organisms in the caecum and colon are passed through the alimentary canal again and so utilised.

Rabbits can be fed on home-produced roughages and green foods, supplemented with cereal grains, such as oats and barley, meal mixtures, or waste bread, but the feeding of such diets does not give rapid growth rates. If home-mixed concentrate meal rations are fed in the form of a dry mash they are not very palatable while the amount of labour required to prepare a crumbly wet mash is too high to make the practice economic.

Commercially bred rabbits are almost always fed on proprietary pelleted foods because these provide a balanced ration, can be fed with a minimum of waste, have low labour requirements, and are digested efficiently. On most rabbit farms a standard all-purpose pellet is generally fed to all classes of stock, the amounts given being adjusted according to the requirements of the different types. Some manufacturers produce two rabbit foods, one for breeding stock and the other for growers, but the benefits are only marginal.

The minimum level of protein in the diet is about 10 per cent, this being adequate for the maintenance of adult non-breeding stock. For a breeding doe during pregnancy and lactation from 16 to 20 per cent is needed, and this level is also adequate for young growing animals. Most rations contain between 15 and 20 per cent of protein, and these levels prove satisfactory in practice.

Although rabbits are less efficient at digesting fibre than cattle or horses they can utilise relatively large quantities of fibrous material. Breeding does and young rabbits need about 14 per cent of fibre in the diet and adult non-breeding stock can be allowed up to 25 per cent. Hay can be fed with pelleted foods to increase the fibre content of a ration, but its consumption is likely to lower the rate of liveweight gain slightly. In order to encourage rapid growth some commercial diets have a fibre content lower than 14 per cent, and hay is particularly useful for recently weaned rabbits on such diets as it helps to reduce the incidence of digestive disturbances. It is generally accepted that rabbits require from 2 to 2·5 per cent of fat for maintenance and 3–5 per cent for lactation, but there is some evidence that higher fat levels produce increased liveweight gains in rabbits being fattened for table.

Mineral and vitamin requirements

Rations containing adequate amounts of green food usually contain sufficient minerals to meet the normal requirements, and compounded diets are almost always fortified with minerals, particularly calcium, phosphorus, and sodium chloride. Salt is commonly included in rations up to a level of 0·5 per cent of the total food intake. Although vitamin deficiencies have been shown experimentally to produce clinical signs in rabbits such deficiencies do not normally occur in animals fed on diets of the type recommended. Compounded rations are usually supplemented with vitamins A and D and the vitamins of the B group are synthesised in the body and so need not be supplied in the food.

Water requirements

A constant supply of clean water must be provided as a restricted intake reduces food consumption and thus the growth rate of fattening rabbits and lowers the volume of milk yielded by suckling does. It is customary to supply water *ad lib.*, but an adult rabbit of the Californian or New Zealand White type will drink about 0·28 l (0·5 pt) per day, a pregnant doe about 0·57 l (1 pt), and a suckling doe may need as much as 3·4 l (6 pt) daily. Obviously the exact amounts will vary with the individual, its size, its age, the type of diet, and the temperature and humidity of the rabbitry.

Table 8.1. General purpose rabbit diets.

Ingredient	Percentage	
	Diet A	Diet B
Barley Meal	—	20
Sussex ground oats	12	20
Maize meal	—	10
Weatings	18	—
Bran	40	—
Fish meal	10	—
Soya bean meal	—	20
Grass meal	20	20
Meat and bone meal	—	10
	100	100

Two suitable formulae for general purpose diets are given in Table 8.1. These rations can be fed as a meal or as pellets. To both rations should be added about 2·5 per cent of a proprietary chick mineral/vitamin supplement according to the manufacturer's instructions.

An indication of the amounts of pellets required per day can be obtained by reference to Table 8.2.

Table 8.2. Approximate daily pelleted food requirements of rabbits of the Californian type.

	oz	g
Growers	3·5–4	99–113 (usually fed *ad lib.*)
Adult bucks or resting does	4	113 (may be increased to 142 g (5 oz) if body condition is lost)
Pregnant does (last week)	7–8	198–226
Lactating does	8–16	226–454 (should be fed *ad lib.*)

The amounts fed to rabbits of other breeds can be adjusted according to body size. Breeding stock must not be allowed to become fat, because excessive fatness can reduce fertility. For fattening rabbits a food conversion of 3·5 units of food to 1 unit liveweight increase is reasonable. Many producers only average 4:1 although some claim to obtain a figure of 3:1.

Breeding

The smaller breeds, such as the Dutch, reach puberty as early as 4 months of age while Californian and New Zealand White does are ready for breeding at about 5 months of age and bucks at about 6 months. The mating of does should not be delayed for too long after these ages are reached or difficulty may be encountered in obtaining conception. Mature does remain in oestrus for long periods and under good environmental conditions can be bred from throughout the year. When ready for service the vulva of the doe is usually purplish in colour and may be slightly enlarged, but these signs are not conclusive as a doe will also mate successfully when the vulva appears normal. For mating the doe must be taken to the hutch of the buck, as introducing the buck to the doe's hutch may lead to fighting. Mating should be observed. After mating the buck normally falls backwards or sideways, possibly emitting a slight scream. If mating does not take place within a few minutes the doe should be removed and returned to the buck later the same day or on the following day. Some does have to be held while mating takes place, one method being described under handling. The ratio of does to bucks varies, but 10:1 is a common figure in practice. Care must be taken not to overwork a buck, as poor fertility can result. The maximum number of matings per buck per day is four and per week twelve. Artificial insemination has been used in experimental laboratories with success, but its use in commercial units is not justified economically because of the high labour requirements.

Does are usually remated when their litters are weaned at 4 weeks of age, so each doe can produce six litters a year. However, some producers remate their does 3 days after the birth of a litter and wean the young at just under 4 weeks of age. The advantage is that it is possible to produce nine litters a year from each doe, but the disadvantage is that the litters tend to be smaller. Only very vigorous strains of rabbits are capable of producing healthy litters continually under this system.

Alternatively, other producers prefer to leave the litter with the dam until the young reach 8 weeks of age and are ready for slaughter. The doe is mated 2 weeks before this stage is reached and so kindles down about 17 days after the litter has been removed. Under this system no extra cages are required and the young do not experience a check in growth through being weaned.

The duration of pregnancy is about 31 days with the possibility of a variation of up to 2 days either way. Pregnancy can be diagnosed by palpation through the abdominal wall between 14 and 16 days after mating. A hand is placed under the body slightly in front of the hind legs and the embryos can be felt as marble-shaped objects as they slip between the thumb and fingers when the hand is pressed against the abdominal wall. The doe must be handled gently for if it struggles the abdominal muscles will be tensed and palpation of the internal organs will be difficult. By the twenty-fourth day of pregnancy there is a palpable increase in the thickness of the mammary glands which gives confirmation of pregnancy.

If a mating does not result in pregnancy, pseudo-pregnancy usually occurs, being indicated by the doe making a nest about 17 days after mating. In this event the doe should be remated immediately as there is every likelihood of a normal conception resulting.

Shortly before giving birth does make nests of hay, lined with fur plucked from their bodies, for the young. The average litter size varies from six to eight, but may range from as few as two to as many as sixteen. Numbers tend to diminish after a doe's eighth litter. The young are born blind, deaf and without any obvious fur. About 24 or 48 hours after parturition most breeders remove the doe from the cage while the litter is examined. Any born dead are removed, and in the case of litters containing over ten young it is advisable to kill one or two weaklings so that the remainder may make a good even weight gain. Alternatively, some youngsters from large litters can be fostered on to does with small litters, preferably within a few hours of birth. The young to be fostered are handled with hands rubbed in some faeces from the foster mother and placed in the nest with the foster mother's own offspring. The nest should be opened from the top so that the sides are not disturbed, and the fur replaced to leave the nest in the same condition as it was originally.

Unlike most mammals, doe rabbits from the second day after giving birth suckle their young only once every 24 hours, and then only for a short period. When a doe jumps on a nest containing active young she lets down her milk, feeds the litter immediately, and then leaps off a few minutes later. Even in well-run rabbitries some does, particularly maiden does, will fail to rear all, or some, of their young. The casualties are caused in a variety of ways. Some does scatter or neglect their litters at an early stage. Individual youngsters may be accidentally lifted out of the nest by the doe if they are attached by their mouths to a teat when the doe jumps out. Others may be injured by the weight of the mother when she leaps on the nest, or may be scattered by such an action and stray and become chilled. Some does will urinate in the nest box, and so cause mortality in the young from chilling. A doe which fails, without an obvious reason, to suckle a litter on two occasions should be discarded. Normally a doe will remain profitable for about 2 years. After this age litter sizes fall, or milking ability decreases, resulting in a high mortality rate of young up to 2 weeks of age or low weaning weights.

As stated earlier, well-grown young rabbits for meat production are usually weaned at 4 weeks of age, by which time they will be eating concentrate food readily, although some are separated from their dams earlier. Any litters, particularly large ones, which have been obviously undernourished should not be weaned at this age but left with the doe for another week or so.

Health

A rabbit should first be observed in its cage as healthy animals are usually active and have a good appetite, and so drowsiness and inappetence are indications of illness. The rabbit should then be handled and examined, starting at the nose and working back to the tail. The nose should be clean and dry as respiratory diseases can be serious, and are characterised by sneezing and a nasal discharge. The eyes should be bright and wide open, as sick rabbits often huddle with eyes half closed. The insides of the ears should be clean for ear canker will cause ear scratching and head shaking. The body should be well fleshed, but not too fat, and evenly covered with sleek, glossy fur without any bare patches. The paws and limbs should be well formed and clean, for if there is a marked nasal discharge the insides of the forelegs will be wet and matted because the rabbit tries to obtain relief by rubbing its nose with its paws. The underside of the hocks should be free from sores. The fur round the anus should not be matted with faeces as this is a sign of diarrhoea, and the genital organs and mammary glands should be clean and free from swellings. The most critical age for a rabbit is soon after it is born, but there is a second critical period between 1 and 3 months during which time rapidly-growing rabbits become specially susceptible to infectious agents. The normal temperature of a rabbit varies between 39 and 40°C (102·2 and 104°F). The pulse rate is not normally used in an assessment of health but averages 200 beats per minute.

Common diseases and their prevention

Snuffles is the common name given to a localised infection of the nostrils with bacteria of the Pasteurella group. The obvious clinical sign is a nasal discharge accompanied by sneezing. In advanced cases the front legs become wet through rubbing the nostrils. Preventive measures include good housing, as draughts and damp are predisposing causes, and the administration of a vaccine which has proved satisfactory on some farms.

Coccidiosis is the most important disease condition affecting young rabbits. The clinical signs are dullness, a pot-bellied appearance, a rapid loss of body condition, diarrhoea, and death. The causal agent is an internal parasite, and there are two main species, one infesting the liver and the other the intestines. Most rabbits are infested with a few coccidia of both types and these do not cause any apparent harm, but heavy infestations of either or both rapidly lead to obvious physical signs. Coccidial eggs are passed in the faeces of infested rabbits and can be spread to other rabbits through contamination of the food supplies or bedding. Housing in cages with wire floors is a useful preventive measure as it reduces the contact between rabbits and infested faeces and so lowers the chance of infestation. The addition of drugs such as sulphaquinoxaline or sulphamezathine to the food or drinking water controls the development of clinical signs.

Ear canker is seen mainly in older rabbits and attention is drawn to it because affected animals shake their heads and scratch their ears. Examination of the insides of the ears reveals a painful inflamed area encrusted with brownish scabs. Parasitic mites, which can be seen with the naked eye in scabs removed from an ear, are the cause. Every effort should be made to prevent the introduction of the mite by examining the ears of newly-acquired rabbits before they are introduced to the rabbitry to ensure that they are not infested, and the disinfection of any secondhand hutches or equipment which may be obtained before use, as the mites can live away from a rabbit for about a month.

Mucoid enteritis occurs most commonly in rabbits between 3 and 8 weeks of age, but occasionally adults are affected. The clinical signs are increased thirst, loss of appetite, grinding of the teeth, and the voiding of large quantities of clear mucoid material from the anus. Palpation of the abdomen produces a splashing sound denoting an increase in the fluid content of the intestinal tract. Death usually occurs in from 3–5 days. The cause is unknown, but it has been suggested that the feeding of hay will reduce the severity of an attack in affected animals and help to prevent the clinical signs appearing in other rabbits.

Handling

Rabbits are best lifted by placing one hand over the ears and grasping the skin of the back behind the base of the ears firmly, but gently, and placing the other hand under the body to bear the weight. Young rabbits may be lifted by grasping them gently round the loins with one hand, the fingers being on one side and the thumb on the other, avoiding pressure in the kidney region. Another method of lifting, sometimes used for pregnant does, is to hold them firmly round the abdomen, with one hand on either side. For inspection rabbits are best placed on a table covered with hessian, on which they will not slip as they do on wood or metal surfaces. The handler can then control the rabbit by placing a hand over the area of the withers.

Sexing

Rabbits over 4 weeks of age can be sexed relatively easily. The rabbit should be held with the right hand by the skin in the withers region so that gentle pressure can be applied on the sides of the reproductive orifice with the finger and thumb of the left hand. This will express an immature penis in a buck, while in a doe a V-shaped slit is revealed. Sexing becomes much easier when rabbits reach the age of 8 weeks.

Restraint during mating

Some does need to be restrained so that the buck can mount. The skin over the shoulders should be held with one hand and the other hand should be placed under the body between the hind legs with the thumb on one side of the vulva and a finger on the other. The hind quarters can be raised slightly and gentle pressure with the thumb and fingers will lift the tail and enable the buck to mate.

Administration of medicines

Oral medication by giving liquids, tablets, and pastes is possible in individual rabbits if thought desirable. By holding a rabbit's ears and neck scruff the head can be tilted back and a hypodermic syringe without a needle or an eye dropper can be inserted in the space between the incisor and molar teeth and the fluid it contains gently released. Tablets, held in a pair of forceps, can be placed at the back of the tongue and pastes can be smeared in the same area. Subcutaneous injections can best be made behind the shoulder, intramuscular injections into the muscles at the top of the hind leg and intravenous injections into one of the ear veins.

General

Vices

Fur chewing

Occasionally groups of young rabbits will chew one another's fur. Overcrowding may be a predisposing cause, but a shortage of fibre is likely to be the exciting cause. If pellets only are fed the provision of hay may stop the habit.

Cannibalism in breeding does

Some does will kill and eat some of their young, usually shortly after giving birth. In the case of an individual baby it is probable that difficulty was experienced at its birth and the doe injured it while attempting to expedite its delivery. Once a baby is injured a doe tends to eat most of the body. When whole litters are killed and eaten it is generally because the doe was frightened or disturbed, although it has been suggested that a shortage of protein during pregnancy or drinking water at the time of parturition may also be causes.

Minor operations

Claw cutting

The claws of adult rabbits kept on wire floors or soft bedding materials may grow excessively long and must be trimmed when necessary. The sharp points are cut with nail clippers, care being taken not to cut too far back and so cause bleeding.

Rabbit meat marketing

In the United Kingdom the consumption of rabbit meat is at its highest during the winter from September to February but, ironically, rabbits breed most freely during the spring and summer. Thus the carcases of many rabbits reaching killing weight during the summer are stored in deep-freeze cabinets before sale. This leads most producers to market their rabbits through large companies with refrigerated storage space, usually sending their rabbits to the marketing firm alive. The rabbits are killed and processed in the firm's packing station, and the firm sells the carcases, skins and offal. Most firms offer a guaranteed price for all rabbits within a stated liveweight range, which is usually from 2 to 2·7 kg (4–6 lb). There is often a price penalty for rabbits outside this range and for rabbits with coloured pelts. Some small-scale producers sell direct to the consumers or to local butchers.

CHAPTER 9
DOGS

Introduction

Dogs have been domesticated for centuries, and have been used by man for a variety of purposes. Dogs are widely kept in the British Isles as pets for they respond to affection and companionship. The breed selected will depend on the taste of the owner, but consideration should be given to the facilities available. Large dogs are not suitable for small flats and dogs of breeds requiring plenty of exercise should not be acquired by persons incapable of satisfying this need. The advantage of acquiring a pedigree puppy is that he should grow into the likeness of the breed and so his owner will know what he will look like when fully grown. A mongrel puppy, on the other hand, is an unknown quantity for an owner has no idea of the likely size, appearance and temperament of the animal when adult.

Many householders now value dogs as guardians of their property and a dog can be left alone for this purpose provided he has a comfortable bed, a constant supply of fresh water and is fed and allowed out to relieve himself regularly. Alsatians are used by the Police and the Services as guards and for the detection and apprehension of marauders.

Dogs of various breeds can be trained to act as guides for blind persons and give valuable assistance in this capacity. Many of the terrier breeds were developed to hunt and kill rats and other vermin, and are still valued for this purpose. Collies are experts at handling sheep and are indispensable on large hill farms, and Cardiganshire Welsh Corgis are still used as cattle dogs in some areas of Wales. Dogs are widely used in sport. There are numerous packs of foxhounds and beagles distributed throughout the country. Coursing hares with greyhounds is an ancient sport and the racing of greyhounds on tracks is now popular. Several breeds have been developed to assist sportsmen when shooting. Pointers are famed for their ability to point to a bird with nose, body, and tail in a straight line;

spaniels find, flush and retrieve game while retrievers are reserved for the task indicated by their name.

Many dog owners find showing an absorbing pastime and travel long distances to attend shows in various parts of the country. The training of dogs to participate in obedience classes is growing in popularity, and some handlers devote considerable time to making their dogs expert in this field. Non-pedigree dogs can be trained for obedience classes.

Dog breeding can be a fascinating hobby with breeders trying to produce animals as near to the Kennel Club standard for the breed as possible. The breeding of dogs is usually not a profitable occupation, but if a kennel owner is successful at dog shows there is likely to be a demand for puppies related to the winning animals which can command a good price. There is also a considerable export trade in pedigree dogs suitable for breeding and showing and every year many high priced dogs are purchased by overseas buyers. The Breeding of Dogs Act 1973 makes it necessary for anyone owning more than two breeding bitches to obtain a licence to breed. The breeder's premises are inspected to ensure that the dogs are properly housed, kept in clean conditions, well fed, and given adequate exercise.

Definitions of common terms

Dog. Adult male.

Bitch. Adult female.

Puppy. Young animal, in law up to 6 months of age and in shows up to 12 months old.

Brindle. The colour obtained from a mixture of dark and light hairs, usually in the form of light hairs on a brown or black background, giving a barred effect.

Merle. A blue colour flecked or streaked with black.

Wheaten. A pale yellow colour.

Tricolour. A dog with a coat of three colours, roughly in the same proportions. The most common combination is black, tan and white.

Feathering. The long fringe of hair on the back of the legs of spaniels and some retreivers.

Undercoat.The soft wooly hair beneath the outercoat of some breeds, which is sometimes of a different colour from the outercoat.

Stern. The tail of a sporting hound, such as a foxhound or beagle.

Docking. The shortening of a dog's tail, usually in the first week of life, to a particular length for the breed.

Principal breeds

Adult dogs vary in weight from about 1·36 kg (3 lb) to 13·6 kg (130 lb).

The Kennel Club classification separates a total of over 110 breeds recognised in the United Kingdom into two main classes, Sporting and Non-sporting, and then divides each of these into three groups. The most popular breeds are described under the six group headings. The height of a dog is usually measured perpendicularly from the top of the shoulders to the ground, and this is the measurement given under individual breeds.

Fig. 9.1. Afghan Hound.

Sporting breeds

Hound group

Afghan Hounds

Originally imported into this country from Afghanistan, where the breed was used for hunting all types of wild animals. One of the largest of the greyhound family standing from 0·61 to 0·74 m (24–29 in) in height with a fairly long tail with a ring at the end. Afghan hounds have a long fine coat. Golden fawn is the most popular colour, but black, black-and-tan, and other colours are recognised. Has a dignified, rather aloof appearance and moves with a smooth springy gait.

Basset Hounds

These hounds have a large head resembling a bloodhound, a long low body and short legs and stand at not above 0·406 m (16 in). The coat may be either smooth or rough and any recognised hound colour is allowed.

Beagles

Have been used for centuries for hunting the hare. Should not exceed 0·406 m (16 in) in height and have a straight tail carried upright. They should have a smooth, dense coat and may be of any of the recognised hound colours.

Bloodhounds

Have long been famed as trackers. The head is characteristic with long ears and skin falling in loose pendulous ridges and folds. In height they are up to 0·69 m (27 in) and are black and tan, red and tan, or tawny in colour.

Dachshunds

Three types of coat are recognised, smooth-haired, wire-haired, and long-haired, and miniature forms of each of these are found. They are very popular as household pets, but were originally bred for sport, particularly badger digging. They stand about 0·25 m (10 in) high, having short legs and long backs. They can be of any colour other than white, but are generally red or black and tan.

Greyhounds

These have been bred for centuries for speed and stamina, and are now used for show and for racing and coursing. The show dogs are heavier and taller than those bred for work. The head is long, the body has a deep chest, and the hind quarters are powerfully developed. They reach 0·75 m (30 in) in height, and have fine, smooth coats.

Favoured colours are black, red, blue, brindle, and fawn, or any of these colours broken with white.

Whippets
A whippet is a greyhound in miniature, and is still used for racing in some parts of the North of England. The height is up to 0·457 m (18·5 in) and any self colour or mixture of colours is accepted.

Gundog group

Irish Setters (Red)
Used for field work to set up and retrieve game and for show. The head is long and lean, the chest is deep, and the body is long and muscular, narrowing towards the hind quarters. The height is up to

Fig. 9.2. Yellow Labrador Retriever.

0·66 m (26 in) and the coat is of moderate length and flattish with a fair amount of hair on the belly. A rich chestnut colour is desired.

Retrievers (*Golden*)

These gentle dogs are popular both at shows and in field trials. The head has a broad skull, the body is short-coupled, and the height is about 0·61 m (24 in). The coat is wavy with good feathering on the back of the legs, and any shade of gold or cream is accepted as the colour.

Retrievers (*Labrador*)

The most popular of the retrievers, they are used for show, as companions, for field work, and as guide dogs for the blind. They are active and powerfully built, with broad skulls, short backs and well-developed hind quarters. In height they are about 0·57 m (22·5 in) and the coat is short and dense without any wave. Black and yellow are the main colours, the blacks being most popular as working dogs while the yellows are very successful in the show ring, although dogs of both colours are used for all purposes.

Spaniels (*Cocker*)

So popular as pets and show dogs that their ability to flush and retrieve game is often overlooked. The head has a well-developed skull and forehead, with long ears. The body is compact and the height about 0·42 m (16·5 in). The coat is flat and silky with good feather. Various colours are found, but black and golden are the most popular whole colours and blue and red roans the most common broken shades.

Spaniels (*Springer, English*)

This breed is of ancient origin, and is widely used for finding, flushing, and retrieving game for the gun. The head is fairly broad with long ears set close. The body is of medium size and the chest is deep. The height is about 0·51 m (20 in), the coat is close and weather resisting and the favoured colours are liver and white and black and white.

Terrier Group

Airedale Terriers

Well known as guard dogs, they stand about 0·61 m (24 in) high. The coat is hard and wiry and the body is black or grizzle in colour with tan markings on the legs. The head is long and flat, with a strong jaw.

Bull Terriers

Were originally bred for bull and bear baiting and dog fighting, but are now kept as smart and faithful companions. They are about 0·56 m (22 in) high, with a short, flat coat, and may be pure white or coloured. Brindle is a favourite colour, but other colours, particularly reds, are seen. The body is strong with wide shoulders and muscular thighs.

Cairn Terriers

One of the smallest of the terriers, being only about 0·25 m (10 in) high, they have a fearless disposition. They are double coated with a hard outer coat and a close furry undercoat. The colour varies from a nearly black brindle, through grey and red, to a sandy shade, all having dark ears and muzzle. The head has a foxy appearance.

Fox Terriers (*Smooth and Wire Coated*)

The wire coated is the most popular type of this breed, which was originally bred as a hunt terrier. The height is about 0·38 m (15 in) and the colour is predominantly white, with black or tan markings.

Scottish Terriers

A sturdy thick set active terrier about 0·28 m (11 in) high. They were originally bred to go to ground after foxes, but are now kept as companions. The back is short and muscular and the chest broad and deep. The outer coat is harsh and wiry with a dense undercoat. The colours are black, wheaten, and brindle.

Staffordshire Bull Terriers

At one time these were used for dog fighting in pits and combine strength with agility. They differ from the bull terrier mainly in the head, which is shorter and broader with well-developed cheek muscles.

West Highland White Terriers

Similar to the Cairn Terrier, from which it is descended, but pure white in colour.

Non-sporting breeds

Utility group

Bulldogs

They were originally used for bull baiting and are the popular symbol of the British people. They stand about 0·41 m (16 in) high and are brindle, red, white, or pied in colour with a black mask. The most striking feature is the large, massive head with broad cheeks. The body is broad and muscular with a short roach back.

Chow Chows

A very ancient breed originally bred in China for human consumption. Chows become devoted to individuals and disinterested in other

Fig. 9.3. Black Chow Chow.

humans, and so make excellent guard dogs. The minimum height is 0·45 m (18 in). The outer coat is coarse, with a soft woolly undercoat. Whole-coloured black, red, fawn, blue, or cream animals are bred, and one unique feature is the purple tongue. The straight hocks, giving a stilted walk, are another breed peculiarity.

Dalmatians

In the days of horse-drawn carriages Dalmatians trotted under the vehicles, on the heels of the horses and had the speed and endurance to keep up. At night they slept in and guarded the carriages, their short coats and the absence of any 'doggy' smell making them particularly valuable for this duty. Their height is about 0·58 m (23 in), and the ground colour is white with black or liver coloured spots varying in size from that of a halfpenny to a ten-penny piece distributed over the body.

Poodles (Standard)

This breed was widely used in France as a gundog, especially for duck shooting, and the hard profuse coat was cut away behind the ribs to facilitate swimming while the other parts were left covered and protected. A head and a tail ribbon were tied so that the dog could be easily seen among the reeds. This method of clipping has now been adapted for show purposes. The head is long and fine, with long ears set low on the side of the face. Strong muscular shoulders slope into a strong, short back. The height is 0·38 m (15 in) or over. Whole colours are required, such as black, white, apricot, liver, or blue.

Poodles (Miniature)

A replica of the standard, but the shoulder height must be under 0·38 m (15 in).

Poodles (Toy)

The same description applies, except that the height must be under 0·28 m (11 in).

Working group

Alsatians

Originally known as German Shepherd Dogs they are widely used as guards, police dogs, guide dogs for the blind, and as companions. They are noted for their watchful alertness, fearlessness, faithfulness and intelligence. The head is long, shows an alert expression, and has rather large ears, placed high on the skull and carried erect. The back is fairly long, with a deep chest, and the height is up to 0·66 m (26 in). The outer coat is smooth and flat with a woolly, thick undercoat. Various colours are recognised, but black and tan is the most popular.

Fig. 9.4. Alsatian.

Boxers

These are descended from the old German bull-baiting dogs, and, being alert and courageous, are used as guards and companions. The head is characteristic with a slightly arched skull and powerful jaws with the lower extending beyond the upper. The body should be short and broad with a muscular appearance. The height is up to 0·61 m (24 in) and the coat is short and lies tightly against the body. The colours are fawn and brindle. White markings are permitted provided they do not occur on the mask and do not exceed one-third of the body colour.

Collies (Rough and Smooth)

For years collies have been used by shepherds for handling sheep flocks. They are alert, intelligent and active, moving smoothly and gracefully. The head is long with small ears thrown back in repose and held semi-erect when alert. The body is long with a deep chest. The height is up to 0·61 m (24 in). The coat in rough collies is very dense with a harsh outer coat and a furry inner coat and there is an abundant mane and profuse hair on the upper parts of the legs. In the smooth variety the coat should be hard, dense and quite smooth. Colour is immaterial, but black and white is usual.

Doberman Pinschers

These are used extensively as police dogs on the continent. The head is long, and the ears are fairly small and set high. The body is square with a short, firm back, and the hindquarters are muscular and well developed. The height is up to 0·66 m (26 in). The coat is smooth, short and hard and the colour is black, brown, or blue with tan markings on the chest, legs, and feet.

Fig. 9.5. Doberman Pinscher.

Great Danes

On the continent these have been used as fighting dogs and wild boar hunters for centuries, but in the British Isles they are kept as companions famed for their gentle natures, their loyalty, and their ease in training. The head is long with a strong jaw and small ears set high. The body is deep and has muscular shoulders and hind quarters. The height is up to 0·76 m (30 in) The coat is short and colours met include brindle, fawn, blue, black, and harlequin. Harlequins have a pure white background and black, or sometimes blue, irregular patches.

Old English Sheepdogs

Originally these dogs, also known as 'Bobtails' because of their short docked tails, were famed for working sheep over rough country without tiring, but are now used mainly for show and as companions. The height is about 0·56 m (22 in) and the coat is profuse, shaggy and free from curl. Regular grooming is essential. Any shade of grey, blue, or blue merle with or without white markings is allowed.

St Bernard

These dogs are famous as life-savers of travellers lost in the Swiss Alps. They are large dogs, about 0·71 m (28 in) or over in height, and have a massive appearance with a distinctive head having a benevolent expression. The coat may be rough or smooth, with orange, mahogany, or brindle patches on a white background. Black shadings occur on the face and ears.

Shetland Sheepdogs

These are miniature collies developed as companions and not for working with sheep on account of their size. They are about 0·36 m (14 in) in height.

Welsh Corgis (Cardigan)

Not as popular as the Pembroke type, from which it can be easily differentiated by the long tail. Any colour, except pure white, is accepted.

Welsh Corgis (Pembroke)

This drover's dog from South Wales has achieved great popularity as a pet, largely because of its intelligence and alert and active disposition. The head is foxlike in shape with medium-sized prick ears. The body is of medium length with strong hind quarters and short legs. The tail is very short. The height is about 0·30 m (12 in). Self colours, such as red, sable, and fawn or black and tan are favoured, but most have some white markings on the legs, chest, and neck.

Toy Group

Chihuahuas (Smooth and Long Coat)

These small dogs, standing only about 0·177 m (7 in) at the shoulder, came originally from Mexico. They have well-rounded skulls, large eyes, a moderately short, pointed nose, and large ears. The back is slightly longer than the dog's height and the hind quarters are

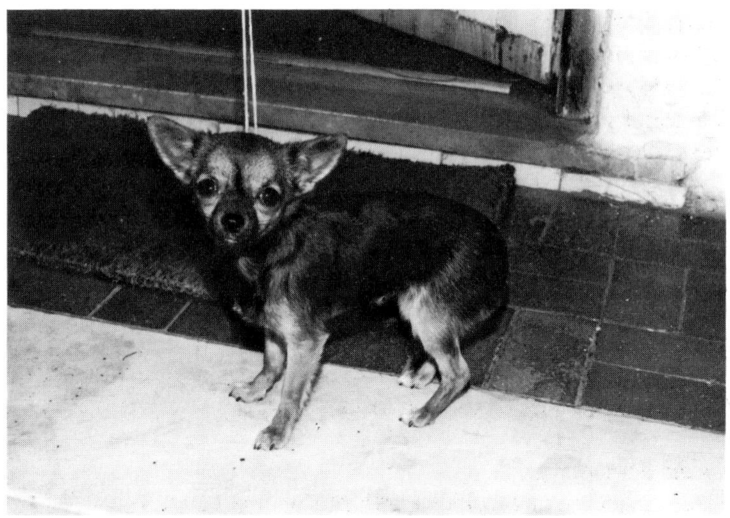

Fig. 9.6. Smooth-coat Chihuahua.

muscular. The coat is either fine and close or long. Any colour or mixture of colours is acceptable.

King Charles Spaniels

Named after King Charles II this breed is divided into four varieties of separate colouring: the King Charles proper, which is black and tan; the Prince Charles, which is a tricolour; the Ruby, a rich red; and the Blenheim, white with red patches. The skull is domed with a square, wide muzzle and large, dark eyes. The body is compact and cobby, and the height about 0·25 m (10 in). The coat is long, silky, and straight.

Cavalier King Charles Spaniels

Now very popular as a pet these Spaniels were extensively used for sport in Stuart times. The muzzle is tapered to a point, and the height is about 0·30 m (12 in). The colours are the same as the King Charles Spaniels.

Pekingese

One of the oldest breeds, having been developed in the Imperial Palace at Peking. The head is broad and massive with long drooping ears and a short, flat nose. The body is lion shaped, broad in front, tapering to the rear, and the height is about 0·23 m (9 in). The tail is turned over the back and well covered with harsh, spreading hair. The undercoat is soft and fluffy with a long, straight topcoat and a profuse frill over the shoulders. Of the whole colours, red is the most popular, but whites and creams occur, and parti-coloured, black and fawn, and tricolours are found.

Pomeranians

At one time this was the most popular toy breed in the British Isles, and it is still favoured because of its bouyant nature. The head and nose are foxy and wedge shaped and the back is short. The tail is turned over the back and carried flat and straight, being covered with long spreading hair. On the body the outer coat is long and straight and the undercoat soft and fluffy. Whole colours, such as white, black, brown, blue, and orange, are found as are parti-coloured animals with the colours evenly distributed in patches.

Pugs

This toy variety of the mastiff family originated in the Far East. It is mainly kept as a pet, but unless well exercised and sensibly fed is apt to become fat and lazy. The head is large and round with a short, blunt muzzle. The eyes are very large and prominent and the ears are small. The body is short backed and cobby, and the height about 0·28 m (11 in). The tail is called the twist and is curled tightly over one hip. The coat is fine and smooth and the colours are black, fawn, silver, or apricot.

Yorkshire Terriers

This is the smallest British terrier, standing about 0·23 m (9 in) high. The head is rather small with a round skull and small V-shaped ears carried erect or semi-erect. The body is compact and covered with moderately long, fine silky hair. The back is dark blue in colour and the chest is bright tan.

Identification

Every dog in Great Britain must, by law, when on a public highway, wear a collar bearing the name and address of the owner. A collar, when fastened, should allow three fingers to be slipped underneath it.

If it is too tight it will cause the dog distress, while if it is too loose it will slip off if the dog pulls backwards.

Registers of the names of pedigree dogs, classified according to breeds and varieties within a breed, are kept by the Kennel Club, which is the authoritative body controlling the showing of dogs. Most pedigree dogs are registered with this body.

Attempts have been made commercially to establish a system of individual registration with the main object of tracing lost dogs. A number, allocated to a particular dog, is tattooed in the animal's ear, or, more probably, in the inner right flank fold. If a numbered dog is found the finder telephones the headquarters of the firm organising the scheme, and a representative informs the owner, who arranges to collect his dog.

Ageing

An examination of the teeth of a dog does not provide an accurate guide to its age because dogs of different breeds vary markedly in their rate of development. Also, the function of a dog's teeth is to cut and tear and not chew food, and so their surfaces do not grind, as, e.g., is the case in the horse, and so there is not the same loss of surface through normal wear during life. However, the following details are useful for making an approximate estimate. The temporary teeth begin to erupt at about 3 weeks of age and are usually all cut by the end of the fifth week. They are softer and sharper than the permanent teeth and are more widely spaced. They are also of a bluish-white colour as contrasted with the ivory-like teeth of an adult dog. The permanent teeth begin to erupt at from 3 or 4 months of age in most medium and large breeds, and slightly later in the toy breeds. The central and lateral incisors are the first incisor teeth to appear, followed at 6 months of age by the corner incisors. The last of the teeth in an obvious position to erupt are generally the canines, but the last molar in the lower jaw may not appear until the dog is 8 months old. By the time all the teeth except this last molar are present the dog is probably 6 months old and by the time this last tooth appears the dog is certainly over 6 months of age, and therefore must have a licence.

Newly erupted incisors have a fleur-de-lys appearance at their upper extremities, which is gradually lost with wear until by 2 years of age the crowns have a level appearance. From then onwards the incisors show increasing signs of wear, but it is not possible to use this change to make more than a rough guess at the age of an animal. For instance, a dog which frequently retrieves stones or habitually gnaws bones will show an abnormal degree of wear while dogs which subsist mainly on soft foods do not exhibit as many signs of wear.

A general sign of old age observed in dogs is the presence of white hairs round the muzzle. This appearance is generally noticed after dogs attain the age of 6 years and is particularly obvious in black animals. Very old dogs, from 8 years onwards, show signs of senile decay, such as emaciation, arthritis, loss of agility and defective hearing and vision.

Housing

House dogs

Indoor accommodation

House dogs should be provided with a comfortable bed. Proprietary models made of canvas or polystyrene filled with granules are available which the dog can mould round his body to form a nest. Alternatively, a fibre glass or wicker-work basket or a shallow, open wooden box with sides 15–30 cm (6–12 in high) which are raised 7·5–15 cm (3–6 in) off the floor can be provided. The size of the box from front to back should be slightly more than the height of the dog at the shoulder. The length should be at least the length of the dog from the tip of its nose to the base of its tail. The bed should be lined with several layers of newspaper, which should be renewed regularly, and placed in a draughtless corner. The floor of the room in which the bed is placed should be covered with linoleum rather than with a carpet. To be effective as a guard a house dog should have the run of the whole house, and, for sanitary purposes, dogs kept in houses should be allowed regular access to the outside.

Outdoor accommodation

Dogs to be kept outside should be accommodated in an outhouse or provided with soundly constructed, well insulated, kennels having a floor area twice the size of the bed described for house dogs. The dog's entrance to the attached run should be at the end of the kennel furthest from the bed and there should be a low bar across the bottom of the opening to minimise the loss of bedding material. The sides

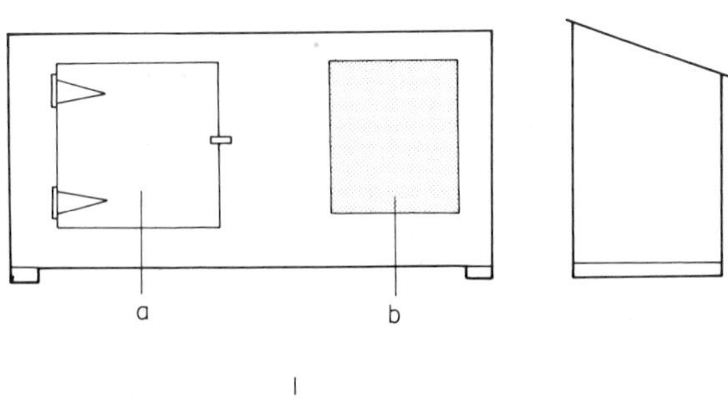

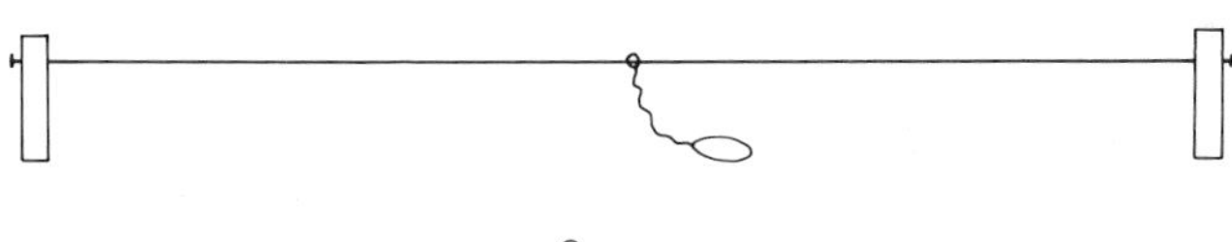

Fig. 9.7. (1) Outdoor kennel, front and end elevations: (a) opening door for cleaning, (b) open entrance for dog. (2) Chain running on a wire.

of the run should be at least three times the shoulder measurement of the dog in height. The run floor should preferably be of concrete, unless it is very large in which case it can be of grass. The kennel should not be sited in a damp position where water could collect.

If it is not possible to provide a run a running chain should be erected. Under this arrangement the dog is tethered to a short chain which has a large ring at its other end. This ring can slide freely along a length of wire cable fixed either 7·5 cm (3 in) above the ground or at double the dog's shoulder height. This enables the dog to pass over or under the wire and so twisting of the chain is avoided. A stop should be fixed near each end of the wire to prevent the chain going round the end posts.

Breeding or boarding kennels

The design will depend on the purpose for which the kennels are required. Although many kennel units are made of wood, when permanent buildings of a substantial nature are being erected the use of timber should be kept to a minimum, especially in those areas which a dog can reach. Wood in accessible positions can be damaged by dogs and it is relatively difficult to disinfect should a disease outbreak occur.

As a general rule large kennels should not be constructed in populous districts, but should be sited where public services, such as electricity and mains water, are easily obtainable. Thus, in most cases they are located on the outskirts of towns. In such areas the barking of dogs kept in kennels may cause annoyance to persons living in neighbouring houses. The noise nuisance may be reduced to some extent by siting the kennels so that they are screened by other buildings, but some form of sound-proofing may be required. Various devices can be employed during the construction of a kennel building which are effective in achieving this, but they add to the cost of erection.

It is usual to provide each dog with an indoor sleeping quarter with an outdoor run attached. A suitable size for the indoor accommodation for a dog such as an Alsatian would be 1·84 m × 0·92 m (6 ft × 3 ft). A smaller area would suit a terrier while an animal of a large breed, such as a Great Dane, would need more space. The runs are normally about twice the size of the sleeping quarters. The most convenient design is to arrange the indoor kennels along each side of a brick building with a passage down the centre. The doors between each kennel and its outside run should be fitted with a catch so that the occupants can be confined either inside or outside.

The walls between the inside areas should be solid, preferably of smooth concrete, up to a height of 0·92 m (3 ft) with wire-mesh divisions above reaching to the ceiling. The floors of the indoor quarters should be of insulated concrete, that is a layer of concrete about 2·5 cm (1 in) thick covering hollow tiles or pipes. A bed should be fitted in each kennel so that the dog need not lie on a cement floor. Each bed should be large enough to allow the dog using it to stretch itself out comfortably. These beds are usually made of wood, but should be movable so that they can be taken out and thoroughly disinfected when necessary. Duplicates should be made so that some beds can be kept unused for a period if this should be desirable. The front of the bed should be fitted with an upright retaining board about 12·7 cm (5 in) high to prevent bedding material from being pushed out, and it is advantageous to have a narrow strip of iron screwed to its upper margin.

Good natural lighting is essential, and most kennel owners find

Fig. 9.8. Beagle bitch and puppies in a box lined with newspaper.

that roof lighting is the most satisfactory form. If windows are fitted in the walls they should be set high up and should be hinged at the base and open inwards with baffle plates at each end so that the incoming air is directed upwards and so does not cause a draught on the dogs. They should not open wide enough to permit a dog to clamber through and escape. Although heating is normally only required during very cold weather some means of artificial heating should be available if needed.

The walls of the exercising runs are often made of chain-link wire fencing with a 5·1 cm (2 in) mesh. They must be at least 1·84 m (6 ft) high, and should preferably be 2·44 m (8 ft). Some dogs are remarkably adept at climbing up fences, and the only way to keep such individuals confined is to fix an additional strip of chain-link fence about 0·61 m (2 ft) wide sloping inwards at an angle of 45 degrees. If wooden posts are used to form the framework of the exercising run walls they should not be less than 5·1 × 5·1 cm (2 in × 2 in) and should preferably be 10·2 × 10·2 cm (4 in × 4 in). It is an advantage to have the ground ends set in concrete. The run floors should be of concrete which need not be insulated, unless the areas are large, when grass is satisfactory.

In large units auxiliary rooms should be included in the design of the unit to serve as an office, food store, bedding store, kitchen and grooming room. There should also be an isolation unit, in case a dog develops an infectious disease, and, possibly, separate sections for whelping and puppy rearing.

Whelping bitches

A bitch kept in a house should be supplied with a box of ample proportions in a room which is comfortably warm. This box should be lined with newspaper, which must be renewed as it becomes soiled with the discharges of parturition. Breeders on a large scale usually erect special kennel units, providing artificial heating, generally in the form of infra-red lamps of the type commonly used for piglets. Such kennels may have a separate bed for the bitch where she can lie out of reach of the puppies if she so desires.

Puppies

Young puppies need a bed which is dry and can be kept at a temperature of about 23·9°C (75°F) for, apart from food, warmth is the most important factor in a young puppy's life. Very young puppies born in a house can be kept in the box in which they were whelped, which should be placed in a kitchen, or any warm room in a house, until they are active enough to leave the nest. After this stage they need to be provided with a dry, weather-proof run in an outhouse or a soundly built kennel with a sheltered run attached. Puppy kennels in specially designed units are provided with outside runs which are linked with the inside areas by a door so that the puppies can be kept in the housed section or allowed out in the run as required.

Bedding

For dogs kept in either houses or kennels newspaper is quite satisfactory and thick layers provide a reasonably warm, soft surface. For kennel units wood wool, when it can be obtained, is good, as it is clean, free from dust, and contains resins which are repellent to fleas and other insects. Hay is not favoured as, although it is soft and warm, it tends to be dusty and contains forage acari which can set up a skin irritation. Oat and wheat straws are satisfactory for dogs of the larger breeds. All these materials have the advantage that they can easily be burnt when they become soiled.

Cushions and blankets are frequently provided for house dogs, but they are relatively costly. Although they can be washed they do not remain clean for long.

Feeding

General

Wild dogs normally have only one meal per day, and this is usually of flesh. To some extent domesticated dogs have become adapted to eating, and deriving nourishment from, other foodstuffs, but domestication has not produced marked changes in the anatomical characteristics of the dog and meat is still an essential constituent of a satisfactory diet for healthy dogs. Because puppies vary so markedly in their growth rates and adults differ greatly in size, activity, and temperament it is difficult to give precise dietary requirements for the various classes of animal.

Puppies

During the early part of their lives puppies live on the colostrum and milk supplied by the bitch, and the quantity and quality of the milk is dependent on the milking ability of the dam and on the diet it receives. At 4 weeks of age, or a week earlier if the puppies appear to be hungry, they should be taught to lap cows' milk fortified with full cream dried milk. This supplementation is necessary because bitches' milk is richer than cows' milk and puppies would have to lap excessive quantities of the latter in order to obtain the desired level of nutrient intake. Soon after the puppies are drinking the fortified milk readily, finely scraped meat can be given. Some bitches regurgitate partially digested food for the puppies to eat at this time. This provision of additional food is necessary in order to make the transition from maternal milk to other foods as gradual as possible, so that the puppies continue to grow steadily.

At 5 weeks of age two milk meals and two meat or fish meals should be given. Proprietary tinned foods can be used to supply the meat and fish. Then starchy foods, such as puppy biscuit, can be included in the diet. Such foods should be eaten in gradually increasing amounts. It is difficult to specify quantities but between 0·05 and 0·08 units by weight of food per unit liveweight is generally allowed per head per day. When using this method of calculation the puppies need to be weighed weekly.

After weaning at about 6 weeks of age meals should be given every

3 hours during the day, which means that the puppies are given five feeds per day. Any food not consumed in 10 minutes should be removed. Fat is of value in puppy foods because of its high energy value, and levels between 7·5 and 20 per cent are used. Excessive amounts are contra-indicated because food intake is reduced and inadequate quantities of the non-fatty nutrients are eaten. Carbohydrates can be included for the provision of additional energy. The protein requirements are important because protein is needed for the formation of body tissue and diets containing less than 17 per cent of good quality protein will not satisfy the needs of young growing puppies. Most diets contain between 20 and 25 per cent of protein, although as the protein requirements diminish gradually as growth continues lower percentages can be used for older animals.

Foods which are suitable for puppies after weaning include milk, meat, fish, and puppy biscuits, as previously advised, plus offal, such as liver, proprietary canned foods, and toasted wholemeal bread. Hard biscuits or bones of a suitable size for the puppies to chew should be provided to encourage tooth development and keep them occupied. The bones supplied should not splinter easily to avoid any chance of injuries to the mouth, or the intestines if fragments are swallowed.

At 10–12 weeks of age the feeding of milk can be discontinued, but a high proportion of good quality meat or fish should still be fed. When the puppies are between 4 and 5 months three feeds a day are sufficient, and at 6 months the adult allowance of one meal, or one main meal plus a light meal, can be started.

Adult dogs

Most dogs need a maintenance diet with little supplementation, as they receive relatively little exercise and are protected from cold weather conditions. The quantities of food required per day vary according to the size, activity, and temperament of the individual, and although size can be accurately measured it is difficult to assess activity and temperament. However, it is generally agreed that an adult animal of one of the medium sized breeds requires daily 0·031 units per unit liveweight and that this needs to be increased to about 0·04 units for individuals of the toy breeds, and can be slightly reduced in the case of the large breeds.

The following approximate quantities per head per day can be used as a guide: dogs about the size of a Great Dane, 1·81 or even 2·04 kg

(4 lb or even 4·5 lb); Alsatian size, 0·68–1·13 kg (1·5–2·5 lb); Fox Terriers and dogs of a similar size, 0·23 kg (0·5 lb); and Pekingese, 113·6–170 g (4–6 oz). Even a guide like this is only approximate, for some large dogs may need 2·27 kg (5 lb) per day while some small animals may require less than 113·6 g (4 oz)

Protein is needed for the replacement of body tissues, and should provide at least 20 per cent of the dry weight of the diet. It is best supplied in the form of meat or fish, either fresh or tinned, but occasionally vegetable proteins, such as soya bean meal, are included in proprietary foods. About 5 per cent of fat needs to be included in order to maintain the skin and hair in good condition, to enable certain body functions to operate, and to help provide the energy required for movement. Most of the common fats, such as beef dripping, are suitable and cereal grains and fatty meats are also sources. Carbohydrates, to provide energy, can form about 50 per cent of the diet. They are usually supplied as wheat biscuits or wholemeal bread, although oatmeal and maize meal may be used. Experiments have shown that purified white flour is not a suitable food for dogs, and white bread or biscuits made from white flour should not be fed over a long period. Some owners feed cooked green vegetables, and small amounts are acceptable but large quantities are to be avoided. Many householders include household scraps in the dog's diet, and, provided these fall within the categories mentioned above, this practice is quite satisfactory.

Regularity in feeding is important and the average healthy adult dog generally receives one main meal per day, possibly supplemented by a second light meal. In a house dog the main meal should be given in the evening, as the dog is then likely to sleep at night, and the light meal at breakfast time. Digestion is comparatively slow in the dog, and some hours elapse after a meal before the alimentary canal can digest a new supply of food. Dogs of the toy breeds, however, are often given two or three small meals in the course of 24 hours.

Bones have no particular food value, but are normally supplied to dogs to give employment to the teeth and keep the gums in a hard, healthy condition. The safest bones are the ribs of cattle, the leg bones of sheep or cattle or other equally large and unsplinterable bones. Small chop bones, rabbit or chicken bones, and fish bones are undesirable as they may become lodged between the canine teeth or, if broken by gnawing, may splinter into sharp spicules which can injure the mouth, oesophagus, or stomach.

Old dogs

With decreasing activity and a lower metabolic rate the requirements for energy diminish and food needs must be adjusted. Unless this is done many old dogs show obesity, and some are markedly overweight at the time of death. The laying down of excessive amounts of fat in an old dog can be prevented and some weight reduction in an excessively obese senile dog can be achieved by reducing the amount of fat and carbohydrate in the diet and substituting carefully regulated quantities of vegetable matter or bran.

Working dogs

The performance of physical work, such as racing, hunting, retrieving game, driving sheep, or acting as guides to blind persons, demands a production ration in addition to a maintenance diet. It is difficult to calculate the amount required because this will depend on the pace and duration of the work. However, up to three times the basic maintenance requirements might be needed by a dog performing about 8 hours a day of really hard work. In some cases, such as hounds which only hunt on certain days, the supplementary food may only be required on these days. To provide the energy required for the performance of work additional amounts of carbohydrate and fat are the essential additions. A slight increase in the protein content will be needed to allow for the replacement of tissues damaged by the work performed.

Breeding bitches

When a bitch comes into oestrus some breeders increase the quantity of food fed by about 5 per cent in the hope that a greater number of ova will be produced so that a larger litter will be born. For the next 3 weeks the level of feeding can drop back to the normal maintenance standard, but from the fourth week onwards an increase is required which is gradual at first, rising to a peak of about one and two-thirds maintenance before whelping. These increases should be in the protein, rather than in the energy, content to provide for the development of the bodies of the fetuses, and at least 30 per cent of the diet should consist of protein, most of it animal protein.

Just before giving birth a bitch's appetite will usually fall. During lactation the amount of food supplied at the end of pregnancy is needed at the start, but at its peak a marked increase is desirable, and levels as high as three or even four times the maintenance level may be required. The actual amount will vary with the quantity of milk being produced which will depend, to some extent at least, on the number of puppies being suckled.

Stud dogs

The total volume of food required by a dog at stud is not much above that needed to provide for maintenance, as stud dogs must not be allowed to become fat. However the quality of the food can, in certain cases, be improved with advantage and the protein level should be about 30 per cent.

Artificial rearing of puppies

The artificial rearing of puppies is an arduous task, but may be necessary in the case of an orphaned litter or when a bitch has an inadequate supply of milk for all the puppies born. Artificial warmth is essential for the first 3 weeks. If at all possible the puppies should be allowed to suck the dam's colostrum before artificial feeding is started. Commercial forms of bitch's milk are available and should be used if possible as bitch's milk is higher in fat and protein than either human or bovine milk. In an emergency a proprietary human baby food of the full cream type may be used, being made more concentrated than is advised for human babies, or cows' milk may be used with some butter added to increase the fat content so that the mixture approximates the 11·8 per cent of fat found in a bitch's milk. The shortage of protein does not appear to be very important. Special puppy feeding bottles, made in different sizes, can be obtained or a premature baby's feeding bottle and teat can be used for puppies of the larger breeds. In an emergency a human eye dropper can be employed.

For the first 3 or 4 days 2-hourly feeding, night and day, is advised, though 3-hourly feeds will be satisfactory for strong puppies. After 4 days 3-hourly feeds should suffice for all puppies until they are about 14 days of age, when the night feeds can be reduced to one. This can be dropped when they are about 3 weeks old. The puppies should be given sufficient to satisfy them without causing noticeable abdominal distension at each feed, and small breed puppies will need from 2 to 3 ml (0·07–0·1 fl oz) each time while larger puppies may take 4 ml (0·14 fl oz). These amounts rise with increasing size. However, because individual requirements vary markedly, it is best to rely on judgment rather than estimated quantities. From the age of

3 weeks the puppies should lap from a saucer, and the feeding of good quality meat essences may be commenced. The diet can then follow the lines recommended for puppies which have been reared by a bitch.

Mineral requirements

The three important minerals are calcium, phosphorus and sodium. Calcium and phosphorus are essential in the diet of puppies for the development of sound bone and not less than 0·10 per cent of each of these minerals should be included in puppy rations on a dry weight basis. Care must also be taken to ensure that the calcium/phosphorus ratio is not wider than about 1·3:1. Meat, fish and cereals are poor sources of calcium, and so rations composed of these foods need to be fortified, and sterilised bone flour added at a level of from 1·5 to 2 per cent of the diet is a suitable supplement. Dogs which chew bones do obtain a certain amount of calcium in this way. The cereal grains contain some phosphorus, but a proportion of this is in a form which is not easily utilised by the dog. Sterilised bone flour also contains some phosphorus and supplies this mineral as well as calcium. Salt at a level of about 0·5 per cent is also required, but this level is attained in most diets without supplementation, as it is present in fish, lean meat and milk. Traces of iron and copper are also required, but these are present in adequate quantities in all the rations fed in practice.

Vitamin requirements

Vitamin A is required in small amounts, particularly by pregnant bitches and growing puppies. A deficiency in the diet of a bitch may cause her to produce weakly, or dead or deformed puppies, whilst a shortage in the growing stage retards development and may cause blindness through nerve damage. Vitamin D is essential for the proper absorption and utilisation of both calcium and phosphorus, a deficiency producing rickets in puppies. Proprietary dog foods are normally fortified with synthetic vitamins A and D and breeders using natural foods usually give cod-liver or halibut-liver oil to provide these vitamins.

Adult pet or working dogs need 88 I.U. per kg liveweight (40 I.U./lb) of vitamin A and 9 I.U./kg (4 I.U./lb) of Vitamin D daily and puppies and breeding bitches are allowed twice the above amounts. Vitamin E will be supplied in sufficient amounts if dogs have a reasonable quantity of fat in the diet. Dogs also need the vitamin B complex, which is present in small amounts in meat offals such as liver, heart, and kidney. Diets deficient in these foods may lead to a suboptimal performance, such as slow growth in puppies, loss of weight in adult dogs, and the production of weak puppies by breeding bitches. If there is reason to believe that a diet is deficient in the vitamin B complex from 2 to 4 per cent of dried brewer's yeast should be added. Vitamin C is produced in adequate amounts in the tissues of healthy dogs, and so there is no dietary need for this vitamin, but sick animals may lose their capacity for forming this vitamin and may benefit from an addition to the diet.

Water requirements

Clean water in unspillable dishes should always be available within the reach of a dog. This is particularly important in the case of hard-working dogs, who can suffer seriously from a water shortage, and lactating bitches, because lack of sufficient water can rapidly bring a lactation to an end. Young puppies need a surprising amount of fluid and they must always be supplied with drinking water even though they are receiving a milk diet before and after weaning.

Breeding

Most bitches come into oestrus for the first time when about 9 months of age, although some come in as early as 6 months and others do not do so until they are a year old. In general a bitch is not mated at her first oestral period should this occur before she reaches the age of 12 months. Exceptions may be made in the case of certain large-headed breeds, which may be mated at the first oestrus so that the litter is born before the pelvic bones set, as this is said to cause a permanent widening of the pelvic passage.

Dogs are not normally used for mating until they are a year old and regular stud work should not be undertaken before the age of about 18 months. The first time a male is used for breeding it is most satisfactory if it is mated with a proven breeder. It is difficult to set a target figure for the number of bitches which a dog can serve during a year, but many breeders consider that 25 bitches each mated twice a year, that is 50 services, is a reasonable maximum. To maintain semen quality dogs should not mate more frequently than every alternate day.

Bitches come into oestrus at approximately 6-monthly intervals, generally in late winter or early spring and late summer or early autumn. Individual variations occur, particularly in animals of the toy breeds. At the start of the oestral period pro-oestrum occurs when there is an enlargement of the vagina and a bloody discharge which persists for from 7 to 12 days but usually lasts for 9 days. The quantity of the discharge varies in different individuals, but may be very copious and have the appearance of pure blood. This is followed by true oestrus which lasts for from 4 to 13 days, but is normally of 9 days duration signalised by a more marked vaginal swelling and a discharge which is not generally bloodstained. Pro-oestrus and oestrus may overlap to some extent and it is difficult to make a clear distinction between the two. By the end of the third week after the first signs of pro-oestrus were observed the animal will have returned to normal.

The bitch is attractive to dogs from the onset of oestrus, and especially after the first week, but will not permit mating. The actual time when the bitch will accept mating varies, but is usually from the first day of true oestrus. The second and fourth days of oestrus, that is about the tenth and twelfth days after the first signs of pro-oestrus, are regarded as being particularly favourable days for mating. Although this is best for the average bitch it must be realised that not all animals are the same. Most bitches indicate a readiness to accept mating by standing rigidly with hindquarters outstretched when approached by other bitches. During mating the dog and bitch become tied together owing to the swelling of the bulbs of the penis, and it is wise to restrain both gently to avoid the bitch moving and dragging the dog along. The actual time taken to mate can be up to 15 or 20 minutes because the dog cannot withdraw its penis until the swelling has subsided. In most kennels bitches are mated again after 48 hours as a routine, but this is not necessary after a normal service. A repeat mating more than 2 days after the first is not advisable as it could possibly lead to the development of embryos of different ages, with the result that some would be born prematurely.

The gestation period lasts, on average, for 63 days, but may be shorter in the case of large litters. In maiden bitches the teats become enlarged and pink by the fourth week. About 6 weeks after conception some teat enlargement is usually noticeable in bitches which have bred before and some abdominal distension may be observed. Experts can feel the developing fetuses by abdominal palpation at this stage. During the last week of pregnancy movement of the unborn puppies can be seen through the dam's flanks. Milk may be present in the teats 1 or 2 days before whelping.

A few hours before the puppies are born the vulva enlarges and the bitch becomes restless and may refuse food. Just before whelping most bitches try to make some sort of a nest and should be provided with some material such as newspaper which can be torn up for the purpose. At term a series of powerful abdominal contractions force the afterbirth of the first puppy outside the vagina. This bursts and the puppy is usually born soon afterwards. A further contraction expels the placenta, which the bitch normally eats after breaking the umbilical cord with her teeth. Once the placenta has been separated the puppy starts to breathe. The intervals between births vary. Several puppies may be born very rapidly and then a period of some hours may elapse before the next puppy is expelled. Such a resting time is normal as it enables the uterus to recover its tone before contracting again. Whelping difficulties are more frequently encountered in some breeds than in others. Bitches of breeds with overdeveloped heads and shoulders, such as Bulldogs or Scottish Terriers, tend to produce puppies which do not pass easily through the pelvis. The delivery of such puppies can often be assisted by gentle traction on the exposed head and legs, but professional veterinary assistance may be needed. A few bitches develop a form of uterine inertia which prevents them from expelling the puppies, and a Caesarean section operation may be required. Most breeders leave the puppies with the mother as they are born, but some prefer to take them away, keep them warm, and return the whole litter together after all the puppies have arrived.

After whelping, bitches normally lick their puppies and turn them over, thus stimulating respiration and circulation. A few, especially of the large breeds, are clumsy and may, unintentionally, injure their offspring. Most bitches have ten teats arranged in two rows with five on each side. However some toy breeds have only eight and occasionally large breeds may have twelve. Healthy puppies soon begin to suckle and bitches usually encourage them to do so. Breeders should not allow strangers to approach a newly-born litter unless they are certain that the mother is not likely to be disturbed by their presence. Many bitches are nervous and are upset if there is any interference with their puppies and may harm or desert them. Very rarely a bitch will neglect one or all of her puppies without any obvious cause, but if these are kept warm, held on a teat while suckling, and returned later the bitch may be persuaded to take to them. If a bitch fails, for any reason, to rear her first litter she should be allowed to breed again, as often the second and subsequent litters will be mothered satisfactorily.

Litter size is frequently related to breed size, as bitches of the larger breeds nearly always produce bigger litters of from eight to ten than do bitches of the smaller breeds who have smaller litters of from two to five. Bitches up to 3 years of age tend to produce larger litters than do older animals. Some bitches will successfully rear large litters but six or seven puppies are probably as many as most bitches can manage satisfactorily. Foster mothers can sometimes be found to take the excess puppies, but these must be very carefully chosen to avoid the introduction of disease.

Occasionally an unmated bitch may show a false pregnancy. At an appropriate time after an oestral period abdominal enlargement occurs and at about the period of whelping milk may be secreted. These signs gradually disappear and the bitch again becomes normal.

Artificial insemination

Artificial insemination is occasionally used in dog breeding. It can be employed when natural matings are difficult due to some physiological or psychological reason or when the dog and bitch are kept at considerable distances apart, e.g. when a mating is wanted between a dog kept in the British Isles and a bitch overseas.

Semen can be collected from a dog by the use of an artificial vagina of a size suitable for the breed. The dog is stood on a table and it is not necessary to allow it to mount a bitch, although it may be introduced to a bitch in oestrus first to arouse sexual excitement. The artificial vagina is similar in design to that used for bulls, but in addition to the water in the space between the outer cover and the inner lining a cylindrical bladder filled with air is fitted The pressure in this space can be raised by means of a hand pump to simulate the tieing effect obtained at a normal mating. Collection takes from 15 to 30 minutes. The volume of the ejaculate averages about 7 ml but can vary from about 2 ml in the case of a small dog, such as a Pekingese, to about 20 ml or even more for a Great Dane. Some operators prefer to use a manual method in which the dog is stimulated by digital manipulation of the penis, which is gripped behind the distended bulbs so that erection of the penis continues and ejaculation starts. The survival time for undiluted semen is about 21 hours, but semen life can be extended up to 6 days by dilution.

For the insemination of a bitch there must be some method of holding the vagina open while a pipette attached to a glass syringe is inserted. A test tube of a suitable size with a small hole blown in the end to allow the passage of the pipette is satisfactory. About 4 ml of diluted semen are injected into the uterus or deposited in the region of the cervix. In the case of the latter method the operator places a finger in the posterior vagina of the bitch so as to stimulate vaginal contractions which aid the passage of the semen along the bitch's genital tract.

In Great Britain the Kennel Club sees no virtue in encouraging the widespread use of artificial insemination in dogs, and does not accept for registration progeny which result from artificial insemination, except in certain cases, each of which is considered on its merits.

Health

A moist nose is generally considered an important sign of good health, but noses can become dry in normal dogs if they lie in the sun or in front of a fire. The hair, whether short, rough, or long, should have a glossy appearance without any bare patches. Healthy dogs normally scratch themselves occasionally, but persistent scratching is indicative of disease. The gums and mucous membranes of the eyes should not show a yellow discoloration and should not be excessively pale. The teeth should be strong and white, although older dogs often have some tartar deposits particularly near the gums, and the breath should not have an offensive smell. The tongue should be unfurred and of a pale pink hue. In most breeds the forelegs should be straight without any abnormal joint enlargements, but in a few breeds they are short and twisted or bowed. The claws should be slightly curved and well worn. The action when moving varies with the breed, but there should be no sign of lameness. A healthy dog will eat its food rapidly, the quantity depending on its size and activity. The drinking of excessive quantities of water can be indicative of disease. The mentality varies both with the breed and the individual. Some are sedate, some boisterous, while others tend to be nervous. Excessive nervousness is a fault as it may lead to gun shyness in breeds used for shooting and to sudden impulsive attacks in breeds kept as pets.

The normal temperature is between 37·8 and 38·9°C (100 and

102°F) being slightly higher in puppies with readings up to 39·2°C (102·6°F). It is also raised in animals of all ages after exercise. The clinical thermometer should be greased or soaped before being inserted into the rectum. It is usual to hold the base of the tail when inserting the thermometer, and it may be necessary to raise the hind legs slightly in dogs which resent the operation. The pulse rate is between 70 and 100 beats per minute, being faster in small and young animals. Thus the rate may be as rapid as 100 beats per minute in a healthy dog of a toy breed and only 70 in a normal Great Dane. The fact that the pulse rate of a healthy dog may rapidly increase under the influence of exercise, pleasure or fear must not be overlooked. The pulse is best felt in the artery which runs near the centre of the hind leg just below the body junction. If this artery cannot be detected the number of times per minute that the heart is beating can be determined by placing a hand on the chest wall behind a dog's left elbow.

Exercise

Exercise is essential for the maintenance of health as insufficient exercise tends to impair both body constitution and breeding efficiency. The amount required varies considerably according to the breed and age of the animal. In general dogs of the larger breeds need more exercise than do those of the small breeds, and the amount of exercise is also regulated to some extent by the purpose for which they are kept. For example, greyhounds, working collies, and sporting dogs need a considerable amount of exercise to develop the physical powers required for their work. Stud dogs and breeding bitches need ample exercise if they are to produce healthy puppies, although for the last few weeks of pregnancy a bitch's exercise should not be too strenuous. Growing dogs require more exercise than mature animals but young puppies must not be overstrained. Soon after weaning voluntary play is sufficient, as, when temporarily exhausted, a puppy will rest. When about 3 months old a puppy should be taken for short walks two or three times a day, the distance being gradually increased according to its development. House dogs should be let out into a garden, or taken out in a road, as early as possible in the morning. Later in the day a further walk is required, if possible with a period of free running.

Common diseases and their prevention

Canine distemper is a specific, highly contagious disease of dogs caused by a virus. Young dogs are more susceptible, but all ages can be infected. The clinical signs are fever, which causes the dog to lie in an apathetic state, refuse food, and become dehydrated and weak. The respiratory tract is usually affected next, with discharges from the eyes and nose and a frequent cough. Vomiting and a slimy diarrhoea follow. A number of dogs proceed to show nervous signs such as convulsions or a tremor of the muscles of the face and limbs. In one form of the disease, commonly known as hard pad, a hardening and thickening of the pads of the feet develop between 1 and 8 weeks after the start of the illness. Recovery from distemper usually confers a lifelong immunity. A vaccine has been prepared which also gives a good immunity, and it is advisable for all puppies to be immunised in this way at 9 weeks of age with a second dose when 13 weeks old. Dogs benefit from a reinforcement dose of vaccine every 12–15 months.

Infectious canine hepatitis is caused by a virus to which dogs of all ages are susceptible but the condition is mainly seen in young dogs from 3–8 months of age. The typical signs are depression, a discharge from the eyes and nose, a red inflammation of the mouth, and swollen tonsils. About a third of the animals affected develop a cloudiness of the cornea from 1 to 3 weeks after the clinical signs have ceased, which usually clears in about a week. New-born puppies, which may have been infected while still in the uterus, may die suddenly with practically no clinical signs. Vaccines are available for the prevention of this disease, and there are several combined vaccines which give protection against distemper and hepatitis. If new-born puppies are affected all the animals in the kennel should be immunised.

Leptospirosis is an acute and often fatal disease caused by microorganisms of the genus *Leptospira*. It is also known as leptospiral jaundice and Weil's disease. Initially an affected dog has a temperature of from 40 to 40·6°C (104–105°F) and is off its food, but then the temperature usually falls and signs of jaundice are shown by a yellowing of the membranes of the eye, the mouth, and even the skin. Pressure over the site of the kidneys may elicit pain, and the animal may have difficulty in urinating and may develop an ammoniacal smell. Rapid wasting and death may follow. The disease is mainly contracted by the consumption of food or liquids contaminated by infected rats, but the organism is easily transmitted from dog to dog by the urine. Thus the control of rats, the prevention of contact between dogs and contaminated urine, and the isolation of affected dogs are of great importance. Vaccination is used as a preventive measure, but needs to be repeated annually. It must be appreciated that human beings can be infected and so rubber gloves should be worn when handling dogs or material which might be infective. Mops or cloths for cleaning up infected urine must be thoroughly disinfected or burnt.

Roundworm infestation in the alimentary canal can retard the development of puppies, especially in the first 3 months of life. After this age puppies become more resistant to the ill effects of a worm burden. The clinical picture is usually that of an emaciated puppy with a distended abdomen, possibly also showing respiratory distress. In some cases the diagnosis is obvious because adult worms are passed in the faeces or vomit. The most common worm is a large roundworm, *Toxocara canis*, with adults varying in length from 40 to 175 mm (1·6–6·9 in). When present in large numbers they may block the alimentary canal and hinder food absorption. The respiratory signs are produced by larvae which hatch from eggs in the intestine and migrate to the lungs. They are coughed up and swallowed, so returning to the alimentary canal to develop into adult worms.

Puppies nearly always acquire the infestation from their mother by ingesting eggs passed in the bitch's faeces which possibly adhere to its teats. Pre-natal infestation is also possible because larvae can develop in the uterus of a pregnant bitch and so enter the embryos. For these reasons bitches should be dosed for worms in the early stages of pregnancy, and puppies should be treated as a routine measure at about 4 and again at about 8 weeks of age to prevent a build up of sufficient worms to cause signs of disease.

The public health aspects of infestations with these worms are important because human beings, particularly children, can become affected. The eggs passed in the faeces of infested puppies are sticky and adhere to their coats and a child handling the puppy and placing his hands in his mouth may become infested himself. The eggs hatch in the child's body and gain access to various tissues. The symptoms produced vary according to the organs affected, but abscesses can occur in the liver and in other internal organs, and growths may develop in the eye, brain, or nervous tissue. Cases of blindness and death have occasionally been reported. To prevent human infestations children should not be allowed to play with a suckling bitch and its puppies or puppies which have not been regularly wormed.

Tapeworms, of which there are many species, are parasitic in the intestines of dogs. Except for the occasional passage of tapeworm sections there are not usually any clinical signs, although the fact that the tapeworm feeds on the food fed to the dog may lead to a lack of body condition. The tapeworm eggs are passed in the faeces and must be ingested by an intermediate host for the next stage of development. The intermediate hosts vary according to the type of tapeworm, but include fleas, rodents such as rabbits, rats and mice, and sheep and other ruminants. One tapeworm, *Echinococcus*, constitutes a public health problem because the intermediate stage, known as a hydatid, forms a cyst, these cysts being found in man, particularly in the liver and lung, as well as in other domesticated animals. The dog is infected by ingesting the intermediate host containing the infective larval stage, and the keeping of dogs free from fleas, preventing them from eating rodents, and the avoidance of food from animals carrying the infective larval stages are the best methods of preventing tapeworm infestation.

Fleas are wingless, laterally compressed insects which feed on blood. Although fleas have certain hosts which they prefer they are rarely host-specific and dog fleas will attack man. Fleas produce injury by wounding the skin and secreting saliva which causes marked irritation. The skin wounds allow bacteria to enter and encourage the development of various forms of dermatitis. The adult fleas can be seen, particularly on the back of an infested animal. Fortunately they are not difficult to control and various dusts, sprays, or solutions in which a dog can be bathed, can be used. As the larvae develop away from the body, the dog's kennel and bed, or other living accommodation, must be treated at the same time as the dog.

Lice are obligatory parasites and are host-specific. The adults can only live away from the body for a few hours, and the eggs are glued to the hair of the host animal. The young lice which hatch are miniature replicas of the adults, becoming adult in about 2 months. Lice infestation can be diagnosed easily if the dog is examined carefully, but as they are relatively small and stay close to the skin they can be missed, particularly in long-haired dogs. It must be appreciated that dog lice can be readily transferred from dog to dog by combs, brushes or similar articles of equipment.

Quarantine regulations

To prevent the introduction of diseases which do not occur in Great Britain from abroad, particularly rabies, imported dogs must be kept in isolation in an approved quarantine kennel for a period of 6 calendar months from the date of landing of the animal, at the owner's expense. The standard required in a quarantine kennel is that no dog must be able to come in contact with any other dog on the premises or escape.

Handling

Training

Puppies which are to live in a house must be house-trained at an early age. In very young puppies defaecation and urination are automatic reflexes and occur after each feed. At first puppies perform these functions in the nest and the bitch licks up all that is passed. This ceases when the puppies receive supplementary food, and the puppies can be put outside in a run immediately after feeding or placed on a tray of sawdust before they have time to soil the kennel floor. When they have relieved themselves they should be allowed back into the kennel or on to the floor. Most puppies are fairly quick to learn and can be house-trained without undue difficulty.

In training a puppy to obey certain commands care must be taken not to discourage it by requiring it to do anything beyond its powers. Observations have shown that early successes lead to greater efforts while failures lead to discouragement. The person doing the training must be patient and kind, but firm. The words of command must be simple, such as 'sit' or 'lie'. A puppy should never be given a meal before a training session and should not be played with during training. When the puppy has performed an exercise correctly it should be spoken to kindly, with words such as 'good dog' and given a tit-bit. Kindness means a great deal to a puppy.

In the house the training of a puppy should commence as soon as it is acquired, as bad habits are difficult to break once established. First, it must be taught to answer to its name and come when called. This can conveniently be started at feeding times when the puppy will come for food, and a fuss should be made of it every time it answers to its name. Next, the house-training, probably started earlier by the breeder, must be continued as a dirty puppy or dog in a house is an abomination. Some puppies need constant supervision and firm methods. A puppy must also be taught not to climb over the furniture, and, possibly, not to climb the stairs or enter certain rooms. Then the puppy must be taught to go to the box or place in the house where it is to sleep on command. The puppy must be shown this place, at the same time being given a command such as 'box'. This exercise must be repeated until the puppy understands. Each time the puppy is in position the handler should move away, facing the puppy the whole time, repeating the word 'box' whenever the puppy moves. Patience is required in teaching this lesson.

When about 9 weeks old the puppy should be taken outside and introduced to a collar and lead and taught to walk at heel. The lead should be held short and the puppy should be made to walk on the left of the handler. At first the puppy will drag behind or pull in front but should soon learn to walk quietly. It is an advantage to train the puppy to stop and stand when required, and this lesson is usually quickly learned. The handler should suddenly stop, pull on the lead and give the command 'stop'. Next comes the sitting position. The handler should hold the puppy's neck with one hand, press on its loins with the other and give the command 'sit'. When this has been learned the handler should slowly walk backwards away from the puppy repeating the word 'sit' whenever the puppy moves. This can be a difficult lesson to learn and a great deal of patience is required. Training for the 'lie down' position comes next. When the puppy is in the sitting position the handler places a hand on its neck and presses down, at the same time pulling the forelegs forward with the other hand and giving the command 'lie'. In most cases this lesson is soon understood. There are a variety of other lessons a dog can be taught, but these few simple examples provide a basic minimum. Dog-training clubs have been started in various parts of the country to which owners needing advice can go with their puppies. Classes on more advanced training methods are usually restricted to animals over 6 months of age.

Restraining strange dogs

Dogs vary greatly in temperament, some being friendly and others vicious. Dogs, particularly nervous individuals, should never be suddenly disturbed, but should be approached quietly, as friendly animals may snap through fear. It is advisable to talk to strange dogs before starting to handle them and then to present the back of a hand first so that the dog does not think that it is being hit. Another advantage is that if a dog does snap less damage is done to the back of the hand than would be the case if the dog grasped the flat of the hand in its mouth and bit into it. Indications that a dog is likely to bite are usually given by a stiffening and lifting of the upper lip and a growl, and a watch should be kept for such signs.

Most dogs can be controlled by holding them by the scruff of the neck, but dogs of the Pekingese and similar breeds must be handled carefully in this way as there is a risk that an eyeball may be displaced if the dog struggles violently. Dogs should not be lifted by grasping the back of the neck only; the other hand should be placed under the abdomen to support the animal's weight.

A dog-catcher can be used to catch vicious individuals. This consists of a pole about 1·22 m (4 ft) in length with a metal attachment at the end shaped to go over the neck of a dog. A piece of thin strong rope is fixed to a hole at the tip of this metal extension and is passed back through a hole at the other end of the metal section and led on to the end of the pole handle. The noose so formed is placed over the neck of the dog and drawn tight. Care must be taken when using a dog-catcher not to strangle the dog.

For a detailed examination the dog should be placed on a table, being controlled by an assistant who holds the scruff of the neck and talks to, and soothes, the animal. If necessary the dog can be laid on its side by grasping the two forelegs with one hand and the two hind legs with the other and gently pulling it over. The dog can be held on its side if its back is kept towards the assistant's chest because one arm can be pressed against the neck to keep the head down and the other can be held against the hind quarters to hold the body down. The mouth can generally be opened by pressing the gums at the side against the teeth. If the mouth has to be held open a thumb can be inserted into the mouth and pressed firmly against the roof. While the animal is held in this way it will not bite, but the thumb must be withdrawn rapidly when the grip is released. If this is not satisfactory pieces of tape can be inserted into the mouth and looped over the upper and lower jaws. The free ends can be grasped by an assistant and the mouth drawn and help open.

If any interference likely to be upsetting or painful is to be made an assistant can hold the scruff of the dog's neck with one hand and place the other hand round the mouth to hold the jaws together, but it is safest to apply a tape muzzle to the dog's jaws. This can be done by using a length of bandage. A half-hitch is made in the middle and the loop so produced is slipped over the dog's nose and pulled fairly tight. The ends are then brought under the jaw, secured with a half-hitch, and finally tied behind the animal's head. These muzzles are

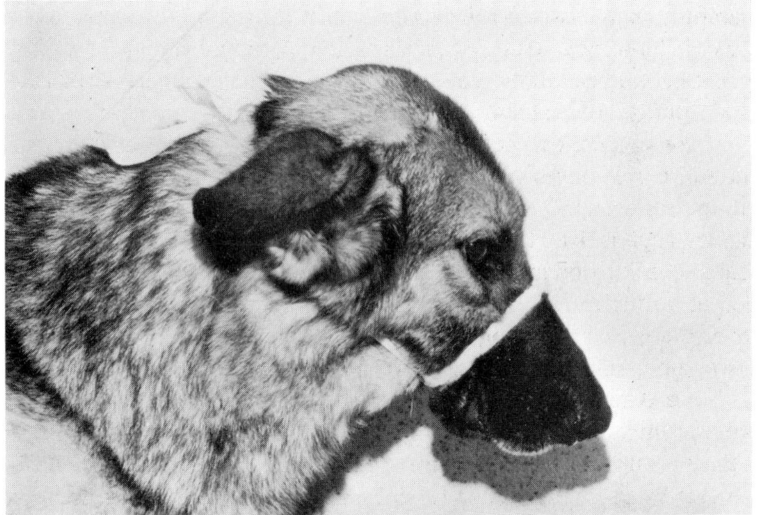

Fig. 9.9. Tape muzzle round the jaws of a dog.

relatively easy to apply to dogs with long noses, but are not so easily fitted in short-nosed breeds such as Bulldogs. The tapes should not be tied too tightly in the case of individuals with short noses, as breathing may be inhibited. As an alternative a leather or wire muzzle can be applied if one is available of the correct size.

Small dogs can be carried by placing one hand under the body with the fingers extending between the front legs, and grasping the scruff of the neck with the other hand to prevent the dog biting. In the case of large dogs both hands will probably be needed to bear the animal's weight, and so a muzzle should be applied if biting is anticipated.

Grooming, stripping and washing

Grooming is important for the preservation of cleanliness by removing dust, dirt, and scurf which can protect skin parasites. It also removes dead hairs. These would eventually drop out, but it is advantageous to remove them at one operation and so avoid the continual annoyance of dead hairs fouling carpets and other objects.

Short-coated dogs normally need very little grooming, but it may be necessary occasionally to remove dried mud and loose hairs. A good rubbing with the hand imparts a glossy appearance to the coat. Wire-coated dogs benefit from a brief daily grooming using a stiff brush with bristles just long enough to reach through the coat. The final brushing is often backwards against the lie of the coat so as to lift the hair and give the coat a rough appearance. Long- or dense-coated dogs need daily grooming to prevent the entanglement of the hairs which so readily occurs. The coat is usually combed first with a steel comb to remove any tangles and then brushed. Particular attention should be paid to the area under the tail as it is possible for a firmly packed mass of faeces and hair to occlude the anal orifice and mechanically prevent the passage of faeces.

Wire-coated dogs are generally stripped or plucked each spring to remove the dead hairs being shed at this time so that the dogs keep cooler during the summer. Stripping is carried out by using a stripping knife. This consists of a cutting blade covered with a finely serated or comb-like edge. The handle of the knife is held between the four fingers and the palm of the right hand, whilst the dog's hair is gripped between the serrated knife edge and the thumb. In plucking the fingers and thumbs are smeared with resin and the dead hairs are actually plucked out. Stripping is a quicker method of improving a dog's appearance, but it is not as efficient as plucking. Wire-coated dogs are trimmed for show purposes according to the standard required for the breed. This does not necessarily mean going over the whole body, but requires the removal of hair from certain parts and the fluffing out of the hairs in other areas to make them look long.

It is sometimes necessary to wash dogs for hygienic reasons, but excessive washing is undesirable, particularly in the case of wire-haired dogs as the hair is softened, although this softening only lasts for a short time. After washing a dog with soap and water he should be rinsed and then allowed to shake himself before being thoroughly dried with a rough towel. Proprietary dog shampoos are excellent as there is no waste and no difficulty in their use. The preparation selected must not contain carbolic or coal-tar derivatives as these can be absorbed through a dog's skin and may cause poisoning.

Administration of medicines

There are a variety of methods which can be used for the administration of medicaments to dogs, and the one selected should be that most likely to be least objectionable to the individual animal. Some medicines are tasty or tasteless and can conveniently be included in a dog's food and the animal will take the prescribed dose without any difficulty. Medicaments, particularly powders, with a slight taste may be hidden in a sardine or a small piece of meat.

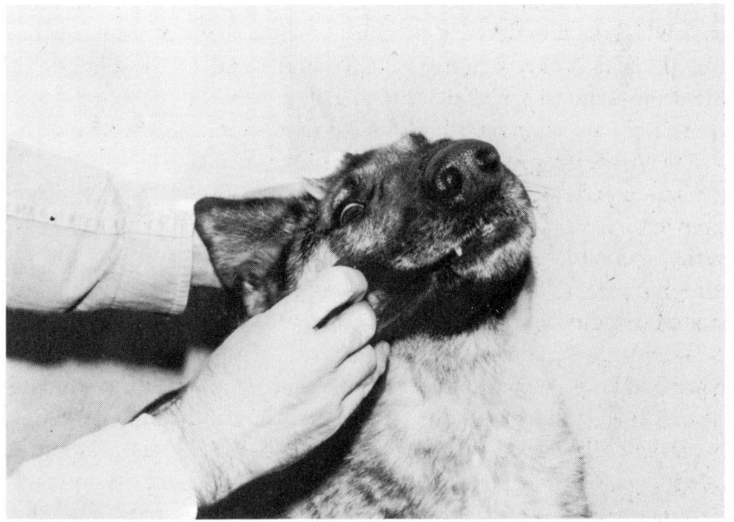

Fig. 9.10. Dog's cheek pulled out to facilitate the administration of a medicine.

Liquid medicines can be given to most dogs fairly easily, although occasionally an animal is encountered which is obstreperous. It is not necessary to hold a dog's mouth open for the introduction of liquids; in fact it is best not to do this as there is a greater risk that some of the liquid will enter the trachea. An assistant holds the dog's head with the nose raised with one hand and with the other pulls out the cheek at one side of the mouth to make a pocket. The fluid can be gently poured into this so that it trickles to the back of the mouth. It is safest to use a spoon or metal container to hold the fluid, as a glass bottle can be broken should the dog make a sudden movement. If the dog does not swallow, its throat can be rubbed gently until it does.

Tablets can usually be given by one person who raises the dog's head, opens its mouth by pushing the gums against the molar teeth and then pushes the gums into the mouth so that if the dog closes its mouth it bites its gums. The mouth can be kept open by pressing on the incisors of the lower jaw with the third, fourth and fifth fingers of the other hand while the tablet is held between the thumb and forefinger. The tablet can then be deposited on the back of the tongue and the mouth closed and held until the dog has swallowed.

Subcutaneous injections are best made in the loose skin of the flank. The common sites for intramuscular injections are the muscles of the thigh or the muscles behind the spine of the shoulder blade. Intravenous injections can be made into the vein on the front of the foreleg or the one just above the hock in the hind leg.

General

Vices

Worrying livestock

A serious vice is sheep worrying. In addition to the sheep which may be attacked and killed or injured, the shock can cause serious harm to in-lamb ewes and new-born lambs, which become separated from their mothers. Other dogs kill poultry and may injure cattle. Most puppies tend to run after sheep and poultry and should be trained, at an early age, not to do so, and dogs being exercised in the country should be kept under strict control.

Dogs which are allowed out without supervision, working sometimes in pairs, can develop a craving for the habit of sheep worrying which can develop into a vice which it is almost impossible to break. Such dogs must never be let out unaccompanied, and must always be kept on a lead when near sheep. If at all possible a home should be found for them in a town away from farm animals and poultry.

A farmer may lawfully shoot a dog that is caught in the act of worrying his livestock if there is no obvious person in charge of it, or if there is no other way of stopping the attack. A court order for the destruction of a dog proved to have injured cattle or poultry, or to have chased sheep, may be issued.

Biting human beings

This also can be a serious vice as it may entail an owner in considerable expense in the payment of compensation to persons bitten. Although dogs do not normally bite people nervous individuals may snap and bite through fear. Such an animal may be safe with people known to him but will run from strangers and bite anyone who tries to handle him. Others become very cunning and give little warning of their intention to bite, with the result that unsuspecting persons fondling the dog are suddenly bitten. The characteristic is usually first noticed soon after weaning and develops with age. Experienced handlers can keep control of such animals and prevent biting, but most pet owners find it necessary to have them destroyed following a serious attack. There is evidence that the excessive shyness which leads to this vice can be inherited and it is not advisable to breed from animals with this weakness.

Fighting other dogs

Some dogs have a pugilistic nature and will attack other dogs with only a scant reason, others will attack dogs which have something of value, such as a bone or piece of meat, while some will suppose it necessary to fight to defend their owner from an imagined attack from some other dog. Dogs which do not normally fight may attack another male when in the presence of a female. Dogs known to be keen fighters should not be allowed their liberty where other dogs are to be found, but should be kept on a lead and prevented from approaching strange dogs too closely.

Caution should be observed in trying to stop a dog fight. A person seizing dogs which are fighting and endeavouring to pull them apart may easily be bitten himself. A bucket of water thrown over the fighters' heads may cause then to desist long enough for their owners to grasp their collars and drag them away. If pepper is available some sprinkled over the noses of the fighters will cause them to start sneezing and stop fighting. Care must be exercised to see that the pepper does not get into the dogs' eyes.

Depraved appetite

Some dogs eat filth of various kinds, particularly horse faeces. It is suggested that mineral deficiencies may be a cause, but as correctly-fed healthy animals often develop this vice, some at an early age, it is not always the reason. Many dogs also roll in faeces and refuse and such a habit appears to be natural in some animals. Prevention is the best cure by keeping dogs away from refuse as far as possible, and ample exercise may reduce the tendency. Some handlers try dusting horse faeces with a powder with a disagreeable taste, such as aloes, and the experience of the unpleasant flavour may teach some puppies to stop the habit.

Excessive barking

A dog which continually barks for no apparent reason can be a very real nuisance to his owners and the occupiers of adjacent properties. Some dogs tied up outside bark because of boredom or discomfort, and these conditions should be remedied if they exist. The habit can develop because of over enthusiastic training when young by the owner in an endeavour to make the animal a good guard. Once developed the vice can be a difficult one to cure, and the dog must be ordered to stop when barking for no known cause.

Minor operations and manipulations

Removal of dew-claws

The dew-claws are the rudimentary fifth digits situated on the inner surface of the forelegs, and, possibly, also on the hind. Dew-claws now serve no useful function and can be a nuisance. In pet dogs they can become caught in furniture and clothing and broken, while working dogs, particularly racing greyhounds, can injure them during the course of their duties. Thus, their removal is recommended, preferably during the first week after birth. An anaesthetic is not required if the amputation is performed before the puppies' eyes are open. In many cases the dew-claw is held only by a fold of skin, but in others there is a bony attachment. In either case the claw can be removed with a pair of curved scissors while the puppies are a few days old, usually on either the second or third day of life, but a dissection operation under a local anaesthetic is required for older animals.

Docking

Certain breeds belonging to the terrier and spaniel groups normally have their tails docked to improve their appearance, and this shortening of the tails is allowed under the Kennel Club Rules. The best time for docking is when the puppy is about 4 days of age when pain and the risk of haemorrhage are reduced to a minimum. An anaesthetic is not required if the operation is carried out before the puppies' eyes are open. As much skin as possible should first be pulled back towards the root of the tail, though in some puppies it may be difficult to pull any skin back. The amputation can be performed with a pair of scissors or with a sharp knife with the tail laid on a pad of cotton wool on a hard flat surface. The cut end should be treated with a mild antiseptic. Alternatively a rubber band can be wound round the tail at the required joint between two vertebrae and knotted. This cuts off the blood supply to the end of the tail so that the end shrivels and drops off after 2 or 3 days. Experience is needed to know how much of a puppy's tail to remove as the required lengths vary with the breed and puppies' tails differ in length when they are born. Thus it is not safe to advise cutting off a set length, such as two-thirds, in a particular breed.

Castration

Castration is not commonly practised in male dogs kept as pets. If performed the operation is generally delayed until the dog is sexually mature, otherwise its character and physical development may be adversely affected. There is considerable variation in the response of male dogs to castration.

Spaying

The spaying of bitches is more frequent because it avoids the twice-yearly inconvenience of controlling the bitch while in oestrus. Some dog owners consider that if bitches are spayed before they reach puberty they lack character and, for that reason, recommend that a bitch should be allowed to have at least one oestral period before the operation is performed.

Ear cleaning

Some breeds of dogs, such as Wire-haired Terriers and Poodles, have extensive hair growth in the ear canal which interferes with the normal expulsion of wax and shed cells. These, when mixed with hair, form a solid plug which interferes with hearing and retains moisture in the base of the canal. If the latter occurs the canal wall may become thickened and infected. This can be avoided by regular ear cleaning with small pieces of cotton wool dipped in surgical spirit twisted round a thin stick or held in a small pair of forceps. Excessive probing should be avoided as this can cause pain.

Nail cutting

A dog's nails should be cut periodically so that they just clear the floor. If allowed to grow long they may cause the foot to spread and may even grow round in a circle and back into the dog's skin. This most frequently happens to the dew-claws which are not in wear, if these have not been removed. The frequency of cutting depends on the type and amount of exercise which the dog is given, but usually once every 2 months is adequate. The cut should be made just outside the pink blood line seen in white nails, while in dark pigmented nails the site for cutting must be judged by noting the length of curvature. The hooklike projection is removed.

Species control

The Kennel Club is the body which controls the showing of dogs and every show in the country now comes under Kennel Club supervision. The club produces codes of rules for the organisation of dog shows and field trials. It also has the power to suspend any person found guilty of malpractices for a specified period. Breeders can register a name prefix or affix which can only be used by the holder and so serves to identify all the dogs they breed. Before any dog can be exhibited it must be registered under a distinctive name with the Kennel Club and if a dog changes hands its registration must be transferred to the new owner by the Club.

Licences

Under the Dog Licences Act 1867, it is necessary for an owner to have a licence costing 37½p for each dog over 6 months of age, except that guide dogs belonging to blind persons and working sheep and cattle dogs do not require licences providing certain formalities are observed. A licence is valid for 12 months from the first day of the month of issue. It is not transferable if a dog changes ownership, but if a dog dies and its owner buys another dog the existing licence is valid for the new dog.

CHAPTER 10
CATS

Introduction

The domestication of cats occurred early in civilisation. In all probability cats were first attracted by the warmth of the camp fires made by the early settlers and by the ease with which food could be obtained around human dwellings. The value of the cats in controlling rodent scavengers would be appreciated by the settlers who would encourage them to stay. Records compiled 5,000 years ago show that cats were highly valued by the Egyptians living at that time.

The majority of people living today either like or dislike cats, there being relatively few who are indifferent. Cats are valued mainly as household pets. Independent by nature, cats maintain their individualities and each animal has its own characteristics and habits but they still make good companions, being willing to live with their owners and show appreciation of their company, while yet remaining remote. Their grace of movement, shown by a supple strength, is also widely admired.

Cats are useful as rodent killers for they are, by nature, hunters, stalking their prey by stealth. Unfortunately, individuals allowed to wander rapidly develop wild instincts and in certain areas there are numbers of stray cats living around warehouses and docks where they sleep during the day and hunt vermin at night. Although they render a useful service by killing large numbers of rats and mice they breed rapidly and their numbers have to be controlled. Unless this is done the cats lose condition because there is insufficient food to support them and diseases can spread rapidly among the weakened animals.

The showing of cats is becoming an increasingly popular hobby. At cat shows the exhibits are confined in cages and brought out individually for inspection by a judge who handles and examines each animal. Considerable skill in breeding is required to produce good show specimens and care has to be taken to ensure that an exhibit is in perfect coat at the time of a show. Pedigree cats bred in Britain are valued in many countries of the world and individual animals are eagerly sought after by overseas breeders. Although there can be no comparison with the number of dogs exported there is a steady demand for British-bred cats in all parts of the world.

Definitions of common terms

Tom cat. Adult uncastrated male.
Stud cat. Adult male kept for breeding.
Queen. Adult female.
Neuter or doctored. Either a castrated male or a spayed female.
Kitten. Young animal, usually up to 6 months of age, but for shows being held under Governing Council of the Cat Fancy rules up to 9 months of age.

Principal breeds

Short-haired

British short-haired

The head should be wide between the comparatively small ears with a fairly short nose and well-developed cheeks. The body should be deep with a broad chest, strong legs and a shortish tail, to give a compact body form. The coat should be short, fine and close. A variety of colours are recognised including blacks, blues, whites with blue or orange eyes, tortoishells, with equally balanced black and red patches, and tortoiseshell and whites with black and red patches on white. There are also cats with markings in a tabby pattern. The stripes should be as dense and as distinct from the body colour as possible. The tabby colours recognised are brown, with a brown background and black markings; silver, with a clear silver ground colour and black markings; and red, with a rich red background and dark red markings.

Manx

The main feature is that they are tailless, with a hollow at the base of the spine where the tail would normally begin. The head is round and fairly large, the body is short and the hind quarters are high. The gait is described as hopping or rabbity. All the colours are recognised.

Fig. 10.1. British short-haired cat—red tabby with white markings.

Fig. 10.2. Seal point Siamese cat.

Manx cats are difficult to breed as in a litter it is possible to have kittens with long tails and some with short tails as well as true Manx.

Siamese

This is the most popular of the foreign type breeds. The head is long and wedge shaped with large ears which are wide at the base. The eyes are a deep blue and are almond in shape, slanting towards the nose. The body is long and the legs are slim, the hind legs being slightly longer than the front ones. The tail is long and tapering and the coat is short and close lying. The body colour is ivory, with the mask, ears, legs, and tail of a contrasting colour. The most common colour for these points is a seal-brown, known as seal, but blue, chocolate, lilac, and red points are among the other colours recognised. All Siamese kittens are born pure white; their points darken as they grow.

Foreign short-haired

The popular colours are white, lilac, and a rich chestnut brown known as Havana. They have a body conformation resembling a Siamese, but are whole coloured.

Burmese

The head should be of a short wedge shape with relatively wide ears and yellow, almond-shaped eyes. The body shape and coat type are similar to the Siamese. Brown is the most common colour, a rich dark seal-brown shading to a slightly lighter colour on the chest and belly being required. Blues and certain other colours are also recognised.

Abyssinian

Slender in build with a long pointed head, comparatively large ears, large green or yellow eyes, and a fairly long, tapering tail. The two popular colours are a ruddy brown with black ticking and a red with dark ticking. There should be no markings, either dark bars or white patches, although a white chin is frequently present. They are not prolific breeders.

Russian blue

The head, neck, body, legs, and tail are all longish. The ears are large and pointed with little fur covering. The eyes are green and the body colour is self blue free from tabby markings.

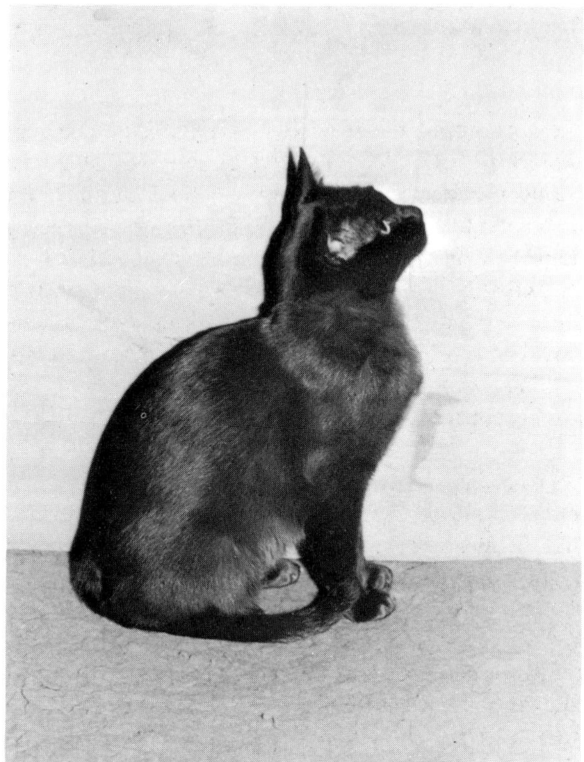

Fig. 10.3. Brown Burmese cat.

Rex

The hair is considerably shorter than in the other short-haired breeds because each individual hair is waved and no guard hairs are visible. This hair type can be introduced into any colour. There are two separate mutations which cause this modification in coat form and these are recognised as separate breeds termed the Cornish Rex and the Devon Rex.

Long-haired

Long-haired, popularly known as Persians

The head should have a wide skull, a short nose, a broad muzzle, and small ears. The body should be cobby and massive and the legs and tail should be short and thick. The coat should be long and flowing, fine in texture and standing out from the body, with a large ruff continuing in a deep frill between the front legs. A variety of colours are recognised, including blacks, blues, creams, whites with either blue or orange eyes, tabbys in browns, silvers or reds, red selfs, tortoiseshells, tortoiseshell and whites, smokes, which are black shading to silver on the flanks, and chinchillas, which have a pure white undercoat with each hair tipped with black to give a silver appearance.

Fig. 10.4. Chinchilla long-haired cat.

Colourpoint

The body and coat resemble the long-haired, but the colour is that of one of the recognised Siamese colours.

Birman

In most respects similar to the colourpoint, but the toes on all four feet are white in colour.

General colour characteristics

Tortoiseshell-coloured cats of any breed are almost invariably females due to a sex-linked factor which is related to this coat colour. Occasionally a tortoiseshell male is born but these are usually infertile. A congenital form of deafness occurs in some strains of blue-eyed white cats.

Identification

Cats are not normally marked in any way for individual identification. When necessary young kittens can best be marked by applying gentian violet under or between the front paws, or cutting some hair from the tip or underside of the tail. An adult can be identified by fitting a special cat collar round the neck with a name disc attached. These collars should be made with an elastic inset so that the cat can pull itself loose if the collar catches on a branch or other projection. Pedigree cats can be registered by name with the Governing Council of the Cat Fancy.

Ageing

The temporary teeth start to erupt during the second and third weeks of life. The incisors erupt first, and, except for the third pre-molar, eruption is complete by the fourth week. The temporary teeth are smaller, sharper and less dense than the permanent ones, and the temporary incisor teeth are only just visible in young kittens. Eruption of the permanent teeth commences at between 15 and 17 weeks of age. The central incisors are the first to be changed, followed about a week later by the laterals and then by the corner incisors. The canine teeth erupt at about 4½ months. All the permanent teeth are through the gums by 6 months and are fully developed by 7 months, at which age the cat can be said to have a full mouth.

Housing

Most cats are kept in the house and a kitten, when first acquired, should be provided with a box or basket lined with a blanket or newspaper in which to sleep. Wicker baskets are still popular, but they do not exclude draughts as well as wood or P.V.C. beds and are harder to disinfect. Many cats prefer a cardboard box, which can be destroyed if necessary. Initially a hot-water bottle should be provided to supply some of the warmth previously obtained from its mother and litter mates, but this can be dispensed with as soon as the kitten becomes adjusted to its surroundings. When accustomed to their new accommodation kittens will often neglect the box or basket supplied as a bed and select another sleeping place. This will vary according to the environmental conditions, and may well be a windowsill on a sunny day or a warm spot near a boiler or radiator at night. However, it is desirable to provide a cat or kitten with its own bed, where it can feel secure. This should be placed in a position where it will not be disturbed.

All house cats should be house-trained, even if they are to be allowed out for toilet purposes, as there may be occasions on which they have to be confined for long periods. A flat tray, such as a baking tin, about 407 mm (16 in) long, 304 mm (12 in) wide and 76 mm (3 in) deep, filled with soil, ashes, or one of the proprietary preparations sold for the purpose, should be placed in a secluded corner. It should always be left in the same position. The kitten will probably have been taught to use such a tray by its mother, and so should only need to be placed on the tray and shown its position, but it may be necessary to move its front paws in a scratching motion so that it realises the purpose for which the tray has been provided.

Some cat owners like to allow their animals complete freedom and so fit a cat door to an outside house door to provide the cat with independent access without allowing the entry of unwanted human beings. A cat door is about 152 mm × 102 mm (6 in × 4 in) in size and swings from a hinge at the top so that it can be opened from either side. Such doors should be of a suitable weight so that they close slowly and do not trap cats' tails. Cat doors do have the disadvantage that stray cats can also gain entry to the house.

All cats need to scratch regularly in order to keep their claws in good condition by pulling off the outer shells of the claws as they become worn leaving a new sharp claw below. A scratching post should be provided for this purpose, this being particularly important in the case of house cats which otherwise will damage furniture, carpets or curtains. A log, complete with the bark, or a solid board upholstered with a piece of carpet can be used. Such a scratching post should be at least 0·61 m (2 ft) long, so that both front and back legs can be drawn along the wood at the same time. Some owners cut the claws of house cats to prevent furniture damage, but this may not be effective as the cats may be stimulated to increase the amount of scratching in an attempt to resharpen the blunted claws.

Breeding or boarding catteries

These are designed along two main lines, indoor and outdoor. An indoor cattery consists of a large room containing pens for individual cats. These pens should have a floor measurement of 0·92 m × 1·22 m (3 ft × 4 ft) and should be 0·76 m (2 ft 6 in) high. The advantages are that such a cattery is relatively cheap to establish and easy to heat. The disadvantage is that infectious diseases can be spread easily. An area should be set aside for exercising the cats or exercising space in an outdoor run should be provided.

Most specially built outdoor catteries consist of small individual sheds with runs attached. Most of the sheds are made of wood but should be well insulated to preserve as stable an environmental temperature as possible. They are usually 1·22 m wide by 1·22 m long by 1·83 m high (4 ft wide by 4 ft long by 6 ft high) at the front. There should be a window facing the run with a shelf about 0·3 m (1 ft) wide, just below it on which the cat can sit or lie. A door through which an attendant can enter should be fitted in the wall leading to the run and a small cat door should be inserted at the base of this door. A catch should be fitted to the cat door so that a cat can be kept confined to the shed, if required. Creosote should not be used as a wood preservative as it is toxic to cats. The floor should be raised about 0·3 m (1 ft) from the ground and covered with vinyl so

Fig. 10.5. Cat house showing separate doors into the run for the attendant and the cat.

Fig. 10.6. Cat house—interior.

that there is no risk of rising damp. Artificial heat may be required in winter in order to maintain an environmental temperature of between 13 and 24°C (55–75°F). Tubular electric heaters are commonly installed, but suspended infra-red lamps are satisfactory. Alternatively, electrically-heated cat beds can be provided. In boarding establishments a space of about 0·61 m (2 ft) is usually allowed between each shed to minimise the spread of infectious diseases.

The runs are the same width as the sheds, namely 1·22 m (4 ft), but the length varies, being normally about 1·83 m (6 ft). The runs should be floored with concrete and have a solid roof so that the cats can sit in the runs without becoming wet. A transparent material can, with advantage, be used for the roof to keep them light. The run walls are best made of wire mesh with 25·4 mm (1 in) holes nailed to posts. Doors should be fitted in the ends of the runs so that any person entering always has to go first into the run and through the run into the shed. It is an advantage to fit a log in the run which the cat can use as a scratching post and to provide a wooden box in which the cat can lie. In boarding catteries a safety passage about 1·22 m (4 ft) wide is normally provided, running down the front of the runs into which all the run doors open. This passage is roofed with wire mesh and fitted with a door at each end so that it forms a cage in which any cat escaping from a run is trapped. In large establishments additional rooms are an advantage which can be used as a kitchen, a store, and an office.

Kittening queens

Whether kept in a house or in a cattery a queen should be provided with a basket or box in which to give birth. Many breeders use a wooden box, painted inside and out so that it can be easily disinfected, about 0·46 m (18 in) square and about 0·41 m (16 in) high, fitted with a hinged lid which can be raised to enable the queen and the litter to be examined. This lid can be turned back if it is felt desirable to provide artificial heat from a heating lamp suspended above the box. At the front there is a solid sill at the base, about 102 mm (4 in) high, to prevent very young kittens from getting out. The remainder of the front may be covered with a curtain which can be drawn back at will.

Stud toms

Cat breeders almost always keep their stud toms in special sheds because mature male cats will usually spray their urine on to curtains and furniture in a house instead of using a tray, and, if allowed free, will wander in search of queens. The shed provided should be high enough to allow the owner to stand up when inside, should be fitted with shelves, at least one of which should be fixed below the window, and provided with an outside run. The roof and walls should be insulated and some form of heating provided for use in cold weather. Artificial lighting will be required. It is also advantageous to have a small shed and run attached in which queens for mating can be placed, so that the male and female can become acquainted without being able to fight. They can be run together when they show an interest in each other. Similar sheds can be provided for breeding queens if the owner does not wish to keep these in the house.

Feeding

Kittens

Kittens start to suckle soon after birth and need only their mother's milk during the early part of life. When between 3 and 4 weeks of age the kittens will require food to supplement their mother's milk, and any of the proprietary human baby milk foods can be given. Alternatively there are specially prepared cats' milk substitutes. If cows' milk is used it should be fortified with dried milk or glucose. Some kittens will lap from a spoon straight away, but others have to be taught by placing a drop of the liquid on their lips or gently dipping their noses in a saucer of milk to make them lick the milk from their own muzzle. After about a week they should all be lapping and the milk supplement can be given in a small saucer. At about 4 or 5 weeks of age baby food cereals can be added to the milk in small amounts. Solid food in the form of cooked fish, scrambled eggs, minced raw beef or minced cooked rabbit will soon be eaten, and weaning from the mother can be completed by about 6 weeks in the case of cats of the British short-hair types and 8 weeks for other breeds. Tough foods can be softened by cooking. Some breeders feed each kitten separately in individual saucers so that they can check that each individual is eating an adequate quantity of food.

It is wise to feed a variety of foods before and after weaning and to see that the kittens eat the different foodstuffs at this age because cats can become faddy and refuse to eat all but one type of food and may starve for a long period before consuming foods they dislike. Once established, such a habit can be difficult to cure, and so it should not be allowed to develop. Suitable house scraps can often be used to provide variety in kitten diets. Once weaned, kittens should be given four meals a day until they are about 4 months old and then the number should be gradually reduced to two.

Growing kittens require adequate supplies of protein for body tissue building and at least 35 per cent, and preferably 45 per cent, of the diet should consist of high quality protein foods. Kittens do well on a high fat ration and up to 20 per cent can be included in the diet. Although kittens do not appear to require carbohydrate they can utilise, and are normally given, about a third of their food in this form. Biscuits or boiled cereals such as rice are acceptable carbohydrate foods, and are usually fed mixed with meat or gravy. Milk can be gradually withdrawn from the ration.

When a kitten is taken from its mother and litter mates it must be appreciated that the move imposes a considerable strain. The stress is very likely to produce an intestinal upset causing diarrhoea. Every effort should be made to minimise the severity of this by continuing to give the same diet as the breeder fed and by under-feeding slightly, and certainly not overfeeding.

Adult cats

Cats are naturally carnivorous but it is usual to provide domestic cats with a diet containing a proportion, usually between a fifth and a quarter, of foods other than meat or fish. The inclusion rate should not exceed a third or the digestive system may be upset. The total quantity of food is generally calculated on the basis of 14·2 g (0·5 oz) of dry matter per 0·454 kg (1 lb) body weight. Under this system the water content of the foods is excluded, which can make a marked difference in the case of liquids such as milk. This allowance scale is satisfactory for adult pet animals. All classes of cat appreciate variety and so different kinds of food should be incorporated in the diet. Cats investigate a meal by nibbling round the edges before eating avidly.

The meat foods include beef, mutton, horse, poultry, and rabbit. Beef and mutton carcase residues, including heads and viscera, such as livers, hearts, lungs and, tripe, are very nutritious but liver is laxative and should not be fed in large amounts. The flavours of horse flesh and portions of fowl carcases, such as necks, gizzards, and livers, are appreciated by most cats, and rabbit is specially liked. Meats can be fed either raw or cooked. Many cats prefer raw meat, but the gravy which is produced by cooking is useful for flavouring the other foods. Fish can be an important item in the diet as it is palatable and makes a distinct change from other foods. Less expensive varieties, such as coley and dogfish or rock salmon are suitable. When eating meat or fish containing bones, cats lick the flesh off large ones, but may chew and swallow small ones. As small bones may become lodged in the mouth or throat many cat owners remove small bones, particularly from rabbit or fish meals. Cats have the ability to digest and utilise fat and about 15 per cent is included in most diets.

The cereal foods given include porridge, bread, flaked maize, and cooked rice or barley, and they are best fed mixed with either milk or gravy. Such foods are not as palatable as meat or fish but a taste for them can be acquired or they can be mixed with protein foods so that the flavour is masked. Milk is not necessary for adult cats, but is often supplied. Diarrhoea may be caused in some cats if milk and raw meat are consumed at the same time.

Many cats are now fed entirely on proprietary foods, either from cans or supplied dry. These foods are nearly always readily eaten and are carefully balanced to provide all the nutrients required. They should be fed according to the makers' instructions.

In addition to their normal diet cats need grass, which when chewed and swallowed, induces slight vomiting, and so prevents the formation of hair balls in the stomach. Cats when cleaning themselves swallow a certain amount of hair which may accumulate in the stomach or intestine. The hazard is obviously greater in long-haired than in short-haired cats. Cats allowed their freedom in the country can obtain grass when required, but cats confined in catteries or houses or kept in built-up areas should be supplied regularly. If grass has to be grown specially in towns, cocksfoot is the most suitable species.

Adult cats are normally given two meals per day, and as cats are creatures of habit the feeding times should be fairly regular.

Breeding cats

Pregnant and lactating queens can be fed a varied diet consisting basically of the types of food recommended for adult cats. A proprietary vitamin/mineral supplement should be added at the level advised by the manufacturer, or a few drops of cod-liver oil and half a teaspoonful of sterilised bone flour given every alternate day. During early pregnancy a queen needs very little extra food, but the amount should be increased from the fifth week onwards so that about twice the normal intake is given at the end of the seventh week. The proportion of high quality protein may be increased during the last 10 days of pregnancy to provide for the tissues of the developing fetuses. At least two meals, and preferably three, should be given daily.

During lactation suckling queens need to be particularly well fed. As soon as a queen has recovered from the upset of kittening she should receive the same quantity of food as she was given during the latter part of pregnancy, namely about twice the normal amount. This should suffice for the first 10 days, but then the amount should be gradually increased until the kittens start lapping at about 3 or

4 weeks of age and the queen's milk yield falls. The actual amount of food needed by individual queens varies, being influenced, particularly, by the number of kittens being suckled, but many queens will eat three times the normal quantity. Milk is a useful food during lactation for those queens which will drink it. The number of meals should be increased to four or even five each 24 hours to encourage maximum food intake, but a queen's food requirements often exceed her appetite and so she has to draw on her body resources for milk production and consequently loses body weight.

Artificial rearing of kittens

This is a difficult task and every effort should be made to obtain a foster mother (see handling section). If artificial rearing has to be adopted a commercial kitten milk substitute can be fed or a human babies' full cream powdered milk food can be used at twice the strength recommended. In an emergency 180 ml (6 fl oz) of cows' milk can be warmed and have beaten into it 3·5 g (0·12 oz) of butter, 45 ml (1·5 fl oz) of egg yolk and one drop of cod-liver oil. It is difficult to administer milk to very young kittens and a hypodermic syringe is useful as the plunger can be pressed lightly as the kitten sucks. Failing this an eye dropper of the type used in human medicine is generally employed. For the first few days the kittens need to be fed every 2 hours throughout the day and night. After this, 3-hourly feeds will suffice until the kittens are about 2 weeks old when only one night feed is required. From 3 weeks of age onwards the kittens should lap from a saucer and they can be fed along the lines recommended earlier for the pre-weaning feeding of kittens reared by their mother.

Mineral requirements

A calcium deficiency may occur in kittens fed almost exclusively on meat, or fish, from which all the bones have been removed, or heart or liver, because these foods are high in phosphorus but low in calcium. This imbalance can produce an effect particularly in kittens of the oriental breeds because they prefer water to milk, and so the calcium shortage is not alleviated in any way through milk drinking. Affected kittens fail to grow, become lethargic and their bones become fragile and fracture easily. Sterilised bone flour is probably the best source of supply for cats or kittens not receiving a calcium-fortified proprietary food. Phosphorus, sodium and potassium are all required by cats for normal development, but the amounts needed are contained in most diets.

Vitamin requirements

Cats require vitamin A, as they cannot synthesise this vitamin from carotene in a similar manner to most other animals and it is in relatively short supply in most of the common natural foods used in cat feeding with the exception of liver. Clinical signs of a deficiency are loss in weight, a scurfy coat, and eye defects such as inflammation and a dislike of bright light. In breeding queens, infertility, abortions at about the fiftieth day of pregnancy, or the production of weak kittens are observed. The vitamin D requirements of cats are relatively low, but a deficiency combined with a low calcium and phosphorus intake does result in bowed legs in kittens. A deficiency of vitamin E has been seen in cats resulting from the feeding of certain types of fish in excess or the administration of large doses of cod-liver oil. It causes the deposition of a yellowish-brown pigment in the body fat, which becomes firm to the touch. Affected kittens have difficulty in moving. The condition is known as yellow fat disease.

Cats need vitamins of the B complex, particularly thiamin and pyridoxine, and fresh meat is a good source of both. A lack of the first can result from the feeding of certain types of fish which contain an enzyme which inactivates this vitamin, and the second can be destroyed by heat during some canning processes. A thiamin deficiency can cause nerve pains and an exaggerated response to touch, and a pyridoxine deficiency can result in a loss of weight and convulsions.

Most proprietary diets are fortified by additions of these vitamins during manufacture.

Water requirements

Clean drinking water must always be available, and should be supplied in an unspillable bowl which should be large enough to hold the quantity of water required by the cats having access to it for a period of 24 hours. The bowl should be refilled with fresh water daily.

It is difficult to calculate the quantity of water required as it varies markedly according to the age of the animal, the weather, and the moisture content of the food. Growing kittens probably need about

28–42 ml per 0·45 kg body weight per day (1–1·5 oz per lb), while adults require about 21–28 ml per 0·45 kg body weight (0·75–1 oz per lb). The requirements of lactating queens are, obviously, very much greater. In the majority of cases over three-quarters of this fluid intake will be derived from the food, and, if liquid milk is given as well, a cat's liquid needs will be satisfied. Although cats on such a ration seldom drink water, others, particularly those of the oriental short-hair breeds, prefer water to milk. Excessive drinking in older cats is usually a sign of kidney disease.

There is some evidence to suggest that the deposition of minute sand-like calculi in the urethra which has been observed, particularly, in cats fed on dry foods, may be linked with an inadequate water intake. These dry foods contain only about 10 per cent of water as compared with the 75 per cent in most other diets. Some cats on dry foods do not drink sufficient liquid to compensate for the lack of fluid in the food, so that their total water intake is reduced. This leads to the production of urine with a relatively high mineral content and these minerals form the basis of the calculi.

Breeding

Tom cats are sexually mature at 6 months of age, but should not be used for breeding until they are 12 months old. Queens come into oestrus when about 6 months old, although Siamese and Burmese may come in at an earlier age. Thenceforth, if not pregnant, oestrous periods occur every 3 or 4 weeks during the spring and summer, although there is usually a quiescent period during the winter. In most queens the oestrous period lasts from 4 to 6 days, but may extend to 8 or 12 days in Siamese or Burmese. As in dogs, the oestrous period is divided into pro-oestrus and true oestrus. The first sign of pro-oestrus is that there is a swelling of the vulva, followed by a slight discharge, although this may not be observed as most queens clean themselves continually. The cat's disposition changes and she becomes more affectionate and rubs round the owner's legs. This lasts for 2 or 3 days. In true oestrus she starts calling with a wailing note, which is particularly loud in Siamese, rolls on the floor and squats in a crouching position with her hind legs in the air. She makes every effort to escape from the cattery or house in order to find a mate. Queens are not normally mated until they are about 10 months old.

A queen will usually refuse to mate during pro-oestrus, and mating normally occurs during oestrus at about the fourth and fifth days after pro-oestrus starts. A female that is free to go out will mate several times on these days with one or a number of tom cats. A female being sent to a stud for mating will be introduced to the male by being kept in an adjoining pen. The two will then be run together and the male will rub round the female before gripping the back of her neck with his teeth. The female will then crouch and they will mate while under observation. They may be separated after mating or permitted to run together for a period during which further matings will occur. If they are separated at least two matings at an interval of about 24 hours should be allowed. A queen may be mated twice each year, preferably in the spring and summer so that the kittens may be reared and weaned while the weather is reasonably good. Artificial insemination can be carried out in the cat, but is not normally practised.

The average gestation period is 63 days, varying from 60 to 66 days. The first sign of pregnancy is a pink shade which shows on the teats 3 or 4 weeks after mating, while abdominal distension will be noticeable 2 weeks later. In long-haired breeds it is advisable to cut away the hair round the teats so that the kittens, when born, will be able to suck more easily. The box in which the kittens are to be born should be placed in a dark, warm, secluded position about 10 days before the kittens are due and the queen introduced to it. If this is not done the queen will probably have her kittens on a bed, or in a cupboard or other inconvenient place. Newspaper is a very suitable bedding material, and the queen will tear this up to make a nest.

About a day before the kittens are due the queen will probably cease feeding. She then shows some discomfort when the involuntary contractions of the uterus start. In the second stage the uterine contractions become more powerful and abdominal straining can be observed. The water-bag of the first kitten to be born next appears at the vulva and breaks and this kitten is soon born. Almost all queens instinctively lick away the fetal membranes from the mouth of the kittens so that they can breathe freely and bite through any umbilical cords which do not break. They eat the afterbirths and membranes produced during parturition. Very occasionally a queen will need help with these tasks. Kittens can be born head or tail first and usually appear at 20-minute intervals, although queens may rest during the birth of a large litter. Provided the queen is not distressed and is not straining for a period without any result it is best to leave her alone. Should a period of prolonged straining occur without the birth of a kitten veterinary advice should be sought. After all the kittens are born and the mother has cleaned them and herself up, the soiled newspaper can be removed and replaced by a clean blanket.

The queen then lies on her side so that the kittens can suck more easily. At frequent intervals during the first 2 weeks of life the queen licks the anal and genital region of each kitten so as to stimulate urination and defaecation.

Health

Healthy cats have sleek coats and normally spend a considerable time each day licking themselves all over and cleaning their fur and paws. A fit kitten is alert and playful, unless sleeping or eating. The eyes should be clear and bright with no sign of a discharge. Cats have a third eyelid rising from the inner angle of the eye and resting on the eyeball. When a cat is ill this third eyelid is usually plainly visible. There should be no sneezing or nasal discharge, as respiratory conditions can be serious. The inside of the ears should be clean without an offensive odour. In a kitten, especially, the abdomen should not be unduly swollen, as this may indicate unthriftiness or worm infestation. There should be no evidence of diarrhoea. Cats are normally independent by nature, and some are excitable, but it should be possible to run a hand lightly over a cat's body without causing pain in any region.

The body temperature of healthy cats is in the range 38·3–38·9°C (101–102°F). The thermometer should be well lubricated with vaseline or soap to facilitate its insertion into the rectum. The pulse rate is normally between 100 and 130 beats per minute and is best counted by feeling the artery on the inside of the hind leg in a manner similar to that advised for dogs. Alternatively, palpation through the chest wall directly over the heart can be employed.

Common diseases and their prevention

Feline influenza is a common disease of cats in the British Isles which can affect animals of all ages, but, although the disease incidence is high, the death rate is low. The main clinical signs are sneezing and coughing with saliva dripping from the mouth which may soak into the hair of the chin and chest. There is also a discharge from the eyes and the animal becomes depressed and loses all desire to eat.

Progressive debility follows. When a cat recovers from the acute state a chronic condition may develop with a nasal discharge and bouts of sneezing, which may persist for a considerable time. The primary cause is a virus but secondary bacterial infections increase the severity of the clinical signs. Resistance following recovery is relatively short. A vaccine is available for injection when kittens are 9 weeks of age with a repeat dose at 12 weeks of age. A booster dose should be given every 6 or 12 months in order to maintain protection.

Feline enteritis is a condition which mainly affects young animals, but older cats may become affected, usually in a milder form The disease often has a sudden onset and healthy kittens rapidly become acutely ill and may die in a short time, usually within 2 or 3 days. Vomiting is a common clinical sign and initially there is bowel stasis without the passage of faeces. If the kitten lives long enough, diarrhoea, often with the production of blood-stained faeces, follows. There is a complete loss of appetite and abdominal pain is shown by a 'tucked-up' appearance and crying when the body is touched or moved. The eyes become sunken and the third eyelids protrude across the eyes. The disease is caused by a virus, which may remain viable on an infected premises for a considerable period. Effective preventive vaccines are available and one of these should be used for protecting young kittens, the actual age of administration depending on the particular vaccine selected. The period of immunity acquired varies from 1 to 2 years, and booster doses should be given when necessary.

Intestinal worm infestation. The clinical signs are more common in kittens than in cats. The adult worms cause loss of condition, and kittens, particularly, become pot-bellied. If the worm burden is large there may be obstruction of the small intestine followed by frequent vomiting and death. Two species of roundworms are found in the intestines of cats, and their life histories are somewhat similar. In both cases the eggs are passed out in the faeces of the host cat and if an infective egg is swallowed by a cat it hatches in the stomach. In one type of worm the larva which hatches makes its way to the liver and after passage through other body organs enters the intestine while a larva of the other worm develops in the intestinal wall. As infestation depends on the ingestion of infective eggs any measures which prevent the contamination of food or water by faeces are indicated to prevent a worm burden being acquired. It is advisable to treat a queen with an effective worm medicine before mating and, possibly, again during pregnancy, to avoid kitten infestation.

Parasitic otitis is caused by a parasite which is a common inhabitant of the ears of cats of all ages. Some cats show few clinical signs of infestation but others scratch their ears and shake and rub their heads. This constant shaking and rubbing removes the hair behind the ears and may leave raw areas which can become infected. An examination of the inside of the ear reveals the presence of a dark brown exudate, over the surface of which it is sometimes possible to see tiny white mites moving. When an affected ear is gently handled scratching movements of the hind leg on that side of the body are stimulated. The infestation is acquired by contact with affected cats. It is, therefore, advisable to ensure that a queen's ears are free from the mite before she kittens to prevent the kittens acquiring the condition from their mother.

Fleas are common parasites on cats of all ages. The severity of the clinical signs varies between cats as some can be heavily infested and yet show few signs, while others scratch and bite themselves vigorously although harbouring very few fleas. In most cases there is a loss of hair and the appearance of skin abrasions in areas within reach of the cat's mouth or claws caused by licking and biting. Individual fleas may be hard to detect but their excreta in the form of small, black, grit-like particles can usually be seen at the base of the hair. Prevention is by keeping cats away from infested cats and from dusty, dirty areas where the fleas lay their eggs from which the larvae hatch.

Quarantine regulations

To prevent the introduction of diseases not known in Britain imported cats, like dogs, are subject to detention, at the owner's expense, in an approved place of quarantine.

Handling

Cats differ widely in temperament. Some are easily handled, while others have a nervous excitable nature and, if upset, can be very difficult to manage. In fact a cat objecting to restraint will fight more than almost any other domesticated animal, and is able to twist its body with surprising agility and attack with both teeth and claws. The golden rule, therefore, for applying restraint to cats is 'the less the better'.

A quiet cat can best be picked up by placing one hand under its body with the middle fingers passing between the front legs. The other hand can be used to hold the skin at the back of the neck. To carry the cat its hind quarters can be held firmly between the handler's body and the elbow of the hand which is placed below the cat's body.

If a cat is brought for examination in a box or bag great care must be taken when the cat is released, as some become upset by the close confinement during transport. The container should be opened in an escape proof room, and the lid should be raised gently so that any sudden attempt by the cat to force its way out can be prevented.

Should a frightened cat get loose in a room it may not allow anyone to approach it. It is possible to use a snare with a running noose similar to, but lighter than, that recommended for dogs, but this is not a very satisfactory method. It is better to throw a blanket or large towel over the cat and grasp the scruff of the neck through the cloth. The blanket can be turned back when the cat is firmly held. For examination a cat should be placed on a table with a hard smooth surface, as this restricts sudden movements. It must always be controlled by being grasped by the neck scruff with one hand to prevent a sudden escape. This is effective with a quiet cat, but spiteful or frightened animals can easily scratch with their hind legs. Although potentially hostile animals need to be held firmly from the start, excessive force must never be used because there is a risk of injury through struggling.

A well-behaved cat can be laid on its side on a table by holding the forelegs with one hand with the forearm over the neck and the hind legs with the other with the forearm over the hind quarters. Alternatively when the cat is on its side the scruff of the neck can be held with one hand and the two underside legs (one fore and one hind) with the other. Some experienced veterinary surgeons with a difficult patient tape the forelegs and hind legs together in pairs. The cat usually stops resisting and, although the restraint may appear to be excessive, it does not prove to be unduly upsetting. Cats are not comfortable if held on their back, and this position should be avoided if possible.

If the head is to be examined mild restraint only in the form of a hand on the back of the neck is best. The mouth can usually be examined without difficulty by placing a hand over the dome of the head and tilting the head gently backwards. This causes the mouth to open slightly and a finger tip of the other hand pressing slightly on the incisor teeth opens it wider. If this is not adequate the cat's body and legs can be rolled in a large towel. The towel is laid out on a table and the cat is placed at one end and then quickly and firmly, but gently, rolled up.

If the hind legs or tail need attention it is possible, after the cat has been rolled over twice, to pull these parts of the body out before the rolling is continued. If necessary the cat's head can be included in the towel as it will be able to breath through the holes in the cloth.

Grooming and washing

Healthy short-haired cats groom themselves efficiently and are not normally groomed by their owners. However, when they are moulting it helps if a hand, curved to fit the body lines, is rubbed along the back, as this brings dead hairs to the surface and removes them. Brushes can be used, but they should not be applied too vigorously.

Long-haired cats and kittens need grooming daily, particularly in spring when the hair is being shed. The coat should be combed with a coarse steel comb to remove any loose hairs and any knots should be teased out. The coat should then be brushed, particular attention being paid to the ruff. A badly matted coat is very difficult to groom and it is frequently necessary to cut the matted lumps of fur away from the skin and allow a fresh coat to be grown.

Cats are not normally washed, but there are occasions when this is necessary in order to remove objectionable or toxic substances which have come in contact with the body, or for the treatment of skin diseases. Most cats resent being washed and object to being stood in water. Thus, it may be easier to stand them in a dry sink or bath and pour warm water over them. An assistant should hold the neck scruff and forelegs while the operator washes the hind part first and then attends to the fore part. The head should be left to the last, as this is the part of the body which cats dislike having washed most. An alternative is to place the cat in a linen bag and draw the string at its mouth round the animal's neck, washing both cat and bag. If a medicament is to be used it is possible to place this in the bag before the cat is inserted. A pillow case can be used instead of a linen bag.

After washing a cat its coat should be carefully dried with a warm towel. An ordinary hair drier can be helpful for the long-haired breeds provided its use does not frighten the cat.

Sexing

Kittens can be difficult to sex. The tail should be raised and the distance between the anus and the genital opening below it should be observed. In female kittens the two openings are close together, while in males there is a space between. In day-old female kittens the distance between the two openings is about 6·3 mm (0·25 in) while in day-old males it is nearer 12·7 mm (0·5 in). Also in females the vagina is slit-like while in the male the opening is round. In litters containing kittens of both sexes it should not be difficult to distinguish between them. As the kittens grow older the spaces between the two orifices obviously increase in length. In adult animals the distances between the two orifices are clearly defined, being about 12·7 mm (0·5 in) in females and 31·7 mm (1·25 in) in males. In male cats pressure round the genital orifice will express the penis and in entire males the testicles can easily be seen in the scrotum.

Fostering

If a queen should die shortly after giving birth or if a cat should produce more kittens than she can rear it is better to find a feline foster mother than to attempt to hand rear them. Every effort should be made to obtain another queen which kittened at the same time as the one which died or produced the large litter. The foster mother selected must be healthy and free from fleas. If she has a litter of unwanted kittens of her own the kittens to be fostered can be mixed with those of her litter during her absence, and, in a short time, when the new kittens have acquired the smell of the others, her own kittens can be removed and painlessly destroyed. If the foster mother's litter has already been removed the kittens to be adopted can be rubbed with some of the foster mother's milk expressed from her teats. If one or other of these measures is adopted it is usually not difficult to persuade the foster mother to accept the strange kittens.

Administration of medicines

Tasteless drugs can be included in the food but this is not a very satisfactory method of administration. Cats normally smell their food before eating and an unusual odour, even if slight, will make them suspicious so that the food is refused. It is not as easy to give a liquid medicine to a cat as it is to a dog as it is not possible to make a suitable pouch by pulling at the side of the mouth. An assistant holds the front legs in one hand and places his other hand on the cat's hind quarters to prevent it from moving backwards. The operator then holds the head with the fingers of the left hand spread wide and twists the head gently backwards. This normally causes the lower jaw to drop and the mouth to open. Then the liquid can be placed on the back of the tongue using a teaspoon or a hypodermic syringe without a needle. Small quantities only should be given at a time to prevent the cat from choking.

Pills are more easily given than are liquids. The mouth is opened as previously described and the pill is dropped on the back of the tongue. If this is not effective the pill can be held in a pair of forceps and placed in position. The mouth is closed immediately and the outside of the throat is stroked to encourage swallowing.

Injections are being used increasingly because they require only a minimum amount of handling. The best position for a subcutaneous injection is in the skin of the flank just behind the shoulder.

General

Vices

Spraying

This undesirable habit is practised by most uncastrated tom cats and occasionally by males castrated after reaching sexual maturity. It consists of ejecting squirts of urine over particular objects to mark a territory, and is not caused by a desire to empty a full bladder. The action is distinctive as the tail is raised, the penis is protruded and the cat directs its hind quarters towards the object and sprays it with urine. In a house curtains and walls receive special attention and the offensive odour of the urine leaves an unpleasant, persistent smell. It is almost impossible to cure the habit and animals which practise it should be kept outside where the spraying does not affect house dwellers.

Bird catching

Hunting and killing are natural instincts in cats, and kittens soon learn to glide with body close to the ground towards a possible prey and then suddenly pounce. The distressing factor is that cats, particularly well-fed animals who do not need to kill their food, torture any birds they catch by worrying them for a considerable time before finally killing them. Cats cannot be taught to give up this habit and the only effective method of prevention is to confine them to a house or shed and run. Some owners try attaching a bell to a light elastic collar so that its ringing when the cats move will warn the birds of the approaching danger.

Minor operations and manipulations

Castration

To prevent the development of problems associated with sexual behaviour in male animals male kittens to be kept as pets should be castrated. These problems include the vice of spraying, previously described, wandering, sometimes for long distances, in a search for female cats, and fighting with other male cats, which can result in distressing injuries. The operation is performed under an anaesthetic, preferably when the animal is from 4 to 5 months old. Males castrated before puberty do not produce objectionable strong-smelling urine and do not develop any of the habits listed above, but some which have been castrated as adults may continue to behave as before.

Spaying

Unspayed female cats can be kept in to prevent them being mated, but they will frequently be in oestrus, will not eat regularly, and will lose body condition. This is particularly true of cats of the oriental breeds. If they escape, unwanted litters of kittens will be born. Spaying prevents these inconveniences. An abdominal operation is required but cats normally stand the interference well and recovery is usually rapid. The operation is best performed when a cat is about 4 or 5 months of age, but the operation may be performed later.

Some cat owners do not favour castration or spaying as they consider that neutering makes cats lazy and fat and reduces their value as vermin killers. In most cases this is not a fact as neuters are good hunters and do not become excessively fat.

Species control

The Governing Council of the Cat Fancy is the body which organises the cat fancy in Great Britain. Its functions are to prepare the rules which govern the exhibition and showing of cats, to act as the central authority controlling pedigree cats, and to arrange for the registration of cats and kittens of all recognised breeds. A breeder can register a prefix which can be used as a distinguishing feature in the names of all the kittens he or she breeds. When a registered cat or kitten is sold it must be officially transferred to its new owner before it can be shown or its progeny can be registered.

CHAPTER 11
FOWLS

Introduction

The poultry industry, which embraces fowls, turkeys, ducks, and geese, is economically the second most important branch of livestock husbandry in the United Kingdom, dairying being the first. Fowls are the most important species of domesticated poultry.

During the last 25 years economic considerations have forced keepers of fowls to make marked changes in their husbandry methods. Intensive housing and labour-saving devices have been developed to reduce labour requirements and improved breeding systems have resulted in the production of more prolific stock. Flocks have been increased in size for the margin of profit per bird is so small that unless the unit is large the net income is too low to enable the owner to survive financially.

There is now a fairly clear dividing line between birds kept for egg production and those kept for meat production, whereas in the past dual purpose-type units were fairly common in which the females were retained for laying and the males fattened for table. Poultry breeding is now largely in the hands of specialists who have used the science of genetics to develop strains or breeds particularly well suited to the production of either meat or eggs. The rapid breeding rate of fowls has been fully utilised by the use of incubators.

Commercial egg or poultry meat production is a highly competitive business which involves a substantial investment of capital and a considerable element of risk. Just at present the prospects for the egg producer are not good as there is a reduction in egg consumption. However, there is an increase in demand for poultry meat, to a large extent because broilers can be marketed at a favourable price when compared with other meats. If the poultry industry is to maintain and improve its position in the national economy, continued improvement in the quality of the poultry and eggs sold is needed and the costs of production must be kept as low as possible. The long-term prospects appear to be good, for both eggs and poultry meat are excellent media for the conversion of vegetable materials into highly acceptable animal protein for human consumption.

Definitions of common terms

Chick. From hatching to 8 weeks. The term 'day-old' chick covers the period from hatching to 72 hours, as chicks do not require to be fed during this period.
Grower. From 8 to 20 weeks.
Pullet. Female from 20 weeks to 18 months.
Hen. Female over 18 months.
Cockerel. Male from 20 weeks to 18 months.
Cock. Male over 18 months

Table birds

Poussin. Killed at about 7 weeks of age at about 0·9–1·0 kg (2–2·25 lb) liveweight.
Broiler. Killed at about 9 weeks of age at about 1·6–1·8 kg (3·5–4 lb) liveweight.
Roaster. Killed at about 20 weeks of age at about 3 kg (8 lb) liveweight.
Capon. A castrated male, usually being reared as a roaster. The testicles can be removed surgically, or synthetic oestrogens can be implanted in pellet form under the skin of the neck to cause testicular regression.
Boiler. A pullet or hen marketed at the end of the laying period.

Principal breeds

Laying fowls

Leghorn

Small, active, quick-growing birds bred for years to a high level of egg production, laying large numbers of good sized, white-shelled eggs. They rarely go broody, but tend to be nervous. Leghorns are found in various colours, white being the most popular. White Leghorn pullets weigh about 1·81 kg (4 lb). Black Leghorns are slightly larger in body size and not quite so prolific, and Brown Leghorns are larger still with a lower average level of production.

Fig. 11.1. Rhode Island Red cock.

Rhode Island Red

Medium in body size, with pullets weighing about 2·26 kg (5 lb), they are good layers of large brown eggs. They are hardy, docile, and adaptable, but the chicks tend to be slow feathering. The birds are chocolate-red in colour with black or greenish-black tail feathers.

Dual purpose Fowls

Light Sussex

A good winter layer of brown eggs but some strains tend to go broody, particularly during the summer months. This breed has good quality white flesh, and certain strains have been developed for their meat producing qualities. In colour the birds have white body feathers with black markings on the neck, tail and wing tips, although an all-white type, often known as White Sussex, has been developed for table bird production. The pullets weigh about 2·50 kg (5·5 lb).

Fig. 11.2. Light Sussex hen.

Plymouth Rock

Of medium body size with pullets weighing around 2·26 kg (5 lb) and laying brown eggs which tend to be small in some strains. For egg production the most popular varieties are those which are barred or buff in colour. The barred birds have dark-coloured bands across each feather. A white variety is used for creating broiler strains particularly in the United States of America and Canada, but is not as popular in Great Britain because it has yellow-coloured flesh.

Table fowls

North Holland Blue

A rapidly growing breed with white flesh but, as it is of a dark barred colour, dark feather stubs remaining after plucking tend to make the carcase unattractive. A white variety, often known as the North Holland white, is now being developed. The pullets weigh about 2·72 kg (6 lb).

Indian or Old English Game

These breeds carry exceptional amounts of breast meat, but have a yellowish skin and flesh which are considered to be unsightly. Game fowls tend to be of low fertility with the hens laying fewer eggs and the cocks producing less spermatozoa than other breeds, but they are pre-potent for meat quality and so are used for crossing. Indian Game pullets weigh about 2·72 kg (6 lb) and Old English Game pullets about 2·26 kg (5 lb).

Cross-breds

At one time the various breeds were mainly bred pure, the object being to improve the genetic values of strains within breeds, but today for commercial fowl production it is usual for some system of cross-breeding to be adopted. The reason is that some pure-breed strains have reached a point at which little further improvement within the strain is possible, but when crossed with other strains or breeds an increased performance can be obtained. First crosses between two breeds, such as a Leghorn cockerel and Rhode Island Red or Light Sussex hen, were popular in early years, the pullet chicks being reared for egg production and the cockerels being reared for table purposes. The increase in specialisation, leading to the production of either eggs or meat, has reduced the popularity of these crosses.

Hybrids

Layers

Modern hybrid fowls for egg production are bred according to carefully planned systems, and the records of the birds used are individually checked. Under one system, four breeds or strains are selected, shown in the following table as A, B, C, and D. For the first three generations brother and sister matings are employed, and the progeny of each pair are culled heavily so that all birds with poor stamina and all females with low egg-production records are eliminated. In the fourth generation the strains are crossed in pairs. This out-crossing of unrelated inbred strains results in the production of birds with greater stamina and egg-production potential, and the best individuals are selected by testing. Selected birds of these two lines are crossed, giving birds with still better performance potentialities. This is the generation which is produced in large numbers and the female birds are sold to egg producers. In general these birds are of a uniformly high standard. The colour of the eggs produced by the hybrids, either white, tinted, or brown, can be determined by the selection of the original breeds.

Table 11.1. A system of hybrid breeding.

Breeding method	Generation	Strains A	Strains B	Strains C	Strains D
Inbreeding	1	Brother and sister	Brother and sister	Brother and sister	Brother and sister
Inbreeding	2	Brother and sister	Brother and sister	Brother and sister	Brother and sister
Inbreeding	3	Brother and sister	Brother and sister	Brother and sister	Brother and sister
Outcrossing	4	AB		CD	
Outcrossing	5		ABCD		

Another system is known as reciprocal recurrent selection, and involves the use of two separate lines, these generally being of different breeds. First, cocks of one line are mated to pullets of the other line and vice-versa and the cross-bred progeny are tested for production efficiency. From the results of these matings it is possible to determine which birds produce the most prolific progeny when crossed. In the following year the best sires are mated to the best females in the same lines to produce pure-bred birds. The following year the fowls so bred are crossed and their progeny evaluated as before. This method leads to the production of lines of birds which produce quality cross-bred fowls. The main advantage is that the closely inbred generations required in the method previously described are not necessary, and for this reason reciprocal recurrent selection is favoured by breeders who operate on a relatively small scale.

There is an increasing tendency to favour small-bodied hybrid fowls weighing only 1·36 or 1·58 kg (3 or 3·5 lb) at point of lay because their food consumption is less than that of heavier birds due to their lower requirements for body maintenance. Early sexual maturity is also being developed because this reduces the cost of rearing to point of lay, although it has to be accepted that the first twenty or thirty eggs laid by birds which mature earlier than normal will probably be below the desired weight of 56·7 g (2 oz). Hybrid fowls can be expected to lay, on average, from 220 to 250 saleable eggs per year, of which at least 90 per cent should weigh 56·7 g (2 oz) or more.

Broilers

The main objects are to increase body weight and growth rate and improve food conversion efficiency, but viability must be maintained and a reasonable level of egg production and good hatchability are necessary for the economical production of large numbers of broiler chicks. As the heaviest birds tend to be poor layers and the best layers usually have poor table qualities it has been found most satisfactory to develop separate strains with these qualities and then cross these to produce the broiler chicks. A three-way cross is popular in which a male from a good table quality strain is mated with females from a good laying strain and the resulting pullets are mated to a male from a second good table quality strain to produce the broiler birds. The advantages are that the female parent has hybrid vigour to improve viability, and that the second cross also gives hybrid vigour as well as improving table quality. Care must be taken to ensure that the birds of the broiler generation have white feathers and white flesh.

Identification

Pedigree chicks being used in controlled breeding programmes are usually marked individually at hatching by the application of numbered aluminium wing bands, manufactured on the safety-pin principle. The pin of the band is forced through the triangle of skin in front of the 'elbow' joint of the wing, and the flap on the numbered section is pressed down on the pin, using a pair of pliers, so that the bird is identified for life. Chicks may also be marked by punching a hole, or holes, in the web between the toes using a toe-punch. There are only sixteen possible number combinations using this method and so it can only be employed for the group marking of chicks from

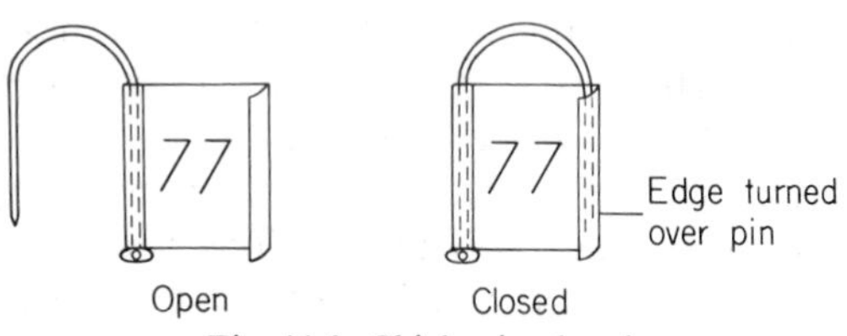

Fig. 11.3. Chick wing band.

certain matings. Another disadvantage is that if the holes are not cleanly punched they may heal across. Chicks were at one time individually identified by the application of numbered leg rings. This method has lost popularity because rings of a size suitable for chicks have to be removed when the birds are about 2 weeks old to avoid leg injuries through pressure. The birds then have to be marked in some other way.

Adult birds being trap nested for pedigree breeding programmes can be fitted with large plastic wing tabs with numbers which can be read easily. Leg rings can also be used as they are very easy to apply. They can be of aluminium with individual numbers and these rings are normally of an adustable design so that they can be made to fit the legs of birds of different sizes. The rings should be attached with the numbers upside down so that they can be read easily when the bird is handled. Coloured spiral rings, often made of celluloid, can be used for group marking to indicate membership of a particular pen or birds of a special strain.

Ageing

There is no reliable method of ageing adult fowls, or any of the other species of poultry described in the next chapter. However, in fowls a hen has more clearly defined scales on the legs, and this may serve as an age indicator. Also, if the tip of the beak is pressed to one side a slight degree of movement will be obtained in a pullet, while the beak of a hen will remain rigid.

Housing

Incubators

The incubation period for fowl eggs is 21 days and the temperature has to be maintained at about 37·2°C (99°F) for the first 19 days and at about 36·1°C (97°F) for the last 2 days. The desired relative humidity is about 60 per cent for the early part of incubation, rising possibly to about 65 per cent just before hatching. The eggs must be turned each day, if possible six or eight times or more, up to the nineteenth day, after which turning is no longer necessary.

A flat type of small incubator holding about 50 or 100 eggs is used by keepers of small breeding flocks. The eggs lie on their sides in one layer and have to be turned by hand. The heat is generated by oil, gas or electricity, the humidity is obtained from trays of water and the ventilation is by natural draught.

On specialist poultry farms or hatcheries cabinet or walk-in type incubators are used. The cabinet machines may take as few as 1,000 eggs, but may hold from 10,000 to 40,000. The walk-in machines are, in effect, specially built rooms in which an operator can walk and wheel a trolley, and each room takes from about 70,000 to 84,000 eggs. In both types the eggs are placed in trays in a vertical position with their broad ends uppermost, and they are ventilated mechanically by fans. The eggs are turned at frequent intervals by tilting the trays through an angle of from 80 to 90°. Each incubator is divided into a large setting and a smaller hatching section, the eggs being transferred from the former to the latter on the nineteenth day of incubation. The advantages are that no turning mechanism need be provided in the hatching section, and that faeces and fluff from newly hatched chicks are confined to that section. On large hatcheries the setting and hatching sections are built separately and housed in different rooms.

Young fowls

General

It is essential for chicks to be provided with artificial heat for the first 5 or 6 weeks of life, temperatures of 35°C (95°F) being allowed for the first week followed by a reduction of about 2·8°C (5°F) per week. It is best not to allow the air temperature to fall below 15·6°C (60°F) during the first 10 weeks of life.

Single-stage intensive method

Most chicks are now reared in groups of from 1,000 up to 5,000 under a single-stage intensive method. This means that the chicks are placed in a large building and allowed 0·47 m^2 (0·5 sq ft) of floor space per head for the first 4 weeks, from 0·47 to 0·93 m^2 (0·5–1 sq ft) for the next 4 weeks, 0·93–1·40 m^2 (1–1·5 sq ft) from 8 to 12 weeks of age and 1·40–1·86 m^2 (1·5–2 sq ft) from 12 to 20 weeks. On some commercial farms the space allocations are less than these. Poussins and broilers are taken out for killing at the required age and pullets for future laying are transferred to their laying accommodation at 20 weeks of age.

The heat required can be supplied by a variety of methods, but the following are the most important. On large farms a central heating system is usually installed. Most are highly automated, thermostatically controlled, and fuelled by electricity, gas, or oil. The warmth may be supplied to the house through hot-water pipes or radiators or as hot air. A series of individual heating units, each surrounded by a hover, may be installed down the centre of the building. The fuel may be electricity, gas, or oil. Low, movable divisions are used initially to keep the chicks near the hovers, and these are moved as the chicks grow to give more floor space. The hovers should be suspended from the roof of the building on pulleys so that the temperature provided for the chicks can be regulated by adjusting their height. This method costs less to install than a central heating system, but is usually more costly to operate. As an alternative infra-red lamps can be suspended above the chicks, so spaced that the whole floor area is heated.

A deep litter system of bedding is generally employed. A new floor should be established under dry conditions by sprinkling chalk at the rate of 0·5 kg per 5·55 m^2 (1 lb per 6 sq ft) of floor space, adding a layer of compost, top soil, or rotted horse manure and applying wood shavings or chopped straw. The various layers are mixed and the bacteria contained in the compost, soil, or manure break down the poultry faeces into a dry powder. Initially there is a need to mix the newly voided faeces with the litter, but later the movement and scratching of the birds renders this unnecessary. About every 2 or 3 weeks additional shavings need to be spread over the floor. Under this system the bedding has only to be changed between batches of chicks. As an alternative the flooring can be of wire or slats, the floor being covered with paper, hardboard, or polythene and a layer of shavings for the first 3 or 4 weeks. The advantage is that the birds do not come in contact with their faeces, which drop through the spaces in the floor, and this limits the spread of enteric diseases.

Adequate trough space must be provided, about 2·5 cm (1 in) per bird being required up to 4 weeks of age, from 5 to 7·6 cm (2–3 in) per bird up to 10 weeks and from 7·6 to 12·7 cm (3–5 in) until the birds are 20 weeks old. It is also important to distribute the feeders over the whole floor area, so that all the birds may feed freely, and to arrange for a sufficient number of drinking points to be provided. Ten 4·5 l (1 gal) drinking fountains are needed per 1,000 birds. Artificial lighting of 10 lux intensity is provided in most cases for 24 hours per day. To avoid cannibalism the light intensity may be reduced at the first sign of feather pecking, or red light, which reduces activity, may be supplied for 1½ hours out of every 2 hours, and white light, to encourage feeding, allowed for the remaining ½ hour. Birds kept for future laying may have the lighting period reduced from about 8 weeks of age in order to slow the rate at which they mature. Details are given later in the section on egg production. As a guide to the optimum ventilation rate it is usual to allow 5·5 m^3 per kg body weight per hour (1 cu ft per minute per lb body weight) for the total number of birds kept in the house when mechanical ventilation systems are employed.

Rearing batteries

For the rearing of future laying birds pullet-rearing batteries are favoured by some poultry farmers. Under this system the birds are kept in tiered wire cages, with wire-mesh floors through which their faeces drop on to solid trays which can be removed for cleaning. Part of the section of each tier used tor the youngest chicks is heated by an overhead heating plate. As the birds grow they are moved on through connecting openings to larger cages, and so the handling is kept to a minimum. The system is efficient, cheap, and requires little labour, but it does involve the housing of birds of different ages in the same building, with a possible risk of disease spread. It is also difficult to provide light patterns suitable for each age group.

Tier brooders

Small numbers of chicks can be reared in tier brooders resembling the first-stage cages described in the previous system. At from 5 to 7 weeks of age they are moved to an alternative form of housing. This could be an extensive system using range shelters or night arks, which are small movable houses in which the birds are confined at night set in a grass field over which the birds can range during the day, or an intensive method such as a deep-litter or wire-floored house. The extensive method is expensive in terms of labour since food and water have to be transported round the fields and the birds have to be shut in the houses each night to avoid losses through foxes.

Adult fowls

Intensive systems

Battery cages

Laying battery cages are the most popular method of housing laying birds. The birds are kept, usually in ones, twos or threes, in metal cages erected in tiers in well-ventilated buildings. The cages are 45·7 cm (18 in) high by 30·5 cm (12 in) deep with frontages of 28–30·5 cm (11–12 in) for a single light hybrid, 35·6–38 cm (14–15 in) for a single heavy or two light hybrids, or 43–45·7 cm (17–18 in) for two heavy or three light hybrids. The floors on which the birds stand are of wire mesh so that their faeces fall through on to a tray below from which they are removed by some type of mechanical cleaning device. The mesh floors slope so that the eggs run down into a collecting trough in front. Water is always delivered to each cage by an automatic watering system, and some form of automatic feeding device is incorporated into most battery cage designs. The birds are placed in the cages just before they come into lay at about 20 weeks of age, and remain in the cages for about 12 months, when they are sold as boiling fowls for table purposes. The level of egg production is high and it is possible to cull easily any bird which stops laying so that food is not fed to unproductive individuals. The eggs keep cleaner than on any other management system, and so require very little cleaning before sale. Because the birds do not come into contact with their own, or other birds', faeces the spread of enteric infections and parasitic infestations is minimised.

Deep litter houses

A deep litter system, with the bedding prepared on the lines given under chick rearing, is also popular, especially for birds being kept for breeding. The house is only cleaned out annually when the birds are removed at the end of the pullet laying year. Drinking founts are distributed throughout the building, and the troughs are often designed so that food pellets are kept constantly circulating through

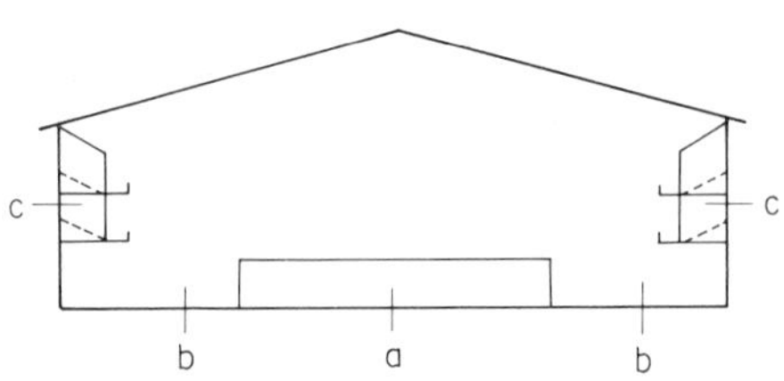

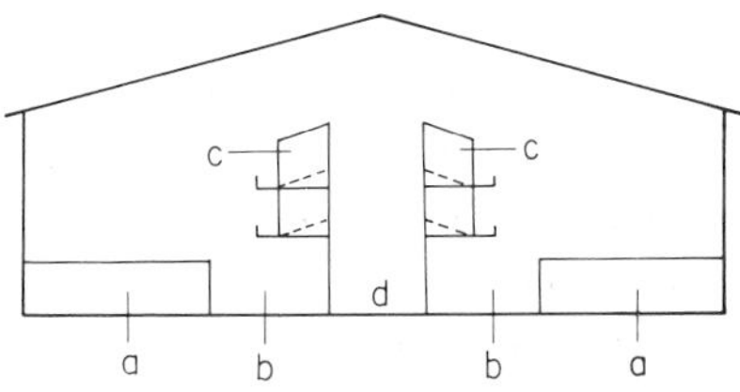

Fig. 11.4. Cross sections of two types of deep litter house. (a) Droppings pit, (b) litter, (c) nests, (d) passage.

the building on a chain-link system. About 15·2 cm (6 in) of trough length are allowed per bird. Perch rails are placed in one area 30·5 cm (12 in) apart, allowing 15·2 cm (6 in) of rail length per heavy-breed bird and slightly less per light-breed bird. As most faeces are voided when the birds are eating or sitting on the perches some buildings have droppings pits covered with wire mesh or slats under the roosting and feeding areas in which the faeces collect. These pits are cleaned out once a year like the rest of the building, but the fact that the faeces so collected do not have to be broken down by the deep litter bacteria means that a smaller floor area per bird can be allowed. A space of 2·8 m^2 (3 sq ft) per bird is provided for deep litter houses, but only 1·86 m^2 (2 sq ft) per bird in houses with droppings pits. On some commercial farms the space allocations are less than these.

Nest boxes must be supplied; these can be of the single type in the form of a 30·5 cm (1 ft) cube, allowing one box per five birds; or of a communal type, providing 11·2 m^2 (12 sq ft) per 100 birds. In some houses the nests are arranged along the side of a passage so that the eggs can be collected from the rear of the nest, thus saving the poultryman the need to enter the house to collect the eggs. On large farms the automatic collection of eggs is arranged by providing rollaway nests in which the eggs roll backwards when laid on to a travelling belt which runs at intervals controlled by a timeswitch, and delivers the eggs to an egg room.

Wire and slatted floor houses

This method of housing poultry is losing popularity, although advantages claimed are that more birds can be kept in a given area, that there are no difficulties with floor litter, and that there is a freedom from intestinal parasites. A space as small as 0·093 m^2 (1 sq ft) per bird has been tried, but as the birds are kept so close together there is always a risk of cannibalism. Wooden slats are made with a 25·4 mm (1 in) gap between them, and the wire floors have a mesh 76·2 × 25·4 mm (3 in × 1 in). Wire floors are favoured as they have a longer life and allow the faeces to pass through more freely. The wire mesh must be carefully made so that there are no projecting edges which might injure the birds' feet. Both types of floor are fitted 0·61–0·76 m (2–2·5 ft) above the base of the house so that the faeces will build up below the wire or slats and only need to be removed once a year. The wire or slats are made in sections so that they can be removed when the building needs to be cleaned out. It is an advantage to provide perches, which can simply be rested on top of the wire or slats, at one side of the building. If the nest boxes are of a satisfactory design the eggs will be cleaner than those produced in a deep litter house.

In order to obtain maximum production from birds kept under the laying battery, deep litter, or wire floor systems, they should not be exposed to environmental temperatures below 4·4°C (40°F) or above 29·4°C (85°F), an optimum of about 12·8°C (55°F) being favoured. Adequate ventilation is also essential and it is customary to allow up to 42·5 dm^3 (1·5 c ft) of air per 0·45 kg (1 lb) of body weight per minute. It has been found that 16 hours of light, at about 10 lux intensity per 24 hours gives the highest egg yields. This light stimulates egg production by acting directly on the pituitary gland, and does not lead to a marked increase in food consumption. The additional lighting required is normally operated by a timeswitch which turns the light on early in the morning, so that the birds go to roost when the light normally fades in the evening. If the artificial light is suddenly switched off during the hours of darkness birds kept on a deep litter system cannot see to find the perches and are forced to sleep on the floor where they were standing.

Semi-intensive and free range systems

Hen yards

This system is mainly used by farmers with redundant buildings, particularly if these are of questionable strength, and with supplies of straw to be converted into manure. Most of the yards are partially covered and an area of 0·14 m^2 (1·5 sq ft) is allowed per bird under cover and about twice this space in the open. Perch units and nest boxes are placed in the covered area and the water troughs are sited in the open section. The food troughs can be fitted either in the open or under cover. Advantages are that there is a low capital cost, but disadvantages are that draughts and cold winds can affect the numbers of eggs laid adversely and a heavy intestinal worm burden can be built up.

Verandahs

The designs vary but they all have slatted or wire-mesh floors raised about 0·75 m (2·5 ft) above the ground so that the faeces accumulate below the buildings. In most cases there is a sleeping area with perches and nest boxes and a covered run in which the food and water troughs are sited. The front walls of the run are always of wire mesh and the side walls may be of the same type of construction, although in exposed areas they may be partially boarded.

Fold units

These are movable units generally made of wood designed to accommodate about 20 laying hens. Each unit is divided into a slatted floored sleeping area 1·52 m × 1·52 m (5 ft × 5 ft) square with a nest box attachment and a run 1·52 m (5 ft) wide and 3·96 m (13 ft) long. The run has no floor so that the fowls can have access to the ground. The units are shifted daily, or at least three times a week, so that the birds are constantly on clean land and their faeces are evenly spread over the whole area of the field. A considerable amount of labour is required to effect this constant moving, but the birds do well provided the land is flat and well drained. As the hens are confined to the unit they are protected from animal predators and so it is not necessary to lock them in the sleeping areas at night.

Fixed pens

The method was popular with pedigree poultry breeders who wished to run small numbers of birds in carefully controlled units. Small wooden houses were placed in lines, each having two grass runs attached. The birds were allowed on to the runs alternately so that while one run was in use the other could be left empty to allow the grass to grow.

Free range

This system is suitable for a general farm, but is going out of favour because of the amount of labour required to travel to the fields to feed the birds and shut them in their houses at night and let them out in the morning. Small houses are provided accommodating from 50 to 150 birds. In slatted-floored houses 0·069–0·093 m^2 (0·75–1 sq ft) of floor area is allowed per bird and in solid-floored houses 0·14 m^2 (1·5 sq ft). The houses should be raised 0·31 m (1 ft) above the ground and should be protected from damage by other farm animals grazing in the vicinity. The birds are allowed to run freely over a field, and 125 birds per ha (50 birds per acre) can be run on grass without reducing its capacity for carrying other stock. If other animals are not running on the field a stocking rate of 375 birds per ha (150 birds per acre) is allowable. It is an advantage to move the houses every 1 or 2 years to prevent the area round each house from becoming fowl-sick. On arable farms it is customary to move the birds on to stubble fields so that they can consume any shed corn.

Disadvantages, additional to the high labour cost, are that food is often eaten by wild birds, hens may be killed by predators, bad weather can markedly reduce egg production, and some hens will lay away from the houses in places where their eggs are not found and collected. The advantage is that there are customers for eggs laid by free range hens who are prepared to pay higher prices than normal in the belief that these eggs are superior in flavour and quality.

Feeding

General

Food accounts for from 70 to 80 per cent of the production costs and so it is essential to feed efficiently if a poultry unit is to be run profitably. Poultry nutrition is necessarily different from that of other animals because of the specialised digestive system. There is an

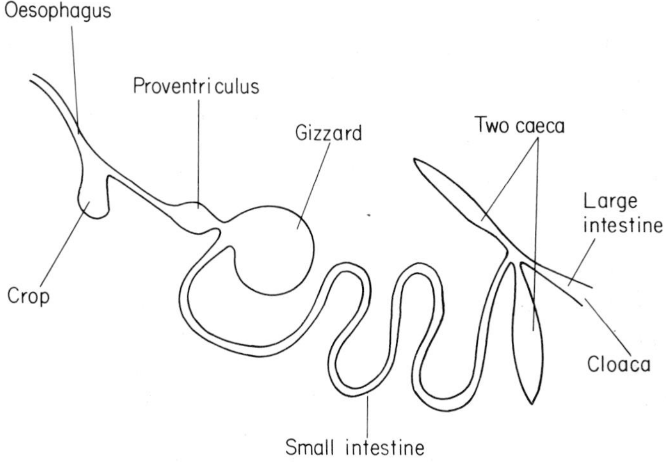

Fig. 11.5. Diagram of the digestive system of the fowl.

absence of teeth, their function being replaced to some extent by the grit in the gizzard; the production of saliva is very limited; the relatively short digestive tract necessitates a more rapid digestion process; and there is little bacterial breakdown of fibre.

There are two main types of grit provided for poultry, soluble and insoluble. Soluble grit, usually limestone or oyster shell, is basically a source of calcium for laying birds to balance their requirements for egg-shell production, and is not normally fed to growing stock. Insoluble grit, usually granite or flint, is supplied to aid the gizzard in the performance of its natural function of food grinding. In general the digestibility of whole grains is increased by 10 per cent and of mashes by about 3 per cent if access to insoluble grit is allowed, and it should be supplied to all types of poultry.

Over 75 per cent of the poultry in the British Isles are fed on rations prepared by compounders or specialist concentrate manufacturers. These firms supply a complete ration, a concentrate which when mixed with a stated quantity of home-grown cereal meals forms a balanced ration, or a diet which will balance certain amounts of whole grain.

Various feeding systems are adopted. In an all mash system the various ingredients are ground and mixed together in the form of a mash. This is generally fed dry in troughs with hoppers holding additional supplies attached so that the troughs are continually kept filled. The birds can eat what they require whenever they so desire. Pellet or crumb feeding is becoming very popular. Under this system the ingredients are ground and mixed together and then compressed into small pellets about the size of a cereal grain for adult birds or smaller particles called crumbs for chicks. The advantages are that there is less food waste and the birds have to consume the whole ration, being unable to select certain ingredients, as is possible when mashes are fed. For growers and adult birds under a grain and dry mash system, a grain balancer formulation mash is always before the birds in troughs and a feed of mixed grain, scattered over the floor of the house, is supplied in the evening. For layers an old-fashioned system, which gave good results in practice and is still used by some 'hobby' poultry keepers, is the grain and wet mash system in which a small feed of grain was supplied in the morning, a feed of wet mash in the early afternoon and a second feed of grain in the evening. This system is very time consuming as not only has the wet mash to be prepared, but the troughs in which it is fed need regular cleaning.

Young fowls

The unabsorbed yolk sac in a newly hatched chick supplies the nutrients required during the first 72 hours of life, and so chicks need not be fed during this period. However, food and water are generally offered at 24 or 48 hours, and some chicks will start pecking at these ages. The food is normally supplied *ad lib.* in mash form at first and this diet form may be continued or replaced by crumbs.

Laying birds

Chicks need a diet containing a high level of energy and from 18 to 20 per cent of good quality protein until they reach 8 weeks of age, when a medium level of energy and a protein percentage of 14 or 15 suffice. The crude fibre content should not exceed 7 per cent up to 8 weeks of age and 8 per cent after that. Birds reared on grass from 8 weeks onwards may be given a grain feed to replace some of the meal, and the grass consumed will reduce the quantity of concentrate food required. The energy foods commonly used in chick starter and grower rations comprise maize meal, which is highly favoured, and coarsely ground wheat, which are included at fairly high levels in starter diets, and barley meal and weatings, which are used particularly for grower rations. The popular protein foods are fish and herring meals, these often being combined with soya bean meal.

Meat and bone meal is sometimes used as a supplement to the above.

Pullets being reared intensively for future egg production are almost always fed entirely on a concentrate ration in meal or pellet form, but broiler breeding stock often receive a meal or pellet and grain diet. This is because birds of the latter type tend to lay down fat excessively and the concentrate feeding level needs to be controlled and the grain allowance can be spread over the floor to encourage scratching and exercise. The change from a rearing to a laying diet should be made before the birds come into lay, when they are between 16 and 18 weeks of age according to the strain, so that a dietary upset is not caused at the start of the laying season.

Table birds

The energy requirements are high because maximum growth rate is usually essential so that the birds can reach market weight as quickly as possible. Because fats have a very high energy content they are frequently included in fattening diets, particularly in broiler rations. In addition to increasing the food value they make a ration more palatable. Various fats are used, tallow and maize oil being favoured. Inclusion levels are normally from 3 to 5 per cent, and should not exceed 10 per cent. Carcase quality is influenced adversely by yellow pigmentation in the fat and skin, and so the content of yellow maize in the diets fed towards the end of fattening is kept low.

Table 11.2. Formulations of some typical rations for young fowls.

Foodstuff	Chick Starter (%)	Grower (%)	Broiler Starter (%)	Broiler Finisher (%)
Maize meal	25	15	25	10
Ground wheat	30	25	41·2	65
Barley meal	15	35	—	—
Weatings	10	10	—	—
Soya bean meal	7·5	7·5	20	12·5
Fish meal	7·5	2·5	10	7·5
Meat and bone meal	2·5	2·5	—	—
Limestone flour	—	1·3	—	—
Fat	—	—	1·3	2·5
Mineral/vitamin supplement	2·5	1·2	2·5	2·5

To provide the rapidly growing birds with sufficient protein to make muscle, protein percentages of between 20 and 25 are usual in broiler starter diets from hatching to 6 weeks of age and between 17 and 21 per cent in finishing diets from 6 weeks to marketing. In the case of capons a lower protein level, 16 per cent, can be used after chemical caponisation, as the birds grow just as well on a diet containing this level of protein as on one containing a higher percentage. The crude fibre content of a broiler starter diet should be 3·5 per cent of crude fibre at the most and a broiler finisher ration not more than 5 per cent.

Examples of typical food formulations for young fowls are given in Table 11.2 and approximate daily food consumption figures in Table 11.4.

Adult laying fowls

High-energy diets are now very popular because they are more efficiently converted into eggs than are those low in energy. Also it has been reported that high-energy diets often lead to the production of larger eggs. It is important that rations high in energy have higher levels of the other nutrients compared with low-energy diets, for birds eat primarily to satisfy their energy requirements and so eat less of a high-energy diet. Birds of the small hybrid strains adjust their food intake very satisfactorily according to the energy level in the food, but medium and particularly heavy-bodied birds tend to overeat and lay down internal fat. Cereal grains are used principally as the source of energy, maize and wheat being favoured in the high-energy diets and wheat by-products, oats and, possibly, barley in the lower-energy rations.

The protein percentages vary from about 17·5 in the high-energy diets fed to the light hybrids down to 15 for the heavy-type layers, with an average range of from 15·5 to 16·5.

On some large poultry farms the protein percentage of the diet is reduced as egg production declines towards the end of a laying period, but such a change must be introduced with care. Fish meal, soya bean meal, and meat meal are the most popular sources of protein for poultry diets. As the digestive systems of fowls are not adapted for the utilisation of fibre the crude fibre content of a laying hen ration should not exceed 9 per cent.

Examples of typical food formulations for laying fowls are given in Table 11.3.

Table 11.3. Formulations of some typical rations for laying fowls.

Foodstuff	High Energy for Light hybrids (%)	Low Energy for Heavy birds (%)
Maize meal	40	20
Ground wheat	25	15
Barley meal	12·5	35
Weatings	—	12·5
Fish meal	10	5
Meat meal	5	6·25
Limestone flour	6·25	5
Mineral/vitamin supplement	1·25	1·25

Mineral requirements

Rations for fowls need to have adequate quantities of calcium, phosphorus, sodium, chlorine, and manganese. A severe deficiency or an imbalance of calcium and phosphorus can cause rickets in growing fowls and chick rations should contain about 1 per cent of calcium and 0·8 per cent of phosphorus and grower diets about 1·3 per cent of calcium and 0·6 per cent of phosphorus. Diets fed to layers must contain adequate amounts of calcium for shell formation, and high-energy diets, which have a low intake level, should contain about 3·8 per cent of calcium, and low-energy diets, which are consumed in larger amounts, about 3 per cent.

A deficiency of sodium and chlorine may cause a reduction in appetite with a resultant decline in growth rate or egg production. Most diets, therefore, have about 0·5 per cent of salt added, but the level of salt in the ration should not exceed 1 per cent or the birds will drink large quantities of water and, as a result, produce wet faeces. A manganese deficiency can lead to an abnormal bone development which can cause the tendon on the hock joint to slip in growing birds. It is normal to add a small quantity of manganese to the ration to prevent this abnormality developing. High levels of calcium in the ration make manganese less available.

Vitamin requirements

Growing fowls react quickly to deficiencies of certain vitamins and in order to obtain optimum growth rates rations for young birds must contain adequate quantities of vitamins. Vitamin A is necessary for growth and the maintenance of a resistance to disease, particularly respiratory diseases. Fowls can convert the carotene in green plants and yellow maize into this vitamin, but most rations are fortified with a synthetic vitamin A preparation or a fish-liver oil. Vitamin D, in the form of vitamin D_3, is required if calcium and phosphorus are to be fully utilised for bone formation. Although fowls can form this vitamin from sunlight most young birds are now reared indoors and so a synthetic preparation or a fish-liver oil needs to be added to the ration. Vitamin E is normally present in adequate quantities in the cereal constituents of the diets in common use, but in certain circumstances it can be destroyed with the result that a disease known as crazy chick disease may develop. Most of the B vitamins are contained in sufficient amounts in the cereals included in rearing rations but deficiencies of some of them can lead to slow growth and poor feather development, and so synthetic forms or dried yeast are normally added as supplements, particularly to diets for young birds, to prevent any risk of shortages occurring. Cases of low hatchability of eggs after incubation have been blamed on a deficiency of vitamin B in the diet, and so the rations for breeding fowls are usually supplemented with vitamins of this complex, particularly riboflavin.

Water requirements

It is important that a supply of fresh water is always available. If nipple drinkers are used care must be taken to see that they are working efficiently at all times and, when birds are being reared in

Table 11.4. Approximate daily food and water consumption figures (per 100 birds).

Type of Stock	Food		Water	
	kg	lb	l	gal
Chick-broiler (1 week)	0·91	2	3·4	0·75
Chick-broiler (4 weeks)	3·14–4·54	7–10	6·8–9·1	1·5–2
Chick-broiler (8 weeks)	5·45–8·18	12–18	9·1–13·7	2–3
Growers-capons (16 weeks)	9·08–13·62	20–30	13·7–18·2	3–4
Adult (light breeds)	11·35	25	18·2–22·4	4–5
Adult (medium breeds)	13·62	30	22·4–27·3	5–6
Adult (broiler breeds)	15·89	35	27·3–31·9	6–7

groups, water troughs should be placed at intervals over the whole floor area so that the birds are never far from a source of supply. A restriction of water intake can lead to a marked reduction in growth rate or egg production and deprivation of water for a few hours can result in serious losses. Thus particular attention should be paid to the water supply during freezing weather, and, on large farms, immersion heaters are frequently fitted in the water supply tanks for use at such times.

Breeding

Some breeders use pullets to produce eggs for hatching but, if possible, it is better to breed from second-year or older birds. Parent birds should be vigorous and have the ability to resist disease, and the fact that a bird has produced a satisfactory number of eggs over 2 or 3 years is evidence of its physical strength. Some method of recording the number and size of the eggs laid, such as trap nesting, is of great assistance in selecting the hens kept for breeding, as a good breeder should be a good layer.

Trap nests differ from the nests commonly provided for laying in that each one is fitted with a trapdoor which a bird shuts itself as it enters. The trap nests are inspected every 3 hours and the birds are released and their leg or wing numbers are noted. Records are kept of the eggs laid. As this task is very demanding in labour it is only used for breeding birds. To reduce the work load some breeders trap nest for the first 3 months of a laying season, to obtain an indication of the level of egg production, and again during the month of August, to show which birds are still laying. This latter record detects the late moulter as, in general, late moulters are the better layers.

Normally cockerels, that is first-season males, are used for breeding as they are more vigorous than cocks and their fertility is usually better. The selection of stock cockerels is difficult and the most reliable indication of the breeding value of a cockerel is obtained by a study of the records of its sisters. Although few breeders retain male birds for two or more seasons some use proven sires whose values have been assessed by progeny testing for the production of pure-bred birds as well as hybrids. Cockerels should be fully mature before being used for breeding, and so should be 6 months old if of a light breed and 9 months old if of a heavy breed. The mating sex ratios which are generally considered optimum are one broiler breed cockerel per ten hens and one laying breed cockerel per fifteen or twenty hens. Some breeders, however, mate larger numbers of hens to their cockerels and obtain a fertility rate of not less than 90 per cent, which is the expected level. Although single-pen mating, in which one cockerel is run with a group of hens, is practised by pedigree breeders, flock mating, where a number of males are run with a large flock of hens, is the common method used on commercial farms. With this system a greater number of cockerels than is necessary may be introduced early in the season so as to allow for some cockerel wastage, as the introduction of new, strange male birds will lead to fighting.

When a pen is first mated fertile eggs are produced on the second day, but it is usual to wait for a week before collecting the eggs for hatching in order to obtain maximum fertility. When a cockerel is removed from a pen fertile eggs will be produced for another week, after which the fertility declines. Artificial insemination is rarely used in the breeding of fowls.

Eggs should not be kept longer than 7 days after being laid before incubation, and during this period should be stored at a temperature below 27°C (80°F) and preferably within the range of 10–15·6°C (50–60°F). On the nineteenth day of incubation, before the eggs are moved to the hatching compartment of the incubator, they can be tested to determine which are likely to hatch. Testing is carried out by candling, that is by placing the eggs on the incubator trays over a strong light which shines through them. An infertile egg is clear as the light shines through it, one containing an embryo which has died at an early stage of development has a small dark area, while a fertile egg is almost filled by the opaque embryo. By removing the eggs which will not hatch more room is available in the hatching compartment for the chicks as they emerge from the egg. The percentage of chickens which hatch varies with the strain of fowl. In light hybrids it is about 80 or 85 per cent while in broilers it is approximately 75–80 per cent.

Sex linkage

With certain matings between breeds it is possible to produce crosses in which the chicks are sex linked to a colour so that the sex can be determined at the time of hatching by differences in the down colour. For example, cockerels of genetically gold breeds, such as Rhode Island Red or Brown Leghorn, when mated to pullets of a genetically silver breed, like the Light Sussex, produce brown pullets and silver cockerels. The reverse cross, for example a Light Sussex cockerel with a Rhode Island Red pullet, gives rise to all silver chicks. Sex linkage also occurs when a non-barred breed male, such as a Black Leghorn, is crossed with a barred female, for example a Barred Plymouth Rock. The barring on the male chicks of the cross is shown in the down colour by a white spot on the back of the head.

In breeding hybrid fowls it is possible to arrange for the final generation to be sex linked by selecting appropriately coloured breeds for the original parent stock.

Egg production

The age at which fowls come into lay varies according to the strain, the level of nutrition during rearing, and the lighting pattern. One recommended lighting programme is to allow about 22 hours of light out of every 24 for the first 8 weeks of life, then decrease the hours by 1·5 per week until the birds are 16 weeks old, and then increase the lighting by 30 minutes per week until 16 hours of light are provided each day. This level is maintained throughout the laying period. Using such a programme the majority of well-fed, egg-producing strain birds will come into lay when between 22 and 24 weeks old and most broiler strain birds when between 26 and 28 weeks of age.

When production commences eggs are laid in sequences. A good producer will lay between four and seven eggs in a sequence, then miss a day and then lay a further sequence. This pattern is repeated. The laying season is usually restricted to about 50 weeks, as during this time the rate of lay is most economic. The birds are kept in the laying accommodation for 50 weeks and then 2 weeks are allowed for cleaning and disinfection before the new birds are introduced. Although individual birds would continue to lay for longer than 50 weeks production would probably fall to a level which might be uneconomic.

The average production of hybrid fowls which lay white eggs should be 250 eggs in the first year of laying. The average number of eggs laid by brown shell laying birds is less, usually about 220 eggs in a season. Broiler breeding birds lay on average between 150 and 160 eggs in a laying season. Pullet eggs are usually smaller in size than are those from older birds, but egg size is largely determined by genetic factors.

Most fowls kept for egg production are slaughtered at the end of their first laying period when they are about 18 months old, because the number of eggs laid in a second season is about 15 per cent less than in the first. However, egg size is generally satisfactory right from the start. Keeping birds for a second season avoids the capital outlay required to rear or buy replacement pullets, but this is offset by the cost of feeding the birds during a moulting period of from 4 to 6 weeks' duration when they are not laying and so are not providing any income. Occasionally birds of special genetic merit are kept by breeders for 3 years or more, but their egg yields diminish each succeeding year.

Moulting

When egg production ceases at the end of a laying season the birds begin to moult. Good layers, therefore, usually moult late and also rapidly. Some of the best layers continue to lay when moulting starts, although at a slower rate, but the continuance of production during the moult retards the rate of moulting. The order in which the different parts of the body lose their feathers is fairly constant, the usual moulting order being the head first, followed by the neck, body, wings, and tail, in that order. The average time taken over a moult is about 8 weeks, but there is a marked variation between individuals.

Birds normally begin to moult in August or September, but most prolific layers do not moult until October or November. If such birds are being retained for breeding they will not be in lay sufficiently early in the season to provide hatching eggs when required, and so force moulting is often practised by breeders. Early in August the birds selected for future breeding are moved to strange quarters, if possible, and the amount of food fed is drastically reduced. Often oats are given in limited quantities as the sole food. Some poultry keepers advise withholding the drinking water for a day, but this drastic measure is not recommended. After a week or 10 days the birds will be heavily in moult and, when the majority of the birds have shed five of the ten flight feathers, the feeding level should be steadily increased, and artificial lighting provided to bring the birds back into lay. Methods of force moulting by the use of drugs are being tested experimentally, but are not widely used in practice.

Broodiness

In modern poultry farming broody hens are not required for sitting on eggs and so broodiness, which is accompanied by a cessation of egg laying, is a serious fault in a laying bird. As it is an inherited character birds which go broody should not be used for breeding. The incidence of broodiness in modern laying strains has been greatly reduced, but if a hen does go broody it should be removed to a broody coop as soon as it is observed. A broody coop is a small cage with a slatted floor placed in the open where the air can circulate. The broodiness generally ceases after about 3 days. The bird can then be returned to the flock, and should be laying again in about 14 days. Drug treatments have been used as an alternative method of curing broodiness, but have only a similar degree of efficiency.

Health

A fowl in good health is alert, with bright prominent eyes. A healthy bird moves actively, and, if given the opportunity, will scratch in a search for food particles. The appetite is normal. Good laying hens usually have ragged, worn plumage, but the skin is soft and pliable. A healthy hen in lay has a bright-red comb and wattles. When the bird ceases to lay the comb and wattles become a dull pink in colour and show some diminution in size. In those breeds with yellow beaks and legs the colour fades in good layers towards the end of a laying period, but returns when the bird stops laying. Early signs of disease are a reduction in food consumption and, in laying birds, a drop in egg production and the appearance of egg-shell defects.

The normal body temperature is about 41·7°C (107°F) and is usually in the range 41–42°C (106–107·5°F). The pulse rate of a fowl is usually from 150 to 300 beats per minute, but is difficult to count and is not normally recorded.

Diseases can cause severe reductions in production efficiency by lengthening the time taken by table fowls to reach slaughter weight and decreasing the rate of egg production in laying birds. Thus every effort must be made to prevent outbreaks of disease, and control any which do occur. The treatment of diseased birds is seldom justified, although it may be attempted in certain cases.

Common diseases and their prevention

Salmonellosis is the term used for infections with organisms of the Salmonella group of bacteria. They are important because they can also cause disease in man, and outbreaks of gastro-enteritis, commonly known as food poisoning, can accompany the consumption of carcases which come from infected birds or which become contaminated while passing through the dressing plant. Outbreaks of food poisoning connected with hen eggs are rare, but have been recorded. Clinical signs are seen most frequently in chicks up to 3 weeks of age, and are dejection, a pot-bellied appearance and diarrhoea with the vent becoming pasted with whitish excreta. The mortality rate can be high, and survivors may continue to excrete the organism for some time. Newly hatched chicks generally become infected by ingesting contaminated food or water. Rats and mice may be carriers and can infect foodstuffs as well as the premises to which they gain access. Control measures include the destruction of wild rodents, the incineration of infected carcases and litter, the thorough disinfection of contaminated utensils and equipment, and the avoidance of contaminated food and water. Eggs from infected birds should not be used for hatching.

Chronic respiratory disease (Mycoplasmosis) can affect birds of all ages, but the greatest losses occur in young birds. Although the death rate may not be high there is a marked loss of condition and a reduction in growth rate in table fowls and a fall in egg production in laying birds. The clinical signs result from an infection of the air sacs, which become thickened, and include nasal discharges and swollen faces. Chronic respiratory disease is often associated with mycoplasma organisms, which are a small primitive type of bacterium without a rigid cell wall. They most frequently cause disease when in combination with certain viruses, such as that causing infectious laryngotracheitis. The best method of prevention is to establish flocks which are mycoplasma free, and this is a practical proposition.

All ages of fowl are susceptible to *infectious laryngotracheitis* and the most obvious signs are sneezing, gasping, and breathing through a partly open beak. Occasionally a loud noise is heard as birds try to dislodge mucous deposits in the respiratory tract. The disease is caused by a virus which is spread by contact. Although the mortality rate is not likely to be high there is a marked loss of production and a proportion of the recovered birds remain carriers. Live vaccines are available which give protection but there is a risk that some of the treated birds will become carriers.

In very young birds *infectious bronchitis* causes coughing, respiratory distress, and death, and in laying birds milder respiratory signs and the production of badly shaped eggs with thin shells with chalky lumps and thin albumen layers known as watery whites. The cause is a virus. Vaccines which can be administered via the drinking water or by aerosol sprays have been developed for the control of this condition.

Newcastle disease (fowl pest) affects fowls, and other poultry species, of all ages. In chicks the clinical signs are gasping, coughing, and depression followed by paralysis of the legs, wings, and head. In laying birds the respiratory signs may be mild or severe but a prominent indication is usually a marked, sudden drop in egg yield. The eggs laid by birds which recover are likely to be imperfectly shelled for a time. A profuse green diarrhoea may also be seen. The mortality rate is often high. The disease is caused by a virus and is controlled principally by vaccines which can be administered by inoculation or in the drinking water. The disease is compulsorily notifiable and all suspected outbreaks must be reported.

Marek's disease (fowl paralysis) usually affects birds between 3 and 5 months of age, although it may be seen in older or younger birds. It can cause serious losses in pullets being reared for egg production. There are two forms, the classical in which paralysis is the main clinical sign, the legs, wings, and neck being most commonly affected, and the acute manifested by widespread tumours affecting the gonads especially. The disease is caused by a virus which can be air-borne or spread by direct contact. Preventive measures include vaccination of day-old chicks, rearing young birds away from adults, planning poultry house ventilation systems to avoid the risk of air transfer of the virus, and the avoidance of the introduction of the virus by newly purchased birds.

Avian leucosis usually occurs in birds over 5 months of age and is observed most frequently in laying hens. Affected birds lose condition and die due to tumours which may develop in all the body tissues, except the nerves, although the liver is the organ most frequently affected. The disease is caused by a virus which may be spread by contact or transmitted through the egg. Some breeds or strains are more resistant genetically than others. An effective vaccine for the prevention of this disease has not yet been produced.

Coccidiosis is an important disease of susceptible fowls. Chicks between 3 and 6 weeks of age are affected by a coccidial parasite which causes disease in the caecum, and older birds by several species of coccidia which produce lesions in the wall of the small intestine. Chicks look dull and listless and void bloodstained faeces. Older birds stand about with closed eyes and drooping wings and rapidly lose weight. Diarrhoea is usually a feature but the faeces are only bloodstained in severe cases.

Most fowls have some coccidia in their intestines, which produce no apparent ill effects. The clinical signs described above are caused when large numbers of coccidia are present. Mature coccidia in the intestines pass minute egg-like bodies which are voided with the chick's faeces, ripen in warm, moist conditions and, when eaten by another chick, develop in the new host's intestine. An effective method of preventing disease outbreaks is to keep the birds from coming in contact with infected faeces and so cage housing systems with slatted floors which permit the faeces to fall through so that the birds do not come in contact with them are very efficient. Deep-litter floors should be kept dry as few oocysts develop in such conditions, while wet litter is ideal for their development. A variety of drugs are available for the prevention of this disease and most proprietary diets for rearing fowls contain preventive drugs at doses which are suitable for continuous administration.

Red mite is a very important external parasite of fowls and can attack birds of all ages. Severe infestations cause anaemia, with the result that young birds fail to grow satisfactorily and egg production in laying birds is seriously reduced. The adult mites are like minute spiders and can be seen with the naked eye. During the day they hide in crevices in the building and by night they attack the birds and suck their blood. Eggs are laid in the building and the mites mature rapidly so that heavy infestations can be built up in a short period of time. The parasite can only be controlled by treatment of the building and this should be undertaken during the day when the mites are in the house and not on the birds. The work can be carried out most efficiently when the house is empty. Special attention must be paid to all crevices. A number of pesticides can be used, including gamma B.H.C. and carbaryl. As a precaution the birds should be dusted with carbaryl, or another product recommended for use on live birds, before they are returned to the house so as to destroy any mites which might have remained on the birds. Another similar mite is the northern feather mite, which differs from the red mite in living on the fowl throughout its life.

Handling

When it is necessary to handle fowls they should be caught with as little disturbance as possible. This is usually done by driving them into a corner and holding them there with a wire-netting frame, or by placing a catching crate, designed to hold from twenty to thirty birds, in a corner or against a door and driving the birds gently into it. For catching individual birds a catching hook, designed on the same principle as a shepherd's crook, is used to catch the birds by the leg. In windowless houses blue light bulbs can be inserted into the artificial lighting system as the birds do not move in blue light. This is the method used in broiler houses when the birds are ready for killing. Broilers should always be caught by the leg and not by the thigh and handled gently otherwise a number of birds may be down graded due to bruising when marketed. It is possible to handle birds of all ages in houses with windows during the hours of darkness, using a flash-lamp delivering a narrow beam of light. Such a light does not cause the birds to leave their perches and they can be easily caught and handled.

To lift a bird from a catching crate or perch a hand should be slipped under the wing close to the body with the thumb over the wing. The bird should not be grasped by a leg as this usually results in a flapping of the wings and struggling. When caught, a fowl can best be held with one hand under its body for support. The thumb is placed outside one leg, the next two fingers between the legs and the other two fingers outside the other leg. Most fowls will sit quietly in this position, but the other hand can be placed over the wings if these are flapped. A bird can be carried by keeping one hand under the body with the head pointing towards the handler's back so that the wings can be held between the upper part of his arm and his body.

It is possible to determine whether a hen is in lay by examining the pelvic bones. In a pullet which has not laid there is only the width of one or one and a half fingers between the pelvic bones, while in a bird in lay three or more fingers can be inserted in the same space. If a bird has stopped laying the width will be reduced so that the space is only one and a half or two fingers wide.

Sexing

When hatched chicks are usually sexed by trained specialists. Sexing can be performed manually by applying gentle pressure to the abdomen to evacuate the faeces and then everting the cloaca. Under a bright light a genital eminence will then be seen just inside the vent in most chicks. It is always present in cockerel chicks and has a knob-like appearance. Many pullet chicks do not have such an organ, and in others it is small. It is essential that a speed of about 800 chicks per hour and an accuracy of about 95 per cent are attained. It is also possible to sex by machine, but the method, though accurate, is not as rapid. The machine consists of a hollow tube, in which a light is reflected, which is inserted into the cloaca. The operator, looking through an eyepiece, is able to see and identify the reproductive organs through the thin intestinal wall.

Administration of medicines

Because of the large number of birds kept together under modern husbandry methods group medication is practised wherever possible. Drugs are frequently included in pelleted food, e.g. most rations for growing birds include a coccidiostat. Vaccines and tasteless medicines can be administered on a mass basis by mixing them in the drinking water. Some vaccines can be sprayed or dusted over a flock of birds with success. Individual treatment is rarely carried out, but liquid medicines can easily be given by holding a bird's comb and opening its beak with a finger and passing an eye dropper or glass pipette down the throat and releasing the fluid. Tablets and capsules can be given by holding the bird's head in a similar manner and pushing the tablet or capsule, held in a pair of forceps, down the throat.

Should individual vaccinations be required most are given subcutaneously or intramuscularly into the muscles of the leg or breast. All vaccines should be given as recommended by the manufacturers, but great care should be exercised in the case of laying pullets not to upset the birds unduly, or egg production will drop markedly or even cease. Intravenous injections can best be made into a vein which runs under the wing. The bird is placed on its side on a level surface where it can be held steady. An assistant takes hold of both legs with one hand and pulls the wing back with the other, thus exposing the vein. An eyedrop vaccination method is occasionally employed.

General

Vices

Feather pecking
A bird will pluck feathers from itself or another bird, causing extensive bare areas. The habit is imitated by other birds.

Cannibalism
This follows feather pecking if blood is drawn. The blood starts the birds pecking at the skin and flesh, leading to extensive wound formation, and, possibly, death.

The causes and prevention of these two vices may be considered together. Both occur most commonly in birds which are kept closely confined, and probably start through boredom. It has been suggested that a deficiency of protein in the diet may initiate these habits, but birds develop these vices when fed on diets without any known nutritional deficiency. Poor feathering is a predisposing cause, and this may be nutritional in origin although it is more likely to be genetic. The selection of rapid feathering strains can reduce the incidence of these vices. In broiler houses they are controlled by adjustments to the light intensity, but debeaking is the normal method of preventing their occurrence in densely stocked laying fowl houses. In small flocks bare areas on fowls can be dressed with Stockholm tar or a commercial preparation which will deter pecking, and efforts can be made to keep the birds occupied by providing fresh green food, such as cabbage or kale, suspended about 0·31 m (12 in) from the floor. These methods are obviously impracticable for large flocks.

Egg eating
The habit usually starts when an egg is accidentally broken and its contents eaten. In many cases it follows the laying of eggs with thin or cracked shells. A taste for eggs is developed by some birds which break and eat sound eggs. Egg eating can be difficult to stop once it has developed. Care should be taken to ensure that ample calcium is provided in the diet to reduce the risk of thin-shelled eggs being produced.

In a battery cage occupied by an egg eater it is possible to fit a guard below the food hopper which allows an egg when laid to roll to the front of the tray and be out of view of the bird which laid it. However, some hens learn to keep their legs together after laying and so prevent the egg rolling forwards. In birds kept in flocks it may be possible to identify an egg-eating bird by observing traces of yolk in the region of the beak. Persons keeping small numbers of fowls sometimes try filling a blown egg with mustard in the hope that the egg eating bird will object to the taste and stop the habit, but this method is of little value. Persistent egg eaters have to be culled.

Minor operations

Debeaking

Debeaking is the normal method of preventing feather pecking and cannibalism in densely stocked laying fowl houses. The upper beak is cut back by approximately a quarter or a third of its length. The usual method is to employ an electric cauteriser which cuts the top off the beak and cauterises any severed blood vessels at the same time. The operation can be undertaken at any age, but is most satisfactorily performed when the birds are between 9 and 16 weeks old. If debeaked at hatching, pullets generally need to be operated on again if cannibalism is to be controlled during the laying period.

Dubbing

Dubbing means the removal of the comb, and the operation is sometimes performed on male birds to avoid frost bite and fighting injuries which can cause pain and so reduce fertility. It is best to remove the comb of chicks within the first 72 hours of life with curved scissors, but the operation can be carried out later using a sharp knife. In the latter case a local anaesthetic should be administered and the raw surface needs to be dressed to prevent excessive haemorrhage.

Caponisation

Chemical caponisation is the method which is now employed. Female sex hormones, or synthetically prepared substitutes, are used and hexoestrol and diethylstilboestrol are the drugs of choice. These cause testicular regression and are supplied in tablets each containing 15 mg. A specially designed injector syringe is used to implant a tablet just below the skin in the upper part of the back of the neck. The needle is directed from the head downwards so that there is no risk of the tablet falling out of the hole made in the skin by the needle when the bird stands upright. The upper part of the neck is used as this will be discarded when the carcase is dressed, and so there will be little risk of any tablet residue being eaten by a human being. One 15 mg tablet should be used for birds up to about 2·27 kg (5 lb) in weight at the time of implantation, but two pellets are needed for larger birds and stock cockerels no longer required for stud purposes.

Chemical caponisation is most profitably used for the production of birds to be killed when between 16 and 20 weeks of age. The birds should be implanted once, 4 weeks before killing. Under the influence of the drug cockerels do not crow, their combs and wattles become small and pale, and their carcases become soft and tender. The effect of the treatment starts to disappear in about six weeks, and the birds then develop their normal sexual characters.

Wing feather cutting

Some birds kept in runs surrounded by wire fences frequently fly over such barriers. This can be prevented by clipping the flight feathers of one wing, so that the bird becomes unbalanced when it rises from the ground. After a moult the new flight feathers will need to be clipped.

Marketing

Egg marketing

A producer may sell his own eggs direct to a consumer on the farm, at a local market, or by door to door visiting, but eggs may only be sold to shops or other traders by registered packing stations. Thus producers who do not sell to customers of the type listed above must sell their eggs to a packing station. Attention must be paid to the quality of the eggs marketed. Externally they must be clean, well shaped, and with smooth, reasonably thick, shells, because stained, dirty, misshapen eggs with rough, thin shells are always discriminated against. Batches should be of a uniform size and of a matching colour. Many consumers favour brown-shelled eggs. Internal quality is assessed by candling, which means the passage of light through the eggs towards the observer in a darkened room. It is more difficult to candle eggs with dark-coloured shells than those with white shells. Characters which can be measured are shell quality, as thin and cracked shells are easily detected, yolk position, which in a fresh egg should remain close to the centre of the egg, the absence of blood spots caused by the rupture of a small blood vessel while the yolk is being formed, and embryo development, which can be avoided if fertile eggs are not used for human consumption.

There are some characteristics which can only be observed when the eggs are broken, and these include odour, flavour, yolk colour, and albumen quality. Yolk colour depends on the diet. Birds on pasture produce eggs with deep-coloured yolks, and such an attractive appearance can also be obtained by including yellow pigments in the concentrate ration. Olive-green or brown-coloured yolks may also result from incorrect feeding. The albumen of a good quality egg will be fairly firm.

Marketing table fowls

Fowl carcase processing and marketing are now highly specialised tasks, and the majority of birds are dressed at poultry packing stations and marketed wholesale by large scale firms. At the packing stations the birds are killed, plucked, eviscerated, cooled, graded according to weight, colour and finish, and packed in boxes containing birds of a uniform quality. The packing stations are equipped with refrigerators and deep-freezing cabinets so that the times of sales can be regulated according to the market demand. Co-operation between a large-scale table fowl producer and a marketing organisation is now essential, so that the birds can be accepted at the packing station when at a suitable weight for slaughter. Small-scale producers can still sell direct to local shops or consumers, and this can prove remunerative.

Welfare code

Newly-hatched chicks have poor control over their body temperatures and so must be housed in environmental conditions which enable them to maintain their normal body temperatures without difficulty. After about 4 or 5 weeks the birds can tolerate a wider range of temperature, but still need protection from excessive heat and cold. If mechanical devices are used to control the heating, ventilation, and lighting of a building, or the feeding and watering systems, care must be taken to see that they are regularly inspected and are working efficiently.

Irrespective of the type of enclosure fowls must be able to stand normally, turn round, and stretch their wings without difficulty. In intensive systems when birds are kept in groups they must have sufficient space to perch or squat without interference from other birds. Birds which are allowed access to land outside a house should not be kept under conditions which are continuously muddy or on ground which has been densely stocked with fowls for a long period and has become 'fowl sick'. Adequate fresh food and water should always be available. During severe winter weather care must be taken to prevent the drinking water freezing.

Recommendations are made that debeaking should only be carried out as a last resort when suffering would be caused in the flock if it were not done, that dubbing of birds older than 72 hours should only be done by a skilled operator on veterinary advice, and that surgical castration, and dewinging, pinioning, notching or tendon severing, which involve mutilation of the wing tissues, should not be undertaken.

CHAPTER 12
TURKEYS, DUCKS AND GEESE

TURKEYS

Introduction

Turkey keeping has developed greatly both in scale and efficiency since the end of World War II. At one time most turkeys were reared in small numbers as a side line on general farms, but now the main producers are turkey specialists keeping large numbers of birds. Unlike fowls, turkeys have not be been bred for any extreme show points by fanciers and all the breeds have utility qualities.

The small-scale producers still keeping turkeys generally undertake the complete task of production including killing, plucking, and dressing. The sale of oven-ready birds at Christmas is their main outlet. The specialists concentrate on breeding or rearing or both. The specialist breeders keep large numbers of breeding birds and take full advantage of the significant advances made in recent years in the science of genetics. Because of the numbers kept a high degree of selection is possible, allowing significant progress to be made in the development of particular attributes. Their aim is to produce eggs all the year round which they attain by careful management of several flocks coupled with the use of artificial lighting. The specialist rearers purchase large numbers of day-old poults which are sold eventually as fattened birds. The aim is to keep the equipment running to capacity for as long a period as is possible during the year and so batches of poults are obtained at regular intervals to replace the birds sold. Turkey production groups are now being formed so that breeding, rearing, and marketing can be integrated. The profits are allocated to those responsible for each of the production stages, and some form of equalisation fund is generally established to ensure that price fluctuations over a period are fairly adjusted.

Although most turkeys are still marketed at Christmas there is an increasing demand for oven-ready turkeys throughout the year, and there is every indication that considerable expansion of this market is still possible. To meet this need turkeys must be produced in the required numbers, offered at prices which the consumers will pay and be available all the year round.

Definitions of common terms

Poult. Young bird up to about 8 weeks old.
Turkey grower. Bird from about 8 to 26 weeks of age.
Turkey hen. Female adult from about 26 weeks of age upwards.
Turkey stag, tom, or cock. Male adult from about 26 weeks of age upwards.

Principal breeds

Broad Breasted White

A white feathered breed with very well-developed breast muscles. The breast width should be at least 89 mm (3·5 in) when measured at a depth of 44·5 mm (1·75 in) at the widest part. The liveweights vary between strains, adult females weighing from 6·35 to 10 kg (14–22 lb) and males reaching weights of 11·3 to 18·2 kg (25–40 lb). The hens

Fig. 12.1. Broad Breasted White turkey hens showing saddles with numbers.

are relatively poor layers, producing from fifty to sixty eggs per season.

Broad Breasted Bronze
Similar but with feathers having a ground colour of black with a terminal edging of white, and copper bronzing on the tail and wings.

American Mammoth Bronze
Similar in colour and body size to the above, but without the well-developed breast muscles.

White Beltsville
A small white variety with a mature weight of 4·54 kg (10 lb) for females and 6·35 kg (14 lb) for males, which was bred to produce the small-bodied bird required by housewives buying for a small family. Egg production is relatively good, being from 100 to 120 eggs per breeding hen per season.

Hybrids
The modern trend is towards some form of hybridisation, using an inbred line strong in one or two factors to supply parent stock to mate with progeny of a second line, inbred for other factors. The object, of course, is to produce poults which grow faster, flesh better and utilise their food more efficiently than the parent stock.

Identification

Poults can be identified individually by numbered wing bands and families can be indicated by the punching of holes in the web between the toes. Leg bands are not suitable for identifying young poults as they grow so rapidly that the bands have to be removed after a few days to avoid leg injuries. Older birds can be marked by using large visible wing tabs or by marking numbers on the canvas saddles used to protect the backs of hens from injury by the spurs of the stags.

Ageing

There are no reliable signs which indicate the age of adult turkeys, but the hair-like tuft attached to the upper part of the breast in stags increases in size with age and may reach a length of from 101·6 to 152·4 mm (4–6 in) in the larger types.

Housing

Incubators

Turkey eggs are incubated in the same types of incubator as are fowl eggs, but more space is required. The incubation period is 28 days, 24 days being spent in the setting section and 4 days in the hatching section of the incubator. The temperature is usually kept at 37·5°C (99·5°F) all through, but the relative humidity is increased from about 62 per cent for the first 24 days to 75 per cent for the last 4 days.

Young turkeys

After hatching, the poults are susceptible to draughts and sudden changes in temperature and are generally kept for 4 weeks in specially designed tier brooders placed in well ventilated, but draught free, buildings. The brooder temperatures are similar to those recommended for chicks, but poults are allowed twice the floor area required by chicks. From 4 to 8 weeks of age the poults can be left in the tier brooders, but may be accommodated in hay box brooders. Such brooders have a covered section with hay between the ceiling above the poults and the roof so that the heat produced by the birds is conserved and artificial heat is not required. Wire floors are fitted well above the ground, as it is possible for young birds to pick up parasitic infestations from contaminated soil. The overall dimensions of a hay box to hold from twenty-five to thirty poults are 0·91 × 2·74 m (3 ft × 9 ft).

When 8 weeks old the poults become stronger and can be kept in pole barns. These are relatively cheap structures of rough-hewn poles with a light roof and back, often of corrugated iron, and wire-netting fronts and sides. They usually have earth floors. Each turkey is allowed 0·28–0·47 m² (3–5 sq ft) of floor space. These buildings are usually fairly wide, 6·1 m (20 ft) or more, which gives shelter from the rain, and most are 3·05 m (10 ft) high in front and 2·42 m (8 ft) at the back. They are frequently constructed face to face, with a wide service track between them. Alternatively the poults may be accommodated in verandahs with floors of slats 50·88 mm (2 in) wide set 25·4 mm (1 in) apart and fitted 0·61 m (2 ft) above the ground. The main advantage is that the faeces fall through the spaces between the slats to the ground below and need only be removed once a year.

Intensive floor brooding is used on some large turkey farms, batches of 1,000 and more poults being brooded on lines similar to those described for broiler chicks. However some farmers consider that poults do better in flocks of 250 or less. For the first 4 weeks, 0·07 m^2 (0·75 sq ft) of floor space per bird are allowed; for the second 4 weeks, 0·14 m^2 (1·5 sq ft); and for the third 4 weeks, 0·23 m^2 (2·5 sq ft). On large commercial enterprises the warmth required is usually supplied by a central heating system, but hovers may be employed. If hovers primarily designed for chicks are being used care must be taken to ensure that the poults have sufficient head room. This system is effective for turkeys to be killed at about 12 weeks of age as they do well and grow evenly, but is too costly for turkey growers to be killed at large sizes, as such birds require a relatively large floor area necessitating a high housing cost.

Occasionally 12-weeks-old poults are kept on free range. The main danger is from attacks from foxes or dogs, and this danger can be minimised by driving the birds into perch yards each night and releasing them in the morning. Perch yards consist of simple fences about 1·8 m (6 ft) high surrounding a series of perches arranged at 0·62 m (2 ft) intervals, allowing 0·31 m (1 ft) width of perch per bird. The perches are placed about 0·76 m (2·5 ft) from the ground. The yard floor is covered with straw, clean straw being added as the old becomes fouled. The yard is cleaned out after each batch of turkeys.

Adult turkeys

On some general farms breeding flocks of turkeys are kept on a range system, being driven at night into a perch yard. A better plan is to bring the turkey flock back to the farmyard each night and confine the birds in a suitable building. The range system is not very satisfactory as birds may be killed by foxes or other vermin and eggs laid in odd places may be missed at collection times.

Most breeding birds are housed in covered straw yards or pole barns, each bird being allowed from 0·37 to 0·56 m^2 (4–6 sq ft) for small-type and from 0·56 to 0·93 m^2 (6–10 sq ft) for large-type birds. The food and water troughs are often fitted outside the pens, the birds passing their heads through the spaces between upright slats in one wall in order to eat or drink. Perches are usually provided so that the birds do not roost on a floor which may be damp. A ladder of poles is sometimes built and is very satisfactory, although the birds do tend to crowd to the topmost perch. Some farmers do not provide nest boxes, allowing the turkey hens to lay their eggs on the

Fig. 12.2. Turkey hen entering a trap nest.

straw, but this practice leads to the soiling of the eggs with faeces and cannot be recommended. If nest boxes are supplied there should be at least one box for every three hens. Each compartment should be 0·46 m (1·5 ft) wide, 0·46 m (1·5 ft) deep and 0·62 m (2 ft) high at the front sloping to 0·46 m (1·5 ft) at the rear. A litter-retaining board about 101·6 mm (4 in) high should be fitted across the front. It is advantageous to site the nest boxes at the side of the enclosure so that they can be opened at the back for egg collection without the attendant having to enter the pen. Suitably designed fronts can be fitted to the next boxes so that trap nesting can be practised if desired. It is difficult to provide effective light control to regulate the laying period, for it is impossible to black-out straw yards or pole barns adequately. However, artificial lighting can be provided after the summer solstice in order to reduce the drop in egg yield which would tend to occur as the daylight hours shortened.

A somewhat similar system giving very good results is the use of slatted floor verandahs, standing about 0·76 m (2·5 ft) from the ground. The verandahs are divided into pens, a space about 5·5 m × 1·56 m (18 ft × 5 ft) being allowed for a stag and ten hens.

On some large farms windowless houses are used so that full use can be made of artificial lighting to stimulate egg production.

Breeders of the broad breasted varieties occasionally keep their breeding birds in individual battery cages and breed by artificial insemination. The cages have a floor area of about 0·37 m^2 (4 sq ft) which is reasonably adequate for the hens but constricts large stags. The food is usually fed in self-filling food hoppers and an automatic system of water supply and a mechanical method of faeces removal are generally fitted. The advantages claimed are that individual records of egg yields can be kept by pedigree breeders without the difficulties associated with trap nesting, and that birds which lay few or poor quality eggs can be easily detected and culled. The disadvantages are foot injuries leading to lower egg yields, a higher proportion of cracked eggs and a tendency to excessive fatness.

Feeding

General

Turkeys need insoluble grit in the same way as fowls, but care must be taken to see that young poults do not eat excessive amounts. Thus chick-size flint or granite grit is usually sprinkled on the food each day for the first 2 weeks and once a week for the next 4 weeks. Poults should not be given grit in a hopper until they are about 6 weeks old, but then hoppers containing adult size grit can be provided.

Poults and fattening turkeys

Poults have poor vision for the first week and if care is not taken to ensure that they find the food and water troughs and start to eat some may die of starvation. Bright strip lights are, therefore, usually fitted over the troughs to attract the birds to the food. Alternative methods not now in general use are to place coloured glass marbles in the food to catch the light and attract the birds or to dye food crumbs different colours to encourage the birds to notice the food. Some turkey rearers place pieces of corrugated cardboard on the floor of the brooder for the first 2 or 3 days and cover these with food so that the poults walk on the food and are tempted to peck at it.

Proprietary foods are generally used and for the first 8 weeks a turkey starter diet high in energy and containing about 28 per cent of protein is given. This diet can be fed in the form of a coarse mash, but better results are obtained if it is prepared in the form of crumbs. A mash consisting of finely ground ingredients should be avoided as it may cause a clogging of the beak and a curling of the tongue during the first 7 days of life. From 8 weeks onwards a turkey rearer diet in pellet form can be given which has a slightly lower energy content and a protein percentage of about 22. This ration is continued up to 16 weeks of age or 3 weeks prior to killing, whichever is the earlier. Some farmers wish to feed grain as part of the diet and grain feeding can be commenced at 8 weeks of age. Two parts of pelleted grain balancer ration containing at least 24 per cent of protein and one part of whole grain is a satisfactory combination. Care must be taken to see that the birds do not consume too much grain and too little grain balancer.

Turkeys run on grass or green crops from 12 weeks of age upwards will eat green food readily and good quality foods of this nature will supply up to 10 per cent of the nutrients required and so reduce the amount of concentrate food needed. Because the green foods contain a high percentage of water a large volume of green food has to be consumed in order to obtain the required nutrient level. Most large scale producers rear the sexes separately from 8 weeks of age upwards because the hen poults, which do not grow as quickly as the stags, can be fed a ration containing slightly less protein. Also the hens rear more satisfactorily if they do not have to compete with the bigger stags for food.

After the rearer ration a special finisher ration is given up to the time of slaughter. Such a ration is high in metabolisable energy and contains about 16 per cent of protein. About 5 per cent of animal fat may be included in finishing rations. Feeders using grain can gradually increase the proportion of grain in the ration until the birds are receiving one part of grain balancer and two parts of grain at 20 weeks of age.

Breeding turkeys

Birds being reared for breeding should be fed as outlined previously until they reach 16 weeks of age. Then they are usually given a pre-breeding ration containing about 15 per cent protein until about 4 weeks before the commencement of laying, when a breeding ration with a protein percentage of about 16 or 17 is started. The energy levels of both these rations are in the medium range so that there is no surplus energy to be converted into fat. Some turkey breeders like to feed grain to their breeding turkeys and use one part of a pelleted grain balancer ration with a protein percentage of 30 to two parts of grain, which gives a protein level in the diet as a whole of about 17 per cent.

Formulations of typical rations suitable for the main stages of development are given in Table 12.1 and approximate figures for daily food and water consumption are set out in Table 12.2.

Mineral requirements

A dietary deficiency or imbalance of calcium and phosphorus in growing poults can cause bone deformities and in adult birds the laying of misshapen eggs. Starter rations containing 10 per cent and grower diets containing 5 per cent of animal protein foods such as fish meal, herring meal, and meat and bone meal should contain adequate amounts of both these minerals, but diets consisting of vegetable protein foods and cereals or their by-products are likely to be deficient. About 2 per cent of calcium and 1 per cent of phosphorus are normally recommended for starter and rearer diets and 2·25 per cent of calcium and 0·75 per cent of phosphorus for breeder rations. Sodium and chlorine are needed for normal growth, and rations which do not contain fish or herring meals need to be supplemented with common salt at a level of about 0·5 per cent. Excessive quantities of salt can be toxic. Manganese is required in small quantities to prevent a slipped tendon condition developing in the hocks of growing poults and for optimum growth and egg production. It is generally considered that about 30 parts per million of manganese are required in turkey rations and manganese sulphate is the compound generally used to supply this element.

Vitamin requirements

Vitamin A is necessary for the maintenance of growth and health, and rations are normally fortified with a synthetic preparation or a fish-liver oil. Fresh green food is a satisfactory source of pro-vitamin A, which turkeys can use. Turkeys require vitamin D_3 for the utilisation of calcium and phosphorus and so a deficiency can contribute to the development of rickets. Birds reared outside can form this vitamin in their bodies from sunlight, but birds kept indoors must have it supplied in their food. Turkeys appear to require greater quantities of vitamin D_3 than do fowls. Vitamin E is normally present in adequate amounts in the cereal part of fowl rations. If there is a deficiency in the ration or if the vitamin present is destroyed by rancid fats, enlarged hocks and muscle dystrophy may develop in young birds and breeding birds may show a fall in the hatchability of their eggs. As the B vitamins have a wide distribution in cereals they are generally present in adequate quantities, but deficiencies of some of them can lead to slow growth and poor feather development and so dried yeast or synthetic preparations are normally added, particularly to diets for young birds.

Table 12.1. Formulation and analyses of typical turkey rations.

Foodstuffs	Starter (%)	Rearer (%)	Finisher (%)	Pre-breeder (%)	Breeder (%)
Maize	25	30	25	25	25
Wheat	25	37·5	31·2	—	19
Barley	—	—	25	20	15
Weatings	10	5	5	40	20
Soya bean meal	22·5	10	5	7·5	10
Fish meal	12·5	7·5	2·5	2·5	5
Meat and bone meal	2·5	5	5	—	—
Limestone	—	—	—	3·7	4·7
Fat	—	2·5	—	—	—
Mineral/vitamin supplement	2·5	2·5	1·3	1·3	1·3

Table 12.2. Approximate daily food and water consumption figures (per 100 turkeys).

Age	Food		Water	
	kg	lb	l	gal
1 week	2·27	5	4·6–9·2	1–2
5 weeks	6·8–9·1	15–20	13·7–18·4	3–4
10 weeks	13·6–18·2	30–40	27·3–40·9	6–9
16 weeks	22·7–27·2	50–60	45·5–54·5	10–12
24 weeks	31·8–36·3	70–80	54·5–68·1	12–15
Adult	36·3–45·4	80–100	63·5–81·8	14–18

Water requirements

It is essential that adequate supplies of fresh water are always available for turkeys of all ages. Failure to ensure this will be reflected in poor growth in young poults and poor egg yields in breeding hens. When poults are being reared in groups, water troughs should be placed at intervals over the whole floor area so that the birds are never far from a source of supply. When birds are kept on free range the water troughs should be moved frequently to prevent the land round the trough becoming wet and muddy.

Breeding

Turkey hens of the small type can be used for breeding when about 30 weeks old and those of the large types at about 36 weeks of age. On most farms the breeding birds are only kept for one laying period because of the cost of maintaining turkeys from one season to another, but when individual birds are trap nested the best will probably be retained for a second year. Care must be taken to see that such birds do not become excessively fat during the period when they are not laying.

Stags can be used for breeding at ages 4 weeks greater than those recommended above for hens. When mated in pens one stag is run with from ten to fifteen hens, but on flock-mating systems from fifteen to twenty hens can be allowed per stag. The birds are mated up at least 3 weeks before the eggs are required for hatching. As the spurs of the stags are liable to lacerate the backs of the hens the latter are usually fitted with specially designed canvas saddles held in position by two canvas loops, one being placed round each wing.

As mentioned under breeds, the number of eggs laid during the breeding season varies from 50 to 120. These figures relate to the first production year because second-season hens can be expected to lay about 20 per cent less. Naturally most of the eggs are produced during the spring and summer, but turkey hens can be brought into production at other seasons by increasing the length of the lighting period through the use of artificial light. It must be appreciated that although artificial lighting will alter the time of the year at which eggs are laid it will not materially increase the number of eggs laid during the season.

Broody turkey hens should be placed in well-ventilated broody coops as soon as they are detected. Turkeys do not cluck when broody as fowls do, but can be observed to remain in a nesting box at night instead of perching.

White plumage colour is favoured because the white stubs which are left after plucking are less obvious and so the dressed carcase has a more pleasing appearance. White is recessive to bronze colour and so if pure breeding bronze birds are mated to white-coloured birds the offspring are all bronze in colour. If these young birds are mated together some white-coloured birds are produced which will breed pure.

Artificial insemination

Stags with markedly broad breasts are not able to mate satisfactorily as they lose their balance before completing the service. In birds of this type, therefore, artificial insemination has to be employed. In most flocks where natural mating is practised the fertility rate usually drops appreciably as the end of the breeding season approaches, and some turkey breeders use artificial insemination during this period only in an attempt to maintain a high level of fertility throughout the season.

The stags to be used for semen production must be kept apart from the hens. They can be housed individually in cages, but are generally run in groups of about six in pens. For semen collection a stag is placed upside down in a metal cone and an operator manipulates the copulatory organ with a thumb and finger until the semen starts to flow, from 0·15 to 0·3 ml being obtained per collection. An assistant touches the semen with the tip of a collecting tube and sucks the semen into a test tube surrounded by water at a temperature of 10–15·6°C (50–60°F) kept in a thermos flask. The semen is diluted and used as soon as possible after, and certainly within 8 hours of, collection. Many breeders try to use the semen within an hour of collection. The hens are best inseminated during the late afternoon when they are likely to have laid as the presence of a hard-shelled egg in the lower part of the oviduct generally results in lowered fertility. For insemination a hen is held by the legs with the head down and the oviduct everted outside the vent by applying abdominal pressure. It is not possible to inseminate hens which are not in lay because the oviduct cannot be everted. The cannula of a syringe is then inserted into the oviduct and a dose of from 0·025 to 0·035 ml of diluted semen injected. The hens are inseminated every 3 or 4 weeks at the start of the breeding season and every 2 or 3 weeks towards the end, although some breeders inseminate once a week.

Health

A normal turkey has an alert appearance, bright eyes and moves actively if disturbed. A turkey should be observed on the ground and made to walk forward as leg weaknesses are found in some turkey strains. The legs should be straight, without a splay-legged attitude, and the knees should not knock together when moving. There should not be any sneezing or gasping, and no discharges from either the eyes or the nostrils. The skin should be clean and healthy. Breast blisters can reduce the value of a carcase, and should not be present. Early signs of disease are reductions in food and water consumption and in feather preening. Egg production is reduced in sick laying hens.

The body temperature is normally about 41·7°C (107°F) and the pulse rate about 150 beats per minute.

Common diseases and their prevention

Aortic rupture causes sudden death in well-grown turkeys over 10 weeks of age, and is commoner in males than females. In most cases the birds are found dead, but may be seen to die suddenly when frightened. Death is preceded by a gradual thinning of the wall of the aorta, which is the main artery running along the spine. A sudden increase in blood pressure, possibly caused by a fright or during fighting, leads to rupture and immediate death. Losses in a group of turkeys in which deaths have occurred can be minimised by disturbing the birds as little as possible, and by adding a tranquillising drug, such as reserpine, to the diet, which reduces the blood pressure.

Salmonellosis occurs in turkey poults, and the description of this disease given under fowls is applicable.

Infectious sinusitis is the commonest respiratory disease of turkeys and is of considerable economic importance in older birds. It is characterised by swellings of one or both the facial sinuses. In the early stages the distended sinuses are soft, but later they become firm. Like chronic respiratory disease in fowls it is caused by a mycoplasma. The establishment of a disease free flock is the best method of prevention.

Newcastle disease (fowl pest). Turkeys may be infected with this disease in a similar manner to fowls, although they appear to be more resistant and the mortality rate is usually low. Fertility and egg hatchability are markedly reduced.

Aspergillosis is a fungal disease affecting young turkeys from a few days of age upwards, being commonly called brooder pneumonia. The clinical signs are respiratory distress followed by death in 1 or 2 days. A chronic form occurs in older birds and shows as a wasting disease associated with breathing difficulties. The disease results from infection with moulds of the aspergillus genus, which are prevalent in mouldy bedding litter and grain. Prevention is by the provision of clean dry litter and wholesome food.

Coccidiosis in turkeys resembles the intestinal form described under fowls, but it is caused by species of coccidia which specifically infest turkeys. Sulphaquinoxaline is the drug normally used for preventive treatment as turkeys will not drink water containing some of the other drugs used for fowl medication.

Blackhead can cause serious losses in young turkeys, particularly in poults between 4 and 8 weeks of age. Affected birds become weak and stand with drooping heads and wings. A characteristic sulphur-yellow coloured diarrhoea develops but the blackening of the head which gave the disease its name is less obvious. The organism causing this disease is a protozoan parasite which primarily affects the wall of the caecum but penetrates this and spreads to the abdomen. The faeces of infested birds contain large numbers of immature parasites and so faecal contamination of food and water is a source of spread, but more important is the fact that the parasite, while in the caecum, can infect a caecal worm. This worm itself does little harm to turkeys, but the immature blackhead parasites enter the worm eggs, which, when passed out with the faeces, protect them and enable them to survive on the ground for long periods. As fowls can harbour this parasite without showing clinical signs young turkeys should be kept away from fowls as well as adult turkeys, and not be run on ground which has recently carried either fowls or turkeys. Rearing turkey poults in buildings with wire or slatted floors excludes the risk of soil contamination. A number of commercial preparations which can be incorporated in the food or water are available for preventive treatment, but it is necessary to give a full preventive dose up to the time of marketing. Attempts to reduce losses by administering anthelmintic drugs to lower the numbers of caecal worms have not proved effective.

Handling

Flocks of turkeys can be driven like sheep, and can be confined in pens prior to catching. Such holding pens should be large enough to accommodate up to twelve birds and should have a door at the top through which the turkeys can be raised. A bird can be caught by the base of a wing at its junction with the body and lifted with one hand on the wing and the other grasping both thighs. This may prove difficult in large birds in which case the right hand may be used to grasp the legs and the left hand placed under the breast at the same time so that the bird is raised by a simultaneous lifting of both hands. If the bird is only seized by the legs the breast will almost certainly hit the ground and be bruised. A bird can best be held with its breast resting on the holder's arm with the hand grasping the upper parts of the legs. If the bird struggles it should be tilted on its crop, with its head towards the holder's body. Turkeys should not be hung upside down by the legs for long periods. As turkeys bruise easily they should be handled carefully, especially before marketing, because bruises reduce the value of a carcase.

A vent examination can be used in a turkey hen to determine whether the bird is in lay as a laying hen has a large moist vent while a hen which is not laying has a dry constricted vent.

Sexing

The procedure in day-old poults is very similar to that employed for the manual sexing of day-old chicks, but the operation in poults is rather more simple. It is not necessary to remove the faeces in poults and the males have two firm genital papillae in contrast to the flabby tissues seen in females. In older birds it is fairly easy to distinguish females from males. In the bronze-coloured varieties the feathers of adult females have white tips which are not present in the feathers of adult males, and the difference can be observed from about 14 weeks of age onwards. In all varieties the males are larger and squarer in shape, their hocks are broader, and their shanks are larger. Most hens do not develop a tuft on the upper part of the breast, but if one is present it is always smaller, shorter, and finer than in a stag. The fleshlike appendage on the head, known as the snood, is larger in males than in females.

Administration of medicines

As adult turkeys are heavy and powerful an assistant will be required to hold a bird while a medicament is given. For the administration of a liquid the assistant should sit, possibly on a bale of straw or a crate, and hold the turkey across his knees, grasping the top of the legs with one hand and holding the back of the head with the other. The operator fills a glass tube about 30 mm (1 ft) long and 10·16 mm (0·4 in) in diameter by suction with the required volume of medicine and places his fingers over the top to prevent the liquid flowing out. The tube is then passed down the throat into the crop and the finger removed from the top to allow the liquid to flow out. Tablets can be given with the bird restrained in a similar position.

An intramuscular injection can be made into the muscles of the leg with the bird held in a similar manner, except that the second hand is used to grasp the upper wing at its junction with the body. For an intravenous injection the vein under a wing is used. The above-mentioned position is suitable, but the assistant pulls the upper wing back to expose the vein.

General

Vices

Feather pecking and cannibalism

Feather pecking, which may lead to cannibalism, is common in turkeys and is associated in many cases with intensive housing methods and overcrowding. Feather pecking a few weeks before slaughter can lead to a reduced market value because, when the carcase is plucked, the bared back has an unattractive appearance. Prevention is by subdued lighting in windowless houses or debeaking. On some farms debeaking is carried out as a routine measure when the birds are about 4 weeks old before they leave the rearing accommodation. Injured areas may be treated with Stockholm tar or a suitable proprietary preparation.

Minor operations

Debeaking

This operation is carried out as recommended under fowls using a debeaking machine which cuts and cauterises.

Toe cutting
To avoid the injuries which stags can inflict on hens during mating, even when the latter are fitted with saddles, the last joints of the inside toes of the stags can be removed. This operation is best performed during the first 3 days of life.

Caponisation
The chemical caponisation of stags can be performed in a similar manner to that outlined under fowls, but is not as commonly practised. Its use is generally confined to birds being marketed at the lower weight ranges. The dose should be that recommended by the manufacturer for the size and age of the bird being implanted. The red fleshy protruberances on the heads of turkeys are not affected by hormone treatment, unlike the combs and wattles of cockerels.

Wing feather cutting
On some farms the feathers of one wing are often cut to prevent flying, but this practice is not advised for breeding stags. Stags with a clipped wing cannot balance when mating and infertile eggs may be laid by the hens with which they are running.

Marketing
There are three main categories of turkey carcase, (1) a small-sized carcase weighing from 2·27 to 3·18 kg (5–7 lb), in which case the birds are killed when about 12 weeks old; (2) a medium-sized carcase weighing between 3·62 and 6·35 kg (8 and 14 lb) from birds killed at about 16 weeks of age; and (3) a large-sized carcase weighing from 6·8 to 13·6 kg (15–30 lb), or even more, produced from birds killed at about 26 weeks of age.

Christmas is still the time of year when there is the greatest demand for turkey meat, and many housewives prefer freshly killed rather than frozen birds. Thus, there is always a big demand for good quality carcases in the three weight ranges at this time. During the remainder of the year catering establishments require large birds and there is an increasing sale through supermarkets of oven-ready frozen carcases in the small- and medium-weight ranges. There is also a sale for portions of cut-up turkeys.

Small producers keeping turkeys as a sideline often sell freshly killed dressed turkeys to meet individual orders, but farmers keeping turkeys as the major source of income and having large numbers of birds need a more organised marketing system. Some cater for the retail trade by installing plucking machinery and deep-freeze cabinets so that they can supply individually packed, eviscerated carcases on demand. Others prefer to send their birds to poultry-packing stations where the birds are killed and the carcases are dressed, graded and marketed.

Welfare code

The code of recommendations for the welfare of turkeys resembles that for fowls. Particular attention should be paid to the environmental conditions provided for newly hatched poults and the stocking densities for birds of all ages.

DUCKS

Introduction

Most ducks are kept for table duckling production, although a small number are kept to produce eating eggs. A large proportion of the ducks in Great Britain are to be found on general farms in relatively small flocks. As ducks are good foragers they find a considerable proportion of their own food if allowed to range freely. They also rid the ground of many insect pests and on marshy land infested with liver fluke they eat the snails which play a part in the life cycle of this parasite. Some ducks are kept intensively on large specialist farms, but small runs are rapidly fouled and if ducks are not able to forage they have to be supplied with large amounts of concentrate food compared with birds of a similar size of other poultry species.

There is a steady demand for the meat of ducklings as it has a distinctive, attractive flavour, and this demand is likely to continue. The carcases of old ducks are difficult to sell and only low prices are obtained for them. Although the majority of the British public do not like duck eggs certain people are keen to buy them provided their quality and freshness are assured. For some forms of cooking duck eggs are superior to fowl eggs. If ducks are to be kept for egg production it is necessary to find a satisfactory market.

Definitions of common terms

Duckling. Young bird, usually under 12 weeks of age.
Duck. Female adult.
Drake. Male adult.

Principal breeds

Table bird production

Aylesbury

Probably the best of the table breeds, having white feathers, a flesh-coloured beak, and orange legs and feet. The birds have a horizontal carriage, the keel being practically parallel with the ground. The adults are large, weighing from 4·18 to 4·54 kg (9–10 lb).

Fig. 12.3. Aylesbury drake.

The ducks are only moderate layers, producing on average about 100 eggs per year, most of which are laid between January and July. Egg fertility is often poor early in the year. The merit of the breed is that the ducklings make rapid growth and develop a large amount of good quality flesh on the breast. The bone is light and the skin is white.

Pekin

The breed is a native of China and the birds have cream-coloured plumage, although a white type is being bred. The body carriage is almost upright, raised in front and sloping downwards behind. The beak, legs, and feet are all orange in colour. The adult weights are slightly less than Aylesburys, varying from 3·64 to 4·18 kg (8–9 lb). The Pekin is a better layer than the Aylesbury and most strains are very fertile. The ducklings make good table birds although they are not as quick growing as Aylesburys and have a yellow skin which is not as popular with consumers.

Muscovy

Some authorities state that the Muscovy is not a true duck, giving as reasons that it is a grazer and eats grass like a goose, and that crosses with other duck breeds are sterile. Unlike other ducks the incubation period is 36 days. The drakes weigh from 4·54 to 5·45 kg (10–12 lb) and are much larger than the ducks which only weight from 2·04 to 2·72 kg (4·5–6 lb). There are several different feather colours, such as white, and black, while parti-coloured birds of black and white and blue and white are common. The face has no feathering as a bright scarlet skin surrounds the eyes and the base of the beak. The ducks are poor layers, averaging only about thirty to forty eggs in a laying season, and, although the flesh of the ducklings is of good quality, meat from adult birds is inclined to be tough and the skin emits a musk-like odour.

Egg laying

Khaki Campbell

The ducks are khaki in colour all over, with the legs matching the body. The beak is greenish-black. During the breeding season the drakes have brownish-bronze heads, necks, and wing bars and khaki-coloured bodies, but from the late spring until the autumn the colour resembles that of the ducks. The birds have compact bodies and a slightly upright carriage. The drakes weigh about 2·27 kg (5 lb) and the ducks 2·04 kg (4·5 lb). The ducks are prolific layers of medium-sized, white-coloured eggs weighing about 70·9 g (2·5 oz),

Fig. 12.4. Khaki Campbell drake.

some flocks averaging 300 eggs per duck per year. Surplus drakes make useful small table birds, weighing about 2·04 kg (4·5 lb) at 10 weeks of age, although the dark coloured skin is not liked by some consumers.

White Campbell

A white-coloured derivative of the Khaki Campbell with an orange beak, legs, and feet and light-coloured skin and flesh. The ducks are nearly as good layers as the khaki-coloured variety and, because of their white plumage and light-coloured skins, the surplus drakes are more highly valued for table purposes.

Indian Runner

A small breed with a long slim body, very upright carriage and active habits. The legs are placed well back so that the birds run and do not waddle like other ducks. The breed name is derived from this characteristic. Various colours are found but the white and fawn and white varieties are favoured for utility purposes. The drakes weigh about 2·04 kg (4·5 lb) and the ducks about 1·59 kg (3·5 lb). The ducks are good layers of white eggs, but are nervous and easily upset

Fig. 12.5. White Indian Runner ducks.

by unusual noises. The surplus drakes are of little value for table purposes.

Cross-breds

A cross between Aylesbury and Pekin ducks produces vigorous, quick growing, early maturing ducklings which give good quality carcases. By using a Pekin drake on Aylesbury ducks the fertility is good, but by mating Pekin ducks to an Aylesbury drake a greater number of eggs is produced. Occasionally flocks of White Campbells are mated with Aylesbury or Pekin drakes for the production of medium-sized table birds because of the larger number of hatching eggs produced, and therefore the increased number of ducklings marketed. Other crosses are not normally favoured.

Hybrids

Hybrid lines for the economic production of high-class table ducklings are now being developed in ways similar to those described under fowls.

Identification

On each leg there are two webs between the three toes. Strains can be recorded by making 'V'-shaped cuts in the front of a web or webs soon after the ducklings hatch. A cut is made by placing the foot of the duckling on a smooth, hard piece of wood and cutting out a 'V'-shaped piece of web with a sharp knife. In addition, there is a small strip of web tissue on the outside of each inner toe which can be used for cutting small additional marks. Individuals can best be identified by the use of wing bands. Adult ducks can be fitted with leg rings similar to those used for fowls, but small rings must be obtained or they will slip off the ducks' relatively fine-boned legs.

Housing

Incubation

The incubation period for the eggs of all breeds other than Muscovys is 28 days. Incubators used for ducks should provide the same conditions as for fowl and turkey eggs, except for relative humidity. Experience has shown that a relative humidity of about 70 per cent is needed during the first 24 days of incubation, followed by 75 per cent for the last 4 days. The capacity of a hen egg incubator is reduced to 85 per cent when used for duck eggs.

Ducklings

No special methods are necessary for rearing ducklings and brooding equipment designed for chicks can be used satisfactorily. However, as ducklings grow very much more rapidly than chicks not more than half the number of ducklings should be placed in a space recommended for chicks. Ducklings need heat for the first 2 or 3 weeks in summer and the first 3 or 4 weeks in cold weather. For the first week a temperature of about 32·2°C (90°F) is required, after which there can be a steady reduction until no heat is provided. However the environmental temperature should not fall to below 15·6°C (60°F) during the first 4 weeks. Brooder houses with solid floors covered with shavings are generally used but the management of the floors requires considerable skill because of the quantity of watery faeces produced by ducklings. Fresh material needs to be added as often as is necessary to keep the floor fairly dry. From 4 weeks of age upwards the ducklings can be moved to large sheds or covered yards. It is possible to keep ducklings being reared for the table in wire-floored tier brooders.

Adult ducks

From two to six breeding ducks mated to a drake are generally kept in a simple, well-ventilated, solid-floored house with a run attached. Unlike fowls, ducks rarely enter the house during the day, even in bad weather. Wire and slatted floors are not a success with adult ducks but have been tried because of the liquid consistency of duck faeces. Sufficient wood shavings or straw should be used as bedding to keep the floor reasonably dry. The house should have a wide door, as narrow entrances may cause injuries because ducks generally enter and leave a house in a rush. Ducks may also be kept in groups of up to 30, this number requiring a yard of about 2·44 m × 3·66 m (8 ft × 12 ft). A small shelter should be provided in which the ducks can lay. The floor of the yard is covered with straw or wood shavings, and additional bedding is added from time to time when necessary. This build up of bedding material forms a kind of drainage mattress. These yards are only cleaned out once a year during August when the birds are moulting and out of production. On some large duck farms the breeding birds are kept in large houses without runs or yards attached.

Often nest boxes are not provided in either houses or yards and the ducks lay their eggs on the floor. If the litter is kept clean and dry this is satisfactory, but the eggs keep cleaner and are less likely to be broken if nests are supplied. There should be one nest for each three ducks, and they should be fitted at floor level. A convenient size is 0·31 m (1 ft) wide by 0·46 m (1·5 ft) deep, with a 76 mm (3 in) high board nailed across the base of the front. The tops are left open.

Laying batteries have been used for housing ducks kept for the production of eggs for human consumption, but this method of accommodation has not proved entirely satisfactory. The ducks' feet become sore through continual standing on a wire floor and it may be difficult to fit water troughs sufficiently deep to allow the birds to immerse their heads. When deep water troughs are supplied the volume of splashed water increases the difficulties of cleaning.

Feeding

General

Relatively little research has been conducted into the nutritional requirements of ducks, and it is generally considered that their needs resemble those of fowls. Special proprietary foods are rarely made for ducks and most producers use those compounded for fowls. Grit should always be available for both growing and adult ducks.

Ducklings

Ducklings should not be fed on a dry mash as they waste a considerable amount trying to wet a meal in the water troughs and may not consume sufficient food for satisfactory growth. Diets for ducklings should always be fed as crumbs or pellets which are not wasted in this way. Table ducklings are usually fed a broiler starter ration containing about 22 per cent of protein for 3 weeks, followed by a finisher diet containing 17 per cent of protein until slaughter. During an 8-week fattening period a duckling will consume a total of from 7·7 to 8·6 kg (17–19 lb) of food. Ducklings being reared for future laying are fed similarly until they are 7 weeks old when a ration with a protein content of 15 per cent and a medium energy level will suffice.

Adult ducks

Whether kept for the production of eggs for hatching or eggs for human consumption a high level of egg production is required, and for this generous feeding is needed. A fowl laying diet in pellet form or grain and a grain balancer ration is generally used. A duck of one of the laying breeds required about 200 g (7 oz) of concentrate food daily if kept confined, but if running on pasture ducks eat a smaller amount of concentrate food as they supplement their diet with grass, slugs, snails, and insects.

Mineral requirements

The major minerals described under fowls are required by ducks. The main clinical signs of rickets in ducklings are a rubbery texture of the beak, which can sometimes be folded back, and a bowing of the legs. Adequate supplies of manganese are required to maintain hatchability.

Vitamin requirements

A deficiency of vitamin A is shown primarily in ducklings by a nasal discharge and later by a general paralysis. In breeding birds there is reduced hatchability and a high mortality rate in newly hatched ducklings, many of which have weak, watery eyes. A vitamin D deficiency contributes to the development of rickets, and a shortage of vitamin E in ducklings is followed by a reduced growth rate and muscle degeneration. Ducklings reared on a diet low in the vitamin B complex grow slowly and a deficiency of nicotinic acid, particularly, results in bowed legs which are not unlike those seen following a vitamin D deficiency.

Water requirements

An ample supply of drinking water is essential for ducks of all ages, and this must be provided in containers sufficiently deep to enable the birds to immerse their heads. If unable to wash their heads they develop crusts round their eyes and, possibly, more serious eye lesions. Some of the water is splashed out of the troughs as the ducks immerse their heads and so ducks require considerably more water than hens do, about 4·55 l (1 gal) being allowed per five adult ducks. Water troughs situated in runs soon become filled with mud and food particles as the ducks wash their beaks, and the surrounding area of ground becomes puddled. Thus the water troughs need to be cleaned frequently and moved periodically.

Breeding

First-year ducks come into lay earlier than older birds and, as an early supply of hatching eggs is desirable, most commercial breeders dispose of their breeding ducks as soon as the hatching season has finished each year. However some breeders use second-year birds for the production of future breeding stock, and most ducks have a useful laying life extending over 3 years or even longer.

It is very easy to trap nest ducks as they lay during the night. Thus one trap nest is provided for each bird, into which it is driven shortly before dark each evening and in which it spends the night. The nests are placed on the floor and are about 0·31 m (1 ft) wide, 0·46 m (1·5 ft) deep and 0·62 m (2 ft) high. The nest tops are in the form of hinged lids to facilitate the removal of the ducks each morning. Because the birds spend their nights in the trap nests these need to be well ventilated and so should have wire-netting fronts.

Ducks of the laying breeds start to produce eggs when about 5 months old and table breed birds when about 6 months of age, but in both cases the season when they were hatched has an influence. In the table breeds a drake is mated with up to four ducks while with the laying breeds a drake can be run with up to six ducks. On a commercial scale flock mating is usually practised, with about twenty table breed and thirty laying breed ducks being run with five drakes. If laying ducks are not released until nine o'clock in the morning almost all the eggs to be produced in the day will have been laid. Generally it is only ducks which are going out of production which lay later. Artificial lighting can be used to bring ducks into lay earlier than would otherwise be the case, and ducks react relatively quickly to light stimulation. It is usual for artificial lighting in the case of ducks to be provided all night, and this practice is commonly started about 10 days before egg production is desired. Another advantage of all-night lighting is that the ducks are less likely to be frightened at night by unfamiliar sounds.

First-year ducks normally moult from about August to October. Most go out of production at this time, although some lay through the moult. Older birds usually lay until about October and then moult and go out of production until the spring. Ducks rarely go broody, and the few birds which do can be brought back into production fairly rapidly if prevented from sitting in their houses.

Health

A healthy duck is alert and should walk without a sign of lameness. The bill should be straight and the eyes bright. The legs should be of medium length, set well apart, and not placed too far back so as to allow the abdomen to extend to the rear of the legs. When in full lay a duck's abdomen will nearly touch the ground. The plumage should be tight and silky with a waxy feel. The wings are carried rather high, but close to the body. The tail should not twist to one side, but should rise slightly from the line of the back. The feathers round the anus should not be caked with faeces through scouring.

Common diseases and their prevention

Ducks are subject to very few of the diseases seen in fowls and turkeys and, if managed carefully, comparatively low losses are suffered. This is a great economic advantage in duck keeping.

Slipped wing is an abnormality which first shows when the birds are about 7 weeks of age and is very noticeable from 10 weeks onwards. The primary feathers on one or both wings are twisted out of position so that they stick out at an angle from the duck's body. The cause is uncertain, although there is some evidence that a lack of adequate ventilation during rearing or long journeys in boxes may be a contributing factor. Although the condition does not appear to be hereditary affected birds should not be used for breeding.

Salmonellosis has been adequately described under fowls, but in ducklings a common clinical sign is that immediately after drinking the affected birds keel (fall) over backwards and die. This has led to the name keel disease being applied to one form of salmonellosis in ducks. Outbreaks of gastro-enteritis in man can follow the consumption of infected duck eggs because duck eggs have more porous shells than hen eggs which the salmonella organisms can penetrate. An additional factor is that ducks often lay on the ground and so their eggs are more likely to come in contact with faeces. Egg transmission can also be the cause of outbreaks of the disease in ducklings.

Duck septicaemia (new duck disease) causes deaths particularly during the period from 4 to 9 weeks of age. Affected ducklings are dull, develop a greenish-coloured diarrhoea, become weak, and show a nervous movement of the head and neck. In some outbreaks the mortality rate is high. The causal organism is a bacterium, but the method of spread is not fully understood. The losses are greatest in ducklings reared under crowded conditions and there is evidence that the organism can gain entrance to the body through minor wounds.

Duck virus hepatitis outbreaks usually start when the ducklings are between 2 and 3 weeks old, but birds under 1 week old may become infected. The mortality rate may be high. The clinical signs are indefinite, but include weakness and a lack of muscular control which causes the ducklings to fall on their sides and kick spasmodically with both legs. The cause is a virus which is spread through the faeces of diseased birds. A vaccine which gives a satisfactory immunity has been evolved.

Handling

Ducks tend to be more nervous than other species of poultry and should not be handled unless it is really necessary. They should be caught as gently as possible to avoid stressing them unduly, for severely upset ducks will go off their legs and be unable to walk, although they quickly recover if placed in a sheltered spot. A group of ducks to be handled should be driven quietly into a corner of a house or run and held there in a catching pen made of two rectangular wooden frames hinged together and covered with 25·4 mm (1 in) mesh wire netting. The area of the space enclosed can be adjusted according to the number of ducks cornered. The best way to catch a duck is for the handler to take hold of the neck just below the head and pull the bird towards him. This causes it to open its wings and the wing bases can be grasped with the other hand. The hand which held the neck is then moved under the body to support the weight and hold the legs. If the duck is to be carried it is held by the hand under the body and the head is tucked under the handler's arm. Ducks should not be grasped by the legs, wings, or tail. Particular care should be exercised when handling ducklings about 6 weeks old as the skin at this time, when they are getting their first full set of feathers, is easily torn. When catching Muscovy ducks they must be held by the neck and wing bases only, as their feet are armed with long sharp claws which can severely cut a human wrist or hand.

Sexing

It is fairly easy to sex day-old ducklings and the examination is best carried out as the birds are transferred from the incubator to the brooder, that is when they are dry but before they have eaten. A duckling is held upside down in the left hand and the tail is bent back with the thumb. Then the thumb and first finger of the right hand are are used to apply light pressure to the sides of the cloaca. This opens the cloaca transversely and forces out the penis in a male, which appears as a tiny glistening projection. In a female there is no such organ and only the inner lining of the cloaca is visible.

From 6 weeks of age upwards there is a sexual difference in sound which can be detected by lifting each bird by the neck and then releasing it. A duck will quack, but a drake only makes a hissing sound. From about 12 weeks of age drakes of all breeds except Muscovys can be detected by the curled feathers, usually three, on the tops of their tails. The ducks have no curls.

General

Vices

These are not found as frequently in ducks as they are in fowls or turkeys. *Quill pulling* is occasionally met in ducks about 6 weeks of age which are being reared in crowded conditions. Blood can be seen on the wings and at the base of the tail where the quills have been pulled out. Allowing the birds more space may stop the habit, but if this is not effective it may be possible to detect the small number of birds responsible and remove these from the flock.

Penis injuries are relatively common in drakes kept for group mating. One cause is pecking by other birds at the time of mating. The injured penis cannot be retracted and protrudes from the vent and may slough off. Treatment is not effective and affected birds are of no value for breeding.

Minor operations

Wing feather cutting

With a pair of scissors about 76 mm (3 in) should be cut from the end of the long, hard wing feathers on one wing. The effect is to throw the bird off balance when it tries to fly.

Marketing

Egg marketing

Duck eggs do not keep as well as fowl eggs because of the porosity of the shells and so must be cleaned soon after being laid and be marketed in a fresh condition. They have a stronger flavour than fowl eggs and are thus not as popular with some consumers, but are used in the confectionery trade.

Marketing table ducklings

Ducklings are generally sold ready for the table, some producers packing the carcases in cellophane bags. Unlike fowls and turkeys, duck carcases have a layer of subcutaneous fat and the flesh is rich in fat. There is a sale for duck feathers, especially white ones, and so efforts should be made to keep the feathers as clean as possible.

GEESE

Introduction

Most geese are maintained in small flocks on a free range system to supply table birds for the Christmas trade and provide general farmers with an additional source of income. Full use is thus made of the special ability of geese to digest and utilise grass very efficiently. There are a few farmers who specialise in goose production and keep them commercially on a large scale. They are trying to reduce production costs by improving the hatchability of goose eggs through better incubator management and breeder selection and by increasing egg production. The goose is a fast-growing bird and breeders are also developing strains which grow increasingly rapidly so that they can be dressed for market at an early age. Thus, very efficient use is made of the food fed. One difficulty is that there is no great demand for geese except at Christmas, and so birds ready to be marketed earlier in the year must be held in cold store until required.

Definitions of common terms

Gosling. Young bird.
Goose. Female adult.
Gander. Male adult.

Principal breeds

Chinese

An upstanding breed with a pronounced knob at the base of the beak. The feather colour can be white or greyish-brown. The birds are relatively small with an adult liveweight of about 4·53–6·35 kg (10–14 lb). The geese are good layers, producing from 80 to 100 eggs per goose per annum. The fertility rate is higher, on average, than it is in the other breeds. The table birds have flesh which is relatively non-fatty and is considered to have a good flavour, but is dark in colour which detracts from its appearance.

Roman

A moderate-sized bird. The geese are fair layers, producing about fifty to sixty eggs per year. The goslings are quick maturing.

Fig. 12.6. Chinese geese.

Fig. 12.7. Embden gander and goose.

Toulouse
Large grey birds with an adult liveweight from 9·1 to 13·63 kg (20–30 lb). The beaks and feet are orange in colour. The geese are relatively poor layers producing about forty eggs per year. The goslings are quick growing and well fleshed.

Embden
Very similar to the above except that the birds are white in colour. The goslings have good breast development. This breed is increasing in popularity to some extent because, like all the white varieties of poultry, any pin feathers remaining on the carcases after dressing are comparatively inconspicuous.

Cross-breds
The majority of the geese seen on farms in Great Britain are crosses of the above mentioned breeds. A popular cross is an Embden gander mated to White Chinese geese. Advantage is taken of the higher egg yields of the Chinese geese, but low fertility may be a problem. The resulting birds have well-formed, clean-looking carcases, which are comparatively easy to pluck. If the cross is made the other way round many less eggs will be laid, but there are unlikely to be any fertility problems. A refinement is to cross a Chinese or Roman gander with Embden or Toulouse geese and mate the geese from this cross with an Embden gander. By this method good table birds are produced from geese having better egg yields than either Embden or Toulouse females.

Identification

Family lines can be identified by punching holes in the webs of the feet of goslings. Nicking the edges of the web is less satisfactory as these nicks may grow out or split. Wing bands can be used for the individual recognition of goslings and metal numbered leg rings are suitable for adult geese.

Housing

Incubation
The incubation period is from 28 to 31 days in length with an average of 30 days. Flat-type incubators are often used by farmers

keeping geese on a small scale, but cabinet machines are favoured by large scale producers. The temperature is kept at about 37·2°C (99°F) initially, being reduced to 36·1°C (97°F) at hatching time. The humidity is very important and the relative humidity should be about 55 per cent for the first 26 days and 85 per cent for the last 4 days. Some goose keepers have found that dipping the eggs in lukewarm water during the last few days of incubation increases hatchability. The humidity is often reduced before the goslings are removed from the incubator, so that their down can dry out. Occasionally goose eggs are hatched under broody hens, from four to six eggs being allowed per hen. Hatchability is improved if the eggs are damped with warm water daily during the last half of the incubation period. Care should be taken to see that the hen is turning the eggs regularly, as some hens find it difficult to move goose eggs because of their size.

Goslings

Goslings only need artificial heat for 2 or 3 weeks with a brooder set at a temperature of 32·2°C (90°F) initially, dropping to 21°C (70°F) at the end of the brooding period. The space requirement rises from 0·07 m^2 (0·75 sq ft) per bird at first to 0·23 m^2 (2·5 sq ft) by the third week. After this they can be kept on grass provided there is an adequately ventilated shed in which they can be shut at night, but they must not be exposed to rain or long wet grass until they are 4 weeks old and their backs have begun to feather. The land may be divided into paddocks so that the goslings can be grazed on the sections in rotation, fences of wire netting with a 50·8 mm (2 in) mesh 0·92 m (3 ft) high being erected for the purpose. It is possible to rear goslings wholly indoors in open-fronted Dutch barns in groups of up to 100, about 0·91–1·22 m^2 (3–4 sq ft) per bird being allowed up to 14 weeks of age, but this method is not popular in the British Isles.

Adult geese

Adult breeding geese are almost always kept on grass and only supplied with a simple shelter, which will give protection from heavy rain and foxes and accommodate the nest boxes. A height of about 0·91 m (3 ft) gives adequate head room. The floor area of the nest boxes should be about 0·45 × 0·45 m (1·5 ft × 1·5 ft). If foxes are a problem the birds should be shut up each night.

The best type of fencing for fields or pens used for adult geese is pig netting, and in most cases a height of 0·91 m (3 ft) is adequate.

Electrified wires fitted at levels of 12·7 cm (5 in) and 30·5 cm (12 in) above the ground have been tried with some success.

Feeding

General

Geese need grit at all times. Chick size grit should be supplied at hatching, but as the birds increase in size any small stones or hard grit will serve to grind the food in the gizzard.

Goslings

A newly hatched gosling will live for 48 hours, and probably up to 72 hours, before having to take food or drink. Then, for the first 3 weeks a chick starter and for the next 2 weeks a chick grower ration can be fed to goslings. Some small producers also give short, fresh lawn clippings to accustom the goslings to feeding on grass. After this period goslings can be reared on grass only during the growing season. Grass is the cheapest crop to grow and is a well-balanced and complete feed for goslings. However only growing, young grass is capable of fully nourishing a gosling. Where no supplementary feeding is given about twenty-five goslings can be run on an acre of grassland during the spring and summer. By dividing the land into paddocks and grazing them in rotation every 2 or 3 weeks the stocking rate can be trebled. Should the grass grow faster than the goslings can keep it down it should be mown. Higher stocking and quicker growth rates can be obtained if grain such as wheat and barley or growers' pellets are given in addition to the grass. At one time fattening geese were fed on grass until the grain harvest and then allowed to glean the stubbles so that they fattened on the shed corn. At Michaelmas (29 September), when, by tradition, there was a demand for dressed geese, the birds were ready for killing without any additional feeding. Now that geese are customarily eaten at Christmas they have to be fattened on concentrate foods from 3 to 5 weeks before that day.

Goslings housed indoors can be fed on a pellet or pellet and grain system. Birds fed intensively in this way are killed at between 10 and 12 weeks of age, by which time they will have consumed between 20·4 and 22·7 kg (45 and 50 lb) of food per head. They yield carcases which are much less fatty than those from older birds reared on grass and the fat has a lighter colour which may be more attractive to consumers. On the Continent geese are overfed to produce enlarged livers from which *pâté de foie gras* is made, but this system is not practised in Great Britain.

Adult geese

Breeding geese are nearly always kept on grass and between the end of the laying season in June and the beginning of January will, in mild seasons, feed mainly on grass. From early in January throughout the laying period a layers' mash, fed wet, or pellets, compounded as recommended for fowls, should be given to stimulate egg production. Home-grown grain, such as wheat, barley, or oats, may be used to replace a part of the compounded diet.

Mineral and vitamin requirements

As far as is known these are similar to those described under ducks.

Water requirements

Fresh drinking water should always be available and geese should be given a sufficient depth of water in the troughs to enable them to immerse their heads completely. Some goose keepers place the troughs for adult birds outside the pen fence so that they can push their long necks through to drink but cannot bath in the water and foul it.

Breeding

Geese are not always successful breeders during their first year, but tend to improve with age up to 10 years and over and will remain in full production for at least 15 years. Geese are mated in sets in the ratio of from three to five geese to one gander, and it is best to allow them to remain in the same breeding sets for life. A gander may take a dislike to one goose and refuse to mate it, and some old ganders ignore all their mates, except one. This can constitute one of the biggest problems in goose husbandry. The flock mating of several ganders to a large group of geese is generally unsatisfactory, but, as it results in a saving of labour, it may be tried. Geese can mate without being allowed access to swimming water, but, particularly for the heavy breeds, it is generally considered that better egg fertility is obtained by providing water for mating. Artificial insemination is possible in geese, but it is not practised commercially.

Geese normally only lay in the spring and summer but artificial lighting can be used to induce them to start laying earlier in the year. When geese are kept in sheds dim lights are left burning all night, but if housed in substantial buildings a time switch can be used to provide light for 8 hours during the pre-laying period, increasing this to 16 hours shortly before the desired laying time.

Geese sometimes go broody, and should be prevented from having access to a nest by way of treatment. This is best effected by placing the bird in a small pen with wire-netting sides. The roof should preferably be solid to give some protection from inclement weather.

Health

Geese are recognised as being the healthiest of all the poultry species commonly kept in the British Isles. Geese should be watched while being gently driven in a flock, as lameness can be observed and physical weakness is shown by an inability to walk normally and keep pace with healthy flock mates. The wings should be examined for, occasionally, one or both wings show signs of the slipped wing condition described under ducks.

Common diseases and their prevention

Staphylococcal arthritis is a condition, caused by a staphylococcus bacterium, which affects other species of poultry as well as geese. It is seen in growing birds, and the predominant clinical sign is lameness. In the acute form the leg and wing joints become swollen, hot and painful, the birds do not move about and many die. Some show a chronic form with thickened joints which inhibit movement. The bacteria normally enter the body through superficial wounds, and so conditions likely to produce wounds should be avoided.

Gizzard worm infestation is one of the commonest causes of death in goslings. The clinical signs of an acute attack are a rapid loss of condition which may terminate in death. The worms are thread-like and about 12·7–25·4 mm (0·5–1 in) in length. They live in the gizzard and by penetrating the wall cause the lining to slough off. As adult geese may have a mild infestation without showing any obvious ill effect goslings should be kept separate from old birds so that they do not become infested.

Coccidia. The kidney is the body organ damaged by the species of coccidia infesting geese. Birds between the ages of 3 weeks and 3 months are most commonly affected, and the clinical signs are weakness and emaciation followed by death within a few days. As the disease is spread in the faeces, goslings should be run on clean land and given uncontaminated food and water. Little information is available about preventive treatments.

Handling

It is advantageous to have a catching cage into which geese or goslings can be driven. The cage should measure about 0·91 m high, 0·91 m wide and 2·74 m long (3 ft high, 3 ft wide and 9 ft long), and should have a door at each end and another door in the centre of the top. The approach should be fenced to form a funnel so that the geese can more easily be driven in. The door in the top is used for lifting the birds out.

If a single goose is to be caught, and a catching cage is not available, it should be driven out of the flock into a corner. A goose should never be seized by the legs, for these are fragile and easily broken, but should be caught by the neck which is a strong feature. A goose can be carried for short distances if held by the neck with its back towards the handler, but over long distances the goose's head should be tucked under a wing and its body carried under the handler's arm.

Sexing

Goslings can easily be sexed at an early age, particularly soon after hatching. Gentle pressure on the cloaca exposes a small fleshy penis in a male, while in a female the genital organ is concave and smooth. In sexually mature birds the penis is large and can be easily seen if the bird is held by an assistant on its back with the head down and tail pulled back and firm pressure is applied to the cloaca. Such an examination is sometimes necessary as in outward appearances males and females resemble one another, although ganders are usually larger than geese of the same age and have slightly longer necks and larger heads.

General

Vices

Vices are not a problem in goose husbandry, although the fighting of ganders being flock mated will reduce fertility. This can be minimised if the ganders are all reared together and the area of land for grazing is large enough for a gander and a few geese to graze away from other similar sets.

Marketing

Most small-scale goose keepers sell dressed carcases to customers in their district or through local butchers. Large scale enterprises usually sell through retailers.

There is still a sale for the down and breast feathers of geese which are used for filling sleeping bags and quilts. White feathers are preferred to coloured ones. At one time the soft feathers on the back, flanks, and breast were plucked from live geese during the late summer or early autumn, but now all the goose feathers which are marketed come from geese being plucked after killing. The down and breast feathers are preserved, care being taken to ensure that they do not become mixed with large, straight feathers.

INDEX

Aberdeen Angus cattle 149
Abortion, contagious—*see* contagious abortion
Abyssian cats 314
Acquired marks, horses 42, 48
Administration of medicines—*see* medicines, administration of
Afghan hounds 267
Aflatoxin 18
Afterbirth—*see also* fetal membranes 126, 178, 328
Ageing
  cats 317
  cattle 104, 126, 162
  dogs 279
  fowls 346
  general 6
  goats 171
  horses 34, 48
  pigs 223
  rabbits 252
  sheep 193
  turkeys 379
Agriculture (Miscellaneous Provisions) Act 1963 93
Airedale terriers 270
Alpine goats 170
Alsatian dogs 273
American Mammoth Bronze turkeys 379
Amino acids 8, 9, 55
Anaemia, piglet 238
Anglo-Nubian goats 168
Angora rabbits 247
Anthelmintics 3, 206, 391
Antibiotics 20, 25
Antibodies 3, 25
Antitoxin 3, 69
Aortic rupture in turkeys 389
Appetite, depraved in dogs 307
Approaching animals 71, 240, 302, 307
Arab horses 44
Arks, movable 254
Artificial insemination
  cats 328
  cattle, 102, 126
  dogs 294
  fowls 362
  general 24
  geese 412
  goats 178
  horses 65
  pigs 236
  rabbits 257
  sheep 204
  turkeys 383, 388
Artificial lighting—*see* lighting, artificial
Artificial rearing
  foals 62
  kittens 325, 334
  lambs 201
  puppies 289
Artificial vagina
  cattle 126
  dogs 294
  pigs 236
Aspergillosis 390
Avian leucosis 367
Aylesbury ducks 395
Ayrshire cattle 98, 153

Bacon 218
Bacon type pig breeds 218
Bandages, horse 78
Banged tails, horses 76
Barking, excessive by dogs 308
Barley 19, 56, 158, 160
Barrow 218
Baskets for cats 317
Bassett hounds 268
Baths, foot 206, 209
  sheep 209

Battery cages
  ducks 400
  fowls 350, 371
  turkeys 383
Beagles 268
Beans 18, 58
Beastings—*see* colostrum
Bedding, deep litter 348
Bedding materials
  cats 328
  cattle, beef 155
    dairy 107, 108
  dogs 280, 282, 284
  ducks 399, 400
  fowls 348
  goats 172
  horses 53
  pigs 231
  rabbits 252, 253
  sheep 194
  turkeys 381
Beds
  cats 317, 320
  dogs 280, 282
Beef 147, 148
  baby 158
  cow 160
  mature 159
  young 158
Beef cattle—*see* cattle, beef
Beef production 147, 158
Beef Shorthorn cattle 150
Beet, sugar—*see* sugar beet
Berkshire pigs 220
Beveren rabbit 250
Bib, horse 92
Billy-goat 167, 177, 182
Biotin 13
Bird catching by cats 335
Birman cats 316
Bitch 266
Biting
  dogs 307
  horses 90
Bitless bridle 81
Bits 79
Blackhead, turkeys 390
Blankets 284, 317
Blaze, horse marking 40
Bloat 131, 162
Bloodhounds 268
Blowfly attack in sheep 206
Blue-eyed white cats 317
Boar 218, 228, 232
Body brush 75
Body colours, horses 36
Body temperatures—*see* temperatures, body
Boiler fowl 340
Bone flour 20, 290, 324, 325
Bones for dogs 280, 286, 287, 290
Border Leicester sheep 186
Bordered horse marking 42
Boxer dogs 274
Bran 19, 56, 58
Branding, freeze 103
Branding horses 48
Breaking horses 82
Breeding
  cats 327
  cattle, beef 161
    dairy 124
  dogs 291
  ducks 402
  fowls 361
  geese 412
  general 22
  goats 177
  horses 65
  pigs 235
  rabbits 257
  sheep 202
  turkeys 387
Breeding of Dogs Act 1973 266
Breeding systems
  cattle, beef 153
    dairy 101
  general 4
  pigs 221
  rabbits 251
  sheep 192
Breeds of domesticated animals
  cats 312
  cattle, beef 148
    dairy 96
  dogs 267
  ducks 395
  fowls 340
  geese 407
  general 1, 3
  goats 168
  horses 42
  pigs 218
  rabbits 248
  sheep 186
  turkeys 378

Brewers grains 19, 56
Bridles 79
Brindle colour 266
British Canadian Holstein cattle 98
British goats 170
British Rabbit Council rings 251
British Saddleback pigs 220
British short-haired cats 312
Broad breasted turkeys
  bronze 379
  white 378
Broiler 340, 345
Bronchitis, infectious in fowls 366
Brood mare 34, 60
Brooders for poultry 349, 380, 399
Broodiness
  ducks 403
  fowls 364
  geese 413
  turkeys 387
Broody coop 365, 387
Browsing by goats 175
Brucella abortus—see contagious abortion
Brucellosis (Accredited Herds) Scheme 133
Brushing in horses 87
Brushing boots 87
Buck rabbit 248
Buckling goat 167
Bulk milk tanks 140
Bull 96
Bull terriers 271
Bullock 148
Bulls—*see* also cattle
  beef from 148
  breeding 124, 125, 161
  feeding 115, 121
  housing 112
  ringing 135, 144
  tethering 113
Bulldogs 272, 293, 303
Bull-dogs or Bull-holders 134
Burdizzo method of castrating 164, 214
Burmese cats 314
Butterfat—*see* milk fat

Cabbage 16
Cages, rabbit 252
Cairn terriers 271
Calcium 11, 64, 122, 130, 160, 176, 200, 234, 255, 290, 325, 359, 385
Calculi in cats 327
Calf—*see also* cattle 96
Californian rabbits 248
Calkins 85, 87
Calves
  casting 136
  diseases 133
  feeding 113, 122, 123, 146, 156
  housing 105, 154
  identification 103
Candling eggs 362, 373
Canine distemper 296
Cannibalism
  fowls 349, 371
  rabbits 262
  turkeys 392
Capon 340, 358
Caponisation 358, 372, 393
Carbohydrate 10, 287, 322
Carotene—*see also* vitamin A 13, 326
Casting cattle 136
Castration
  calves 163
  cats 336
  dogs 309
  horses 92
  lambs 208, 214
  pigs 240, 242
Cat door 318
Catching cage, geese 414
Catching crate, fowls 369
Catching hook, fowls 369
Cats
  ageing 317
  artificial insemination 328
  breeding 327
  breeds 312
  definitions of common terms 312
  diseases 329
  feeding 322
  handling 331
  health 329
  housing 317
  identification 317
  medicines, administration of 334
  operations, minor 336
  vices 335
  water requirements 326
Catteries 318
Cattle, beef

Cattle (*cont.*)
- artificial insemination 162
- breeding 161
- breeds 148
- definitions of common terms 148
- diseases 162
- feeding 156
- handling 162
- health 162
- housing 154
- identification 154
- operations, minor 163
- water requirements 161

Cattle, dairy
- ageing 104
- artificial insemination 126
- breeding 124
- breeds 96
- definitions of common terms 96
- diseases 130
- feeding 113
- handling 134
- health 129
- housing 105
- identification 102
- medicines, administration of 141
- milk production 127
- milking 137
- operations, minor 142
- vices 142
- water requirements 124

Cavalier King Charles spaniels 277
Cereals 19, 56, 158, 323, 358
Charolais cattle 151
Chestnuts on horses' legs 35, 48
Cheviot sheep 190
Chianina cattle 152
Chick—see also fowls 340
Chihuahua dogs 276
Chinese geese 407
Chlorine 11, 122
Choline 13
Chow Chow dogs 272
Claw cutting in rabbits 262
Cleveland Bay horses 43
Clipping horses 75
Clothing for horses 76
Clun Forest sheep 189
Clydesdale horses 42
Cobalamin—see vitamin B_{12}
Cobalt 11, 124, 200
Coccidiosis
- fowls 367
- geese 414
- rabbits 260
- turkeys 390

Cock 340
Cock turkey 390
Cockerel 340
Cod-liver oil 20, 123, 290, 324
Colbred sheep 192
Colic, horses 55, 70
Collars
- cat 317
- dog 278
- goat 173
- horse 81, 82

Collie dogs 274
Colony system of keeping rabbits 253
Colostrum 25, 113, 133, 146, 173, 202, 236, 285, 289
Colours, horse 36
'Colour marking' of calves 151
Colourpoint cats 316
Colt 34
Concentrate foods 14, 17, 56, 115, 158, 160, 174, 175, 202, 217, 231, 254
Connemara ponies 46
Contagious abortion 133
Contagious diseases 28
Copper 11, 200, 234, 290
Copper sulphate 20, 21, 206
Coprophagy in rabbits 254
Cottage type piggeries 230
Cow—*see also* dairy cattle 96, 116
Cow kennels 109
Cowsheds 105
Crate, handling 163
Crazy chick disease 360
Creep feed for piglets 232
Creosote 319
Crest, horse 35
Crib-biting, horses, 53, 91
Cross-bred animals
- cattle, beef 151
  - dairy 100
- ducks 398
- fowls 343, 344
- geese 409
- general 3, 5
- pigs 221
- rabbits 250
- sheep 191

Crude fibre—*see* fibre

Crush 136
Cryptorchid horse 34
Cubicles, cow 108, 155
Curry comb 75
Cutting in horses 87

Dachshunds 268
Dairies 112
Dairy cattle—*see* cattle, dairy
Dairy Shorthorn cattle 100
Dales ponies 46
Dalmatian dogs 273
Dam 2
Dandy brush 75
Danish piggeries 229
Dartmoor ponies 46
Debeaking 371, 372, 375, 392
Definitions of common terms
  cats 312
  cattle, beef 148
    dairy 96
  dogs 266
  ducks 395
  fowls 340
  geese 407
  general 2
  goats 167
  horses 34
  pigs 218
  rabbits 248
  sheep 186
  turkeys 378
Deep litter bedding 348
Deep litter houses 350
Dehorned 3
Dehorning 143, 163, 177
Dental star on a tooth 49
Dentition—*see* ageing
Depraved appetite, dogs 307
Devon cattle 150
Dew-claws, removal in dogs 308
Diarrhoea
  calves 133
  cats 323, 329, 330
  dogs 297
  ducks 404
  fowls 366, 367, 368
  general 26
  goats 180, 181
  pigs 239
  rabbits 260
  sheep 205
  turkeys 390
Dipping, sheep 211
Dipping bath for sheep 209
Disbudded 3
Disbudding 142, 163, 177, 181, 183
Disease
  contagious 28
  infectious 28
  prevention of 28
  signs of 27
Diseases
  cats 329
  cattle, beef 162
    dairy 130
  dogs 296
  ducks 403
  fowls 365
  geese 413
  general 28
  goats 180
  horses 69
  pigs 238
  rabbits 260
  sheep 205
  turkeys 389
Doberman Pinscher dogs 275
Docking
  lambs 208, 214
  pigs 241, 243
  puppies 267, 308
Doctored cat 312
Doe rabbit 248
Dog 266
Dog-catchers 302
Dog Licences Act 1867 310
Dogs
  ageing 279
  artificial insemination 294
  breeding 291
  breeds 267
  definitions of common terms 266
  diseases 296
  exercise 296
  feeding 285
  handling 300
  health 295
  identification 278
  licences 310
  medicines, administration of 305
  operations, minor 308
  vices 306
  water requirements 291
Dorsal band, horse marking 42
Dorset Horn sheep 189, 203

Down sheep breeds 187
Down-calver, cow 96
Drake 395
Draught horses 33, 42, 56
Drainage in buildings 7
Drenching
  cow 141
  goat 183
  sheep 213
Dried grass—*see* grass, dried
Driving horses 33, 81
Dry period in cows 128
  goats 178
Dual purpose breeds
  cattle 100
  fowls 341
  pigs 220
Dubbing 372, 375
Duck 395
Duckling 395
Ducks
  breeding 402
  breeds 395
  definitions of common terms 395
  diseases 403
  feeding 400
  handling 404
  health 403
  housing 399
  identification 399
  operations, minor 406
  vices 405
  water requirements 402
Duck septicaemia 404
Duck virus hepatitis 404
Dutch rabbit 249

Ear canker, rabbits 260
Ear cleaning, dogs 309
Ear notching 193, 222
Ear tags 104, 154, 171, 193, 222, 251
Ear tatooing 104, 154, 171, 193, 222, 251, 279
Early weaning—*see* weaning, early
Ears, lop, horse 35
Eating faeces
  dogs 307
  horses 91
Echinococcus infestation—*see* tapeworms in dogs
Egg eating 371
Egg production
  ducks 394, 402
  fowls 339, 363
Eggs, storage 362
Electric fencing 3, 116, 195, 228, 411
Electuaries 88, 241
Embden geese 409
Enterotoxaemia 180, 205
Entire—*see* stallion
Environmental temperatures—*see* temperatures, environmental
European Economic Community 185, 247
Ewe—*see also* sheep 186
Exercise
  boars 228
  dogs 296
  general 28
  horses 68
Exmoor ponies 46

Face, dished, horse 35
  white, horse marking 40
Faeces
  appearance of 26, 68
  eating of
    dogs 307
    horses 91
Farriers 85
Farriers (Registration) Act 1975 87
Farriery 83
Farrowing crates 224
Farrowing pens 225
Fat 9, 10, 55, 255, 286, 287, 322, 357, 384, 385
Fat, milk—*see* milk fat
Fat-soluble vitamins 13
Feather on horses 42
Feather pecking 349, 371, 392
Feathering on dogs 266
Feeding
  beef cattle 158
  bulls 121
  calves 113, 156
  cats 322
  dairy cows 116
  dogs 285
  ducks 400
  fowls 354
  geese 411
  general 8
  goats 175
  heifers 115
  horses 54

Feeding (*cont.*)
  kids 173
  pigs 231
  rabbits 254
  sheep 195
  turkeys 383
Feline enteritis 330
Feline influenza 329
Fell ponies 46
Fences 173, 195, 227, 283, 410
Fetal membranes 24, 67, 328
Fetus 24, 61, 66, 121, 196, 232, 288, 292, 324
Fibre 10, 55, 234, 255, 356, 358
Fighting in dogs 307
Filly 34
Fish, for feeding 323, 326
Fish liver oils 13, 20
Fish meal, white 18
Fixed pens, fowls 354
Fleas 299, 331
Fleeces 210, 215
Flemish Giant rabbit 249
Flesh marks, horses 42
Flies 69, 89
Floors, slatted 108, 155, 194, 351, 352, 353, 382, 400
  stable 52, 53
  wire 252, 351, 352, 400
Flushing
  ewes 198
  mares 61
  sows 232
Flute bit 91
Foal 34
Foaling—*see* parturition
Foaling boxes 52
Foals—*see also* horses
  feeding 62
  orphan 62
  weaning 63, 67
Fold units 353
Food conversion ratio 3, 248
Foot bath for sheep 206, 209
Foot rot in sheep 205
Force moulting 364
Foreign short-haired cats 314
Fostering
  foals 62
  kittens 334
  lambs 213
  rabbits 258
Fowl paralysis 367
Fowl pest 367, 389
Fowls
  ageing 346
  breeding 361
  breeds 340
  definitions of common terms 340
  diseases 365
  feeding 354
  handling 369
  health 365
  housing 346
  identification 345
  incubation of eggs 346
  medicines, administration of 370
  operations, minor 372
  vices 371
  water requirements 360
Fox terriers 271
Free range housing
  fowls 354
  turkeys 381
Free-martin 96
Freeze-branding 103
Friesian cattle 98, 102, 153, 159
Fullered shoes 85
Fur chewing in rabbits 262
Fur, rabbit 247, 251

Galloway cattle 150
Galvayne's groove on a tooth 49, 50
Game fowls 343
Gander 407
Gastro-enteritis
  in man 365, 404
  in pigs 239
Geese
  breeding 412
  breeds 407
  definitions of common terms 407
  diseases 413
  feeding 411
  handling 414
  health 413
  housing 409
  identification 409
  vices 414
  water requirements 411
Gelding 34
Gestation—*see* pregnancy
Gilt—*see also* pigs 218
Gimmer 186
Gizzard worm in geese 413

Glucose 59, 63, 322
Goats
  ageing 171
  breeding 177
  breeds 168
  definitions of common terms 167
  diseases 180
  feeding 173
  handling 181
  health 179
  housing 171
  identification 170
  medicines, administration of 182
  milking 182
  operations, minor 183
  vices 183
  water requirements 176
Goatling 167, 171
Goose 407
Gosling 407
Governing Council of the Cat Fancy 312, 317, 336
Grass
  beef cattle 158, 159
  cats 324
  dairy cows 115, 116
  ducks 401
  fowls 354, 356
  geese 407, 410, 411, 412
  goats 174, 175
  horses 57, 60, 61, 63, 70
  pigs 232
  rabbits 254
  sheep 196, 199
  turkeys 384
Grass, dried 15
Grass meal 59
Grass staggers 130
Grazing
  rotational 159
  strip 116, 131
  zero 116
Great Dane dogs 275, 294
Greyface sheep 191
Greyhounds 268
Grit for poultry 355, 383, 401, 411
Grooming
  cats 333
  dogs 304
  horses 74
Grower
  fowl 340
  rabbit 248
  turkey 378
Growth-promoting substances 20
Guards, dogs as 265, 280
Guernsey cattle 99, 153
Guides, dogs as 265, 310
Gullet strap 91
Gundog group of dog breeds 269

Hackamore—*see* bitless bridle
Hackney horses 43
Hacks 45
Hair balls in cats 324
Half-brothers and half-sisters 2
Halibut-liver oil 20, 290
Halters 72, 134
Hampshire Down sheep 187
Hand, horse measure 35
Handling
  cats 331
  cattle, beef 162
  dairy 134
  dogs 300
  ducks 404
  fowls 369
  geese 414
  general 29
  goats 181
  horses 71
  pigs 240
  rabbits 261
  sheep 207
  turkeys 391
Harness
  horses 79, 82
  sows 227
Harness galls 69
Harness room 54
Harnessing
  horses for driving 81
  horses for riding 79
Hay for feeding
  cattle 115, 117, 121, 160
  general 15
  goats 174, 175
  horses 56, 58, 63
  rabbits 255, 261
  sheep 196, 198
Hay box brooders 380
Hay nets 53
Hay racks 51, 53
Headcollars 72, 181
Health, signs of
  cats 329

Health (*cont*.)
  cattle, beef 162
    dairy 129
  dogs 295
  ducks 403
  fowls 365
  geese 413
  general 26, 29
  goats 179
  horses 67
  pigs 238
  rabbits 259
  sheep 204
  turkeys 289
Heavy hogs 244
Heat period—*see* oestrus
Heifer—*see also* cattle 96
  breeding 124
  feeding 115
Hen 340
Hen turkey 378
Hen yards 353
Hereford cattle 149
Herring meal 19
Highland cattle 150
Highland ponies 46
Hinny 34
Hock boots 78
Hog 218
Hogg or hoggett 186
Hogging manes 76
Hoof structure 83
Hoof trimming 83, 143, 181, 183, 215
Hoof-pick 75
Hoofs, colour 42
Horse—*see* stallion
Horse Breeding Act 1958 93
Horse—rugs 76, 92
Horseshoes 85
Horses
  artificial insemination 65
  ageing 48
  breaking 82
  breeding 65
  breeds 42
  clipping 75
  clothing 76
  colours 36
  definitions of common terms 34
  diseases 69
  exercise 68
  feeding 54
  grooming 74
  handling 71
  harnessing 79
  health 67
  housing 51
  identification 47
  measuring 73
  licences 93
  medicines, administration of 88
  operations, minor 92
  shoeing 83
  vices 89
  water requirements 65
Hound group of dog breeds 267
House cats 318
  dogs 280
Housing
  cats 317
  cattle 105, 154
  dogs 280
  ducks 399
  fowls 346
  geese 409
  general 1, 6
  goats 171
  horses 51
  pigs 223
  rabbits 252
  sheep 194
  turkeys 380
Hoven—*see* bloat
Humidity 7, 162, 410
Hunters 44
Hutches, rabbit 252
Hybrid
  ducks 398
  fowls 343
  general 3, 5
  horses 34
  pigs 221
  rabbits 250
  sheep 191
  turkeys 379
Hybrid vigour 5, 100, 345
Hypocalcaemia—*see* milk fever
Hypomagnesaemia 130

Identification of animals
  cats 317
  cattle 102, 154
  dogs 278
  ducks 399
  fowls 345

Identification (*cont.*)
geese 409
general 6
goats 170
horses 47
pigs 222
rabbits 251
sheep 193
turkeys 379
Inbreeding 5, 344, 379
Incisor teeth—*see* ageing
Incubation periods 346, 362, 380, 399, 409
Incubators 346, 362, 380, 399, 409
Indian Runner ducks 397
Indoor systems of sow housing 224
Infectious bronchitis in fowls 366
Infectious canine hepatitis 297
Infectious diseases 28
Infectious laryngotracheitis in fowls 366
Infectious sinusitis in turkeys 389
Infra-red heaters 195, 226, 284, 320
Infundibulum on a tooth 49
Inhalations 88
Injections 30, 88, 142, 183, 241, 262, 306, 335, 370, 392
Injuries 28, 69, 87, 244
Insemination, artificial—*see* artificial insemination
Intersexes in goats 177
Iodine 11
Irish setters 269
Iron 11, 234, 238, 290

Jersey cattle 99, 153
Jersian cattle 100

Kale 16
Kennel Club 279, 295, 308, 310
Kennels
cow 109
dog 281
Kerry Hill sheep 189
Khaki Campbell ducks 396
Kicking 89, 135, 182
Kid 167
Kids—*see also* goats
feeding 173
handling 181
housing 171
Killing out percentage 3, 164
King Charles spaniels 277
Kitten 312
Kittens—*see* cats
Knee caps 78

Lactation 25, 127, 129, 137, 179
Lactation tetany—*see* hypomagnesaemia
Lamb—*see also* sheep 186
Lamb creep 199
Lamb dysentery 205
Lambing pens 195
Lameness
dairy cows 131
general 27
Landrace pigs 219
Large Black pigs 220
Large White pigs 218
Laryngotracheitis, infectious in fowls 366
Laying batteries 350, 400
Laying ducks 396
Laying fowls 340, 356, 358
Leghorn fowls 340
Lehmann system of pig feeding 233
Leptospirosis in dogs 297
Ley 3
Lice 180, 299
Licences
dogs 279, 310
general 31
stallions 93
Lifting legs
cows 136
horses 72
Light Sussex fowls 341
Lighting, artificial in poultry houses 349, 352, 382, 383, 387, 403, 413
Limb markings, horses 42
Limousin cattle 152
Linebreeding 5
Litter 2
Liver fluke disease 206, 394
Lockjaw—*see* tetanus
Long-haired cats 315
Longwool sheep breeds 186
Loose boxes
cows 105, 112
goats 171
horses 51, 52
Lowland sheep 197
Luing cattle 150
Lungeing horses 68

Lux 3, 252, 349

Maggots in sheep 207
Magnesium 11, 64, 122, 130, 200
Maintenance requirements
  cows 117
  dogs 286
  goats 175
Maize 19
Mane, horse 76
Manganese 11, 359, 385
Manipulations—*see* operations, minor
Manx cats 312
Mare 34, 60
Marek's disease 367
Mark on a tooth 49
Marker straps for cows 104
Marketing
  down and feathers, goose 415
  eggs, duck 406
    fowl 373
  general 30
  meat, beef 164
    duck 406
    fowl 374
    goose 415
    pig 244
    rabbit 263
    sheep 185
    turkey 393
  milk 144
  wool 215
Markings, horse 37
Martingales 81
Masham sheep 191
Mastitis 132, 142
Mating
  cats 328
  cattle 124, 161
  dogs 292
  ducks 402
  fowls 361
  geese 412
  general 23
  goats 177
  horses 66
  pigs 236
  rabbits 257, 261
  sheep 202
  turkeys 387
Mating, age at first
  cats 327
  cattle 124
  dogs 291
  ducks 402
  fowls 361
  geese 412
  general 23
  goats 177
  horses 65
  pigs 235
  rabbits 257
  sheep 202
  turkeys 387
Measuring horses 73
Meat
  beef 164
  duck 406
  fowl 374
  goose 415
  pig 244
  rabbit 247, 263
  sheep 185
  turkey 393
Medicines, administration of
  cats 334
  cattle 141
  dogs 305
  fowls 370
  general 30
  goats 182
  horses 88
  pigs 241
  rabbits 262
  sheep 213
Merle colour 266
Metabolic disturbances 29
Middle White pigs 220
Milk—as a food for
  cats 322, 325
  dogs 286, 289
  humans 25, 95, 179
Milk
  cooling 139
  composition, cows 95, 128
    goats 179
  production, cows 95, 117, 127
    ewes 198
    goats 167, 178
    mares 62
Milk fever 130
Milk tanks, bulk 140
Milk and Dairies (General) Regulations 1959 105, 141
Milk fat 95, 98, 129

Milk Marketing Board 144
Milking
  cows 137
  goats 182
Milking bail 105, 111
Milking equipment, cleaning 141, 182
Milking machines 138, 182
Milking parlours 105, 109
Milking stand for goats 173
Mineral requirements
  cats 325
  cattle, beef 160
    dairy 122
  dogs 290
  ducks 401
  fowls 359
  geese 412
  general 11
  goats 179
  horses 64
  pigs 234
  rabbits 255
  sheep 200
  turkeys 385
Mixed horse marking 42
Molasses 17
Molassine meal 56, 59, 202
Monorchid horse 34
Mordax cogs 85
Moulting in poultry 364, 403
Mountain sheep 189, 196
Mucoid enteritis, rabbits 261
Mule 34
Muscovy duck 396
Mustard 16
Mutton 186
Muzzles 59, 90, 303
Mycoplasmosis 366
Myxomatosis 247

Nail cutting in dogs 310
Nailing on a shoe 86
Nails, horseshoe 86
Nanny-goat 167
National Foaling Bank 62
National Milk Records Scheme 101, 144
Near side of a horse 35
Neck, ewe, horse 35
Nest boxes 253, 351, 381, 400
Neuter cat 312
New duck disease 404
New Forest ponies 46
New Zealand White rabbit 248
Newcastle disease 367, 389
Nicotinic acid 13, 402
Nipple-type waterers 235, 253, 360
Non-sporting breeds of dogs 272
North Holland Blue fowls 342
Nose, Roman, horse 35

Oats 19, 56, 58
Obesity in dogs 288
Oestrus
  bitches 292
  cats 327
  cows 125, 129
  ewes 203
  general 23
  mares 65
  nanny-goats 177
  rabbits 257
  sows 235
Off side of a horse 35
Oil—*see* fat
Oil-seed cakes and meals 17, 56
Old English Sheepdogs 275
Operations, minor
  cats 336
  cattle, beef 163
    dairy 142
  dogs 308
  ducks 406
  fowls 372
  general 30
  goats 183
  horses 92
  pigs 242
  rabbits 262
  sheep 214
  turkeys 392
Orphan animals, rearing
  foals 62
  kittens 325, 334
  lambs 213
  puppies 289
Outcrossing 5
Outdoor system of rabbit keeping 254
Outdoor system of sow housing 227
Overreaching 87

Pacers, horses 44
Pantothenic acid 13

Paper for bedding 54, 280, 284, 317, 328, 348
Parakeratosis in pigs 234
Parasitic gastro-enteritis 180, 206
Parasitic otitis in cats 331
Parlours, milking—*see* milking parlours
Parturition
  bitches 293
  cats 328
  cows 126
  ewes 203
  general 24
  mares 66
  nanny-goats 178
  rabbits 358
  sows 236
Patch, horse marking 40
Peas 18
Peat moss for bedding 54
Pecking, feather 349, 371, 392
Pekin duck 396
Pekingese dogs 277, 294, 302
Penis injuries in drakes 406
Percheron horses 43
Perches 351, 353, 381
Performance testing
  beef cattle 153
  general 4
  pigs 221
  sheep 192
Permanent pasture 3
Persian cats 315
Phosphorus 11, 64, 122, 160, 176, 200, 234, 255, 290, 359, 385
Pig catcher 241
Pig handling boards 240
Pig runs 244
Piglet 218
Piglet anaemia 238
Pigs
  ageing 223
  artificial insemination 236
  breeding 235
  breeds 218
  definitions of common terms 218
  diseases 238
  feeding 231
  handling 240
  health 238
  housing 223
  identification 222
  medicines, administration of 241
  operations, minor 242
  vices 242
  water requirements 235
Pine, nutritional in sheep 201
Plymouth Rock fowls 342
Pneumonia in calves 162
Poisonous plants 28, 181
Poisons 28
Pole barns for turkeys 380
Polled 3
Polo ponies 45
Pomeranian dogs 278
Ponies—*see also* horses
  breeds 45
  feeding 60
Poodle dogs 273, 309
Pork 218
Pork-type pig breeds 220
Potassium 11, 122, 325
Poult 378
Poultry industry 339
Poussin 340
Pregnancy
  bitches 292
  cats 324, 328
  cows 121, 126, 128, 161
  ewes 196, 203, 208
  general 24
  mares 66
  nanny-goats 176
  rabbits 258
  sows 225, 232, 236
Pregnancy, false 294
Pregnancy testing
  bitches 292
  cows 125
  mares 66
  rabbits 258
Progeny testing
  cattle, beef 153
    dairy 101
  general 4
  pigs 221
Proprietary foods 56, 118, 254, 324, 355, 383
Protein 8, 17, 55, 56, 62, 118, 159, 254, 287, 322, 356, 358, 383, 384, 401
  animal 9, 18
  vegetable 9, 17
Pseudo-pregnancy in rabbits 258
Puberty
  colts 65

Puberty (*cont.*)
- fillies 65
- general 23
- rabbits 257

Pug dogs 278
Pullet 340
Pulp cavity on a tooth 49
Pulse
- cats 329
- cattle, beef 162
  - dairy 130
- dogs 296
- fowls 365
- general 27
- goats 179
- horses 68
- pigs 238
- rabbits 260
- sheep 205
- turkeys 389

Puppy—*see also* dogs 266
Purebred 2
Pyridoxine—*see* vitamin B_6

Quarantine regulations 300, 331
Queen cat 312
Quidding food 69
Quill pulling in ducks 406

Rabbits
- ageing 252
- artificial insemination 257
- breeding 257
- breeds 248
- definitions of common terms 248
- diseases 260
- feeding 254
- handling 261
- health 259
- housing 252
- identification 251
- medicines, administration of 262
- operations, minor 262
- vices 262
- water requirements 255

Rabies 300
Racehorses—*see* horses
Raddle for rams 202, 204
Ram 186, 200, 202
Rams
- breeding 192, 202
- feeding 200
- vasectomised 203

Rape 16
Rearing batteries, fowls 349
Rearing in horses 90
Reciprocal recurrent selection in fowls 344
Red mite infestation 368
Red Poll cattle 100
Relative humidity 8, 230, 347, 380, 399, 410
Reproductive cycles—*see* oestrus cycles
Respiratory conditions in horses 69
Respiratory disease, chronic in fowls 366
Retriever dogs 270
Reuff's method of casting cattle 136
Rex cats 315
Rex rabbits 251
Rhode Island Red fowls 341
Riboflavin—*see* vitamin B_2
Rickets
- dogs 290
- ducks 401
- pigs 235
- turkeys 385

Riding horses 33, 44, 79
Rig horse 34
Ringing
- bulls 144
- pigs 243

Rings, leg
- ducks 399
- fowls 346
- geese 409
- rabbits 251

Rings, nose
- bulls 144
- pigs 244

Ringworm 70
Roaster, fowl 340
Roman goose 407
Romney Marsh sheep 186
Roots for feeding
- cattle 115, 117
- general 16
- goats 175
- horses 57
- pigs 233
- sheep 198

Roughages 14, 15, 56, 57, 115, 158, 160, 217
Roundworms in
- cats 330

Roundworms (*cont.*)
  dogs 298
Rubbers, stable 75
Rugs, horse 77
Runt, piglet 218
Russian blue cats 314
Rye 19

Saanen goats 170
Saddles
  horse 79, 82
  turkey 379, 387
Salmonellosis 365, 389, 404
Savaging in pigs 242
Sawdust for bedding 54, 107, 108
Scottish Blackface sheep 190
Scottish Blue Grey cattle 151
Scottish Half-bred sheep 191
Scottish terriers 271, 293
Scour, white—*see* white scour
Scratching post for cats 318, 321
Semen 24, 291
  dilution 25, 127, 178, 237, 295, 388
  frozen 25, 127, 178
Service—*see* mating
Sexing
  chicks 370
  ducklings 405
  goslings 414
  kittens 334
  rabbits 261
  turkeys 391
Sex-linkage 362
Shavings, wood for bedding—*see* wood shavings
Shearing sheep 210
Shearling sheep 186
Sheep
  ageing 193
  artificial insemination 204
  breeding 202
  breeds 186
  definitions of common terms 186
  dipping 211
  diseases 205
  feeding 195
  handling 207
  health 204
  housing 194
  identification 193
  medicines, administration of 213
  operations, minor 214
  shearing 210
  spraying 212
  washing 211
  water requirements 201
Sheep handling unit 208
Sheep scab 207
Shepherd's crook 207
Shetland ponies 47, 93
Shetland sheepdogs 276
Shire horses 42
Shoeing horses 83
Short-haired cats 312
Shorthorn cattle, beef 150
  dairy 100
Shortwool sheep breeds 189
Shows 4, 266, 310, 311, 337
Shying in horses 90
Siamese cats 314
Sibs 2
Silage for feeding
  cattle 108, 115, 117, 121, 160
  general 17
  goats 175
  horses 56, 57
  sheep 198
Simmental cattle 152
Sire 2
Slap marking pigs 222
Slatted floors—*see* floors, slatted
Slink veal 148
Slipped wings
  ducks 403
  geese 413
Snip, horse marking 40
Snuffles, rabbits 260
Sodium 11, 122, 325
Sodium chloride 64, 234, 255, 290, 359, 385
Solari type piggeries 230
Solids-not-fat in milk 95, 98, 129
Southdown sheep 189
Sow 218
Spaniel dogs 270
Spaying
  cats 336
  dogs 309
  mares 34
Special foods 14, 20
Speedy cutting in horses 88
Spermatozoa—*see* semen
Sporting breeds of dogs 267
Spot, horse marking 40
Spraying in cats 335

Spraying sheep 212
Springer cow 96
St. Bernard dogs 276
Stable rubber 75
Stables 52
Staffordshire Bull terriers 271
Stag turkey 378
Stallion 34
Stallions—*see* also horses
  breeding 65
  feeding 62
  licence 93
Stalls for
  cows 106
  horses 51
  sows 223
Staphylococcal arthritis in poultry 413
Star, horse marking 37
Steaming-up, dairy cows 121
Steer—*see* bullock
Sterilised bone flour—*see* bone flour
Stern, hound 266
Stomach tube 88
Store cattle 148
Straw 15
  barley 54, 115, 118, 160
  oat 54, 115, 160, 284
  wheat 54, 284
Straw, for bedding 54, 107, 160, 172, 194, 230, 284, 400
Straw for feeding
  cattle 115, 117, 155, 160
  goats 172
  horses 57
Strip cup 132, 138
Strip grazing 116
Stripe, horse marking 40
Stripping dogs 304
Stud cat 312
Stud cat accommodation 321
Subsidies, Government 165, 216
Succulent foods 14, 16
Suckling calves 156, 157
Suffolk horses 42
Suffolk sheep 189
Sugar beet pulp for feeding
  cattle 118
  general 19
  goats 175
  horses 57
Sugar beet tops 16, 200
Sulphur 11
Sussex cattle 150
Swaledale sheep 190
Sweat-box piggery 230
Swill 234
Swine—*see* pigs

Table ducks 395
Table fowls 342, 357
Tail biting in pigs 242
Tail, horse 76
Tail savers 78
Tamworth pigs 220
Tape muzzles for dogs 303
Tapeworm infestation in dogs 299
Tattooing
  ears 104, 193, 222, 251, 279
  horses 48
  udders, cows 104
Tearing clothing, horses 92
Teats, supplementary 143, 184
Teeth—*see* dentition
Temperatures, body
  cats 329
  cattle, beef 162
    dairy 130
  dogs 295
  fowls 365
  general 27
  goats 179
  horses 68
  pigs 238
  rabbits 260
  sheep 205
Temperatures, environmental
  calves, veal 155
  cats 319, 320
  dogs 284
  ducks 399
  fowls 347, 352
  geese 410
  general 7
  pigs 229, 230
  rabbits 252
  sheep 201
  turkeys 380
Terrier group of dog breeds 270
Tetanus 69
Tethering
  bulls 113
  dogs 281
  goats 173
Thermometers, clinical 27, 68, 296, 329

Thiamine—*see* vitamin B_1
Thoroughbred horses 44
Tier brooders
  fowls 349
  turkeys 380
Toe cutting in turkeys 393
Toggenburg goats 170
Tom cat 312
Tom turkey 378
Tooth cutting in pigs 243
Tortoiseshell coloured cats 316
Toulouse geese 409
Toxin 3
Toxocara canis—see roundworms in dogs
Toy group of dog breeds 276
Trace elements—*see* also mineral requirements 11
Training dogs 300
Trap nests 346, 361, 382, 402
Tricolour 266
Trotters, horses 44
Tup—see ram
Turkeys
  ageing 379
  artificial insemination 388
  breeding 387
  breeds 378
  definitions of common terms 378
  diseases 389
  feeding 383
  handling 391
  health 389
  housing 380
  identification 379
  medicines, administration of 392
  operations, minor 392
  vices 392
  water requirements 387
Tusks 49, 224, 243
Twitches 73, 241

Udder 97, 168
Udder, inflammation of—*see* mastitis
Udder kinch 135
Udder sucking 142
Udder tattooing 104
Undercoat 266
Urine 26, 298
Utility group of dog breeds 272

Vaccine 3, 297, 330, 366, 367, 370
Vasectomized rams 203
Veal 148
Veal production 154, 157
Ventilation of buildings 7, 52, 105, 155, 171, 230
Verandahs for poultry 353, 382
Vices
  cats 335
  cattle 142
  dogs 306
  ducks 405
  fowls 371
  geese 414
  general 30
  goats 183
  horses 89
  pigs 242
  rabbits 262
  turkeys 392
Virus pneumonia in pigs 239
Vitamin
  A 13, 64, 123, 161, 176, 201, 234, 290, 326, 359, 385, 401
  B_1 13, 326
  B_2 13, 360
  B_6 13, 326
  B_{12} 13, 124, 201, 235
  B complex or group 14, 123, 201, 234, 255, 291, 326, 360, 386, 402
  C 13, 14, 291
  D 13, 14, 64, 123, 161, 176, 201, 234, 290, 326, 360, 385, 401
  E 13, 64, 123, 201, 290, 360, 385, 402
  K 13
Vitamin requirements
  cats 326
  cattle, beef 160
    dairy 123
  dogs 290
  ducks 401
  fowls 359
  geese 412
  general 13
  goats 176
  horses 64
  pigs 234
  rabbits 255
  sheep 201
  turkeys 385
Vitamins 12

Wall-eye in horses 40

Washing
  cats 333
  dogs 304
  sheep 211
Water bowls
  cattle 106, 156
  goats 173
  horses 51, 53
  pigs 223, 235
Water brush 75
Water requirements
  cats 326
  cattle, beef 161
    dairy 124
  dogs 285, 291
  ducks 402
  fowls 360
  geese 412
  general 21
  goats 176
  horses 65
  pigs 235
  rabbits 255, 259
  sheep 201
  turkeys 387
Water-soluble vitamins 13
Weaning
  foals 63, 67
  kids 174
  kittens 322
  lambs 197
  pigs 228, 236
  puppies 285
  rabbits 257, 259
Weaning, early
  calves 114, 156
  pigs 228, 233
Weatherby 92
Weaving, horses 91
Web punching 345, 379, 399, 409
Wedder—*see* wether
Welfare codes
  cattle 145, 165
  fowls 374
  general 31
  pigs 244
  turkeys 394
Welsh Corgi dogs 276
Welsh Half-bred sheep 191
Welsh Mountain sheep 191
Welsh pigs 219
Welsh ponies 45
Wensleydale sheep 187
West Highland White terriers 271
Wether 185, 186
Wheat 19
Wheaten colour 266
Whelping—*see* parturition
Whey 233
Whippets 269
White Beltsville turkeys 379
White Campbell ducks 397
White scour in calves 133
Whorls on horses 42, 48
Wind sucking, horses 91
Wing bands 251, 345, 379, 409
Wing feather cutting 373, 393, 406
Wire floors—*see* floors, wire
Wood shavings for bedding 54, 107, 172, 253, 348, 400
Wool 185, 215
Work, feeding for
  dogs 288
  horses 55, 56, 57
Working group of dog breeds 273
Worms
  cats 330
  dogs 298, 299
  geese 413
  goats 180
  horses 70
  pigs 239
  sheep 206
  turkeys 390
Worrying livestock, dogs 306

Yard, collecting 163
Yard fattening of cattle 160
Yards
  cattle 105, 107, 155
  ducks 460
  goats 172
  hens 353
  pigs 224, 230
  sheep 194
  turkeys 381
Yearling horse 34
Yeast 13, 291
Yellow fat disease in kittens 326
Yolk colour in eggs 374
Yorkshire coach horses 43
Yorkshire terriers 278

Zebra marks, horses 42
Zero grazing 116
Zinc 11, 234